Light Propagation in Linear Optical Media

Light Propagation in Linear Optical Media

Glen D. Gillen and Katharina Gillen

California Polytechnic State University
San Luis Obispo, California, USA

Shekhar Guha

Wright-Patterson Air Force Base
Ohio, USA

CRC Press
Taylor & Francis Group
Boca Raton   London   New York

CRC Press is an imprint of the
Taylor & Francis Group, an **informa** business

CRC Press
Taylor & Francis Group
6000 Broken Sound Parkway NW, Suite 300
Boca Raton, FL 33487-2742

First issued in paperback 2017

© 2014 by Taylor & Francis Group, LLC
CRC Press is an imprint of Taylor & Francis Group, an Informa business

No claim to original U.S. Government works
Version Date: 20130918

ISBN 13: 978-1-138-07632-7 (pbk)
ISBN 13: 978-1-4822-1094-1 (hbk)

Library of Congress Cataloging-in-Publication Data

Gillen, Glen D., 1972-
    Light propagation in linear optical media / Glen D. Gillen, Katharina Gillen, Shekhar Guha.
        pages cm
    "A CRC title."
    Includes bibliographical references and index.
    ISBN 978-1-4822-1094-1 (hardcover : alk. paper)
    1. Light--Transmission. 2. Electromagnetic waves. I. Gillen, Katharina, 1977- II. Guha, Shekhar. III. Title.

QC389.G55 2014
535'.32--dc23

2013034925

Visit the Taylor & Francis Web site at
http://www.taylorandfrancis.com

and the CRC Press Web site at
http://www.crcpress.com

To Lani
— from Glen and Kat Gillen

To Rahul, Rakesh, Subhalakshmi, Meghna, and Preetha
— from Shekhar Guha

Contents

7   Scalar and Vector Diffraction Theories   177

8   Calculations for Plane Waves Incident upon Various Apertures   209

Preface

The main aim of this book is to describe light propagation in linear media by expanding on diffraction theories a little beyond what is available in the optics books that we love and admire (such as Born and Wolf, Jenkins and White, Goodman, Hecht, Brooker, or Akhmanov and Nikitin). This book can possibly be used as a technical reference book by professional scientists and engineers interested in light propagation and as a supplemental text for upper-level undergraduate or graduate courses in optics. The first two chapters of the book, written at a more introductory level, are included for the sake of completeness and perhaps can be used as a refresher to the subject.

The general flow of chapters and topics in the book is illustrated in Table 0.1. Chapters 1 and 2 provide background information and descriptive explanations of the fundamentals of light and physical optics. Chapter 1 provides a background of the origins of light, our historical understanding of it, and a brief explanation of both classical and quantum mechanical models for the creation of light. Chapter 2 describes light as electromagnetic waves and provides a mathematical description of the fundamentals of linear physical optics. It is assumed that the light is traveling through a single, homogeneous, isotropic medium. Chapter 3 continues the topic of light traveling within a single unbounded medium. However, the medium is now an anisotropic crystal. Calculations of walk-off angles in both biaxial and uniaxial crystals, which are of particular importance in nonlinear optical applications, are emphasized in this chapter.

Chapters 4 and 5 present mathematical models for what happens when light encounters planar boundaries between different media. In Chapter 4 the behavior of the electromagnetic fields when they encounter a single planar boundary between two different optical media is described, whereas Chapter 5 presents light propagation within a slab waveguide bounded by two different media. In Chapter 6 we derive and explain the propagation of Gaussian beams through homogeneous media, through slabs and lenses, and mathematical models for higher order Hermite and Laguerre-Gaussian beams.

In Chapters 7–10 we discuss various diffraction models for the propagation of light through regions where sizes of apertures and optical components can affect the propagation of the light via diffraction. In Chapter 7 we present various scalar and vector diffraction theories for light fields within and beyond a planar aperture, with the results of calculation for several scenarios illustrated in Chapter 8. Chapter 9 expands the vector diffraction theory to propagation

Chapter number(s)	General topic(s)
1, 2	Light and EM waves in isotropic media
3	EM waves in anisotropic media
4	EM waves encountering a single planar boundary
5	EM waves in waveguides
6	Gaussian beams
7, 8, 9	Diffraction of plane waves
10	Diffraction of Gaussian beams
11	Trapping atoms using linear light-atom interactions

TABLE 0.1
General flow of chapters and topics.

across curved surfaces. In Chapter 10 we describe the diffraction of Gaussian beams by spatially limiting apertures or obstacles.

In Chapter 11 we present an application of diffraction theory: spatially confining (trapping) cold atoms within localized light intensity patterns. We present a few methods to trap atoms using a variety of propagating light fields including a scalar light-atom interaction, a vector light-atom interaction, and projecting a near-field diffraction pattern to trap atoms located in what would be considered to be a "far-field" location.

Significant portions of this book have been compiled from several sources: research work performed by Shekhar Guha and Glen Gillen at the Air Force Research Laboratory, research performed by Glen and Katharina Gillen at California Polytechnic State University, and adaptations of in-class lecture notes of Glen and Katharina Gillen from a variety of courses in physics and optics they have taught.

Acknowledgments

Glen D. Gillen would like to thank his college professors at Denison University, Granville, OH, for providing him the academic and analytical training that has served as a strong foundation for a scientific and academic career. Glen would also like to thank Shekhar Guha for the wonderful opportunity to have worked with him at the Air Force Research Labs, Dayton, OH, right out of graduate school, and all of his guidance and support while pursuing a career in academia. Last, but certainly not least, Glen thanks his family and loving wife for their love and support throughout the years.

Katharina ("Kat") Gillen would like to thank Shekhar Guha for giving her the opportunity to collaborate on research and to participate in the writing of this book. She would like to acknowledge useful discussions with Ivan Deutsch, Marianna Safronova, and her friend and colleague Thomas Gutierrez. Most importantly, she would like to thank her research students for all their hard work and enthusiasm. In particular, she appreciated helpful discussions with Travis Frazer, and the joint research project with Bert D. Copsey, the results of which have been included in this book. On a personal level, Kat is thankful for all her mentors along the way, from her high school math and physics teachers Jutta Bremken and Ludwig Weisgerber, who believed in her talents, to her Ph.D. advisor Greg Lafyatis, to her faculty mentor Paula Turner, who helped give her a good start into the college teaching world. Finally, she would like to express her deepest gratitude for her friends and colleagues at Cal Poly, and her friends and family back home for supporting her and inspiring her. Last, but not least, she thanks her collaborator Glen Gillen for being patient when she wasn't writing the book, and her husband Glen for when she was. She could not have accomplished this without the love and support of her family.

Shekhar Guha would like to take this opportunity to thank his collaborators, Profs. Glen and Katharina Gillen for embarking on this venture and for skilfully bringing it to completion despite demanding academic and family responsibilities. Also he would like to gratefully acknowledge his high school physics teacher, Anjan Baran Das Gupta, for instilling in him the love of learning physics, for continuing to provide guidance and inspiration throughout his college days, and for providing many precious books as gifts. Working on this book reminded him of those days of long ago. Shekhar would also like to thank his colleague Dr. Leonel Gonzalez not only for the everyday discussions and keen insight that must have permeated this book but also for reviewing and helping with much of this manuscript. He also thanks the Air Force Research Laboratory for providing him the facilities and the stimulating work environment that made this book possible. Lastly and mostly, he thanks his family for their patience, support, and love.

About the Authors

Glen D. Gillen is an Associate Professor in the Physics Department at California Polytechnic State University, San Luis Obispo (more affectionately known as "Cal Poly").

Dr. Gillen grew up in south Florida and moved to Ohio in 1990 to begin his academic journey at Denison University, Granville, OH, where he earned a B.S. in physics with a minor in mathematics in 1994. From 1994–1997, he attended Miami University, Oxford, OH, where he received an M.S. in physics and an MAT in secondary education. In 1997, he continued his post-baccalaureate education at The Ohio State University, Columbus, where he received another M.S. in physics in 2001 and a Ph.D. in physics in 2002. His first M.S. in physics was in the field of experimental molecular spectroscopy, and his doctoral work was in the field of experimental femtosecond high-intensity laser-atom interactions.

After completing his academic training, Dr. Gillen joined the Infrared Materials Research Group of the Air Force Research Labs at Wright-Patterson Air Force Base, Dayton, OH, where he studied the linear and nonlinear optical properties of infrared materials. Under the guidance of Dr. Shekhar Guha he also studied theoretical and computational light propagation methods. In 2006, he joined the faculty at Cal Poly to pursue his lifelong passion for teaching undergraduate physics.

Katharina ("Kat") Gillen is also an Associate Professor in the Physics Department at Cal Poly, San Luis Obispo. Kat is originally from Germany. After three years of undergraduate studies in physics at the Rheinische Friedrich-Wilhelms-Universität Bonn, she came to the U.S. as an exchange student in 1999. After deciding to extend her stay indefinitely, she received an M.S. in physics in 2000 and an Ph.D. in physics in 2005, both from The Ohio State University, Columbus. Her research field is cold atomic physics, with an affinity for neutral atom quantum computing. During her time at Ohio State, Kat discovered her passion for teaching undergraduate physics. Following the completion of her Ph.D., she taught at Kenyon College, Gambier, OH, for one year before taking her current position at Cal Poly in 2006.

Shekhar Guha obtained a Ph.D. in physics from the University of Pittsburgh and did post-doctoral work at the University of Southern California, Los Angeles. He has been working at the Air Force Research Laboratory since 1995. His research interests are in the field of nonlinear optical materials.

1

Electromagnetic Fields and Origin of Light

In this chapter we review the concepts of

- electric and magnetic fields, **E** and **H**
- the vector and scalar potentials, **A** and ϕ
- the Hertz vector potential, **Π**

We describe how to use the potentials to find **E** and **H**, discuss some properties of electromagnetic waves, and review a few fundamental concepts of quantum mechanics pertaining to the generation of light from moving charges and atoms.

1.1 Introduction

The importance of light to us is self evident. Our vision, arguably the most important of all our senses, is enabled by light. The energy needed for life to flourish on earth travels from the sun mainly in the form of light; light from other stars gives us the information needed to understand the universe; and investigation of properties of light absorbed or emitted by flames led to the understanding of the structure of atoms that everything is made out of. Scientists have tried to understand the nature and properties of light at least since the times of the ancient Chinese and Greek civilizations. The understanding that light is an electro-magnetic wave came about in the nineteenth century after the laws of electromagnetism were consolidated by James Clerk Maxwell into the set of "Maxwell's equations" and since then most of the characteristics of light propagation have been described by using the solutions of Maxwell's equations. The invention of laser in 1960 made it possible to easily obtain coherent and high intensity light, which accelerated advancements in many fields including physics, chemistry, and engineering, and ushered in the field of nonlinear optics with its many practical applications.

In many optics experiments, light generated by one or more lasers or some other sources travels through air, then through a set of optics, such as mirrors, prisms, polarizers, filters, gratings, lenses, etc. as well as through other linear or nonlinear optical media that can be solid, liquid, gaseous, liquid crystalline, plasma, or of some other form, in bulk, waveguide, thin film, or some other configuration, and is reflected, absorbed, or scattered by that medium or by some other object and detected by one or more light sensor. In this book, we attempt to describe and compile the formalisms and equations that describe the propagation of light after it leaves the source and before it is detected by some sensor.

In this book we will describe in detail light propagation through linear optical media, mainly through various solutions of the wave equation derived from Maxwell's equations. In this chapter we recapitulate the laws of electricity and magnetism that are consolidated as Maxwell's equations, from which the existence of electromagnetic radiation is predicted. Since classical consideration of radiation emitted by moving charges cannot explain the observed properties of light emitted by electronic motion in atoms, we will briefly delve into the quantum mechanical origin of light from atoms.

1.2   Electric Fields

From the observations that glass rods rubbed with silk can attract hard rubber rods rubbed with fur, and that two glass rods both rubbed with silk repel each other, scientists in the eighteenth century deduced that electric charge (denoted by the symbol q) is a property of matter that causes two bodies having charges to exert a force on each other. Observing that charges are of two kinds that can cancel each other, Benjamin Franklin named the two kinds positive and negative. The region surrounding a charge distribution can be thought of as being permeated by an electric field (denoted by the vector $\mathbf{E}$), which is defined at any position to be the force experienced by a unit charge placed at that position.

From the experimental observation (known as Coulomb's Law) that the field strength due to a charge q at a distance r from it is proportional to q/r^2, it can be deduced that the electric field ($\mathbf{E}$) in a material medium with a free charge density ρ must follow the relationship:

$$\boldsymbol{\nabla} \cdot \mathbf{E} \propto \rho, \tag{1.1}$$

where $\boldsymbol{\nabla}$ is defined in terms of partial derivative operators as

$$\boldsymbol{\nabla} = \frac{\partial}{\partial x}\hat{x} + \frac{\partial}{\partial y}\hat{y} + \frac{\partial}{\partial z}\hat{z}, \tag{1.2}$$

where $(\hat{x}, \hat{y}, \hat{z})$ are the unit vectors in the directions of the Cartesian coordinates x, y, and z. The divergence of the electric field can then be written as

$$\boldsymbol{\nabla} \cdot \mathbf{E} = \frac{\rho}{\epsilon_o}, \tag{1.3}$$

where ϵ_0 is the experimentally measured constant (called the permittivity constant) with the value of 8.85×10^{-12} Coulomb^2 s^2/kg/m^3.

Two electric charges of equal magnitude (say, q) and opposite sign, separated by a distance (say denoted by the vector **d**) constitute an electric dipole, with a dipole moment $\mathbf{p} = q\mathbf{d}$. Common solid, liquid, or gaseous materials are made out of neutral atoms and molecules, which can be considered entities with overall equal and opposite charges. The distribution of these charges can either completely cancel each other out leaving no dipole moment, or can result in a net displacement between the positive and negative charge distributions yielding a permanent dipole moment. But in either case, the molecules acquire an induced dipole moment in the presence of electric fields that are associated with electric charge. The macroscopic amount of the induced electric dipole moment per unit volume is called the polarization (**P**). In vacuum, **P** of course goes to zero.

If the medium is polarizable the total net effect of the incident electric field, **E**, and the induced polarization, **P**, can be expressed as one vector defined to be

$$\mathbf{D} \equiv \mathbf{P} + \epsilon_0 \mathbf{E}, \tag{1.4}$$

where the vector field **D** is referred to as the displacement vector. Including the polarization of the medium, the observed behavior of the displacement vector is described by

$$\nabla \cdot \mathbf{D} = \rho. \tag{1.5}$$

A medium is defined to be linear if **P** is linearly dependent on **E**. In an isotropic medium, the vectors **P** and **E** are parallel and the relationship is written as

$$\mathbf{P} = \epsilon_0 \chi \mathbf{E}, \tag{1.6}$$

where χ, the linear susceptibility, is a constant. In general, for anisotropic media, χ is a tensor relating each component of **P** to the components of field **E**. The dielectric permittivity of the medium is defined to be the tensor

$$\epsilon \equiv \epsilon_0 (1 + \chi). \tag{1.7}$$

For an isotropic medium, ϵ is a scalar and the **D** and **E** vectors are parallel, with

$$\mathbf{D} = \epsilon \mathbf{E}. \tag{1.8}$$

1.3 Magnetic Fields

The attraction and repulsion of magnets and deflection of magnetic needles by current-carrying wires can be described by assuming that the space surrounding magnets or moving electric charge contains a *magnetic field*, characterized by a magnetic field vector **B**, which is called the *magnetic induction.* **B** is defined in the following way: If a charge of amount q moving through a point P with velocity **v** experiences a force **F** that is perpendicular to **v**, then a magnetic induction **B** is present at P given by the relation

$$\mathbf{F} = q \ \mathbf{v} \times \mathbf{B}. \tag{1.9}$$

The distribution pattern of **B** at different distances from magnets and electric currents led to the conclusion that single poles of magnets do not exist and even down to electronic and nuclear levels elementary particles are magnetic "dipoles." From the non-existence of isolated magnetic poles, the Gauss's law for magnetism

$$\nabla \cdot \mathbf{B} = 0 \tag{1.10}$$

was derived.

1.4 Electromagnetism

From the observation (by Michael Faraday) that an electric field is produced by a magnetic induction changing with time (t), the Faraday's law

$$\nabla \times \mathbf{E} = -\frac{\partial \mathbf{B}}{\partial t} \tag{1.11}$$

was deduced. An electric dipole (with moment **p**) placed in an external electric field (**E**) experiences a torque given by

$$\boldsymbol{\tau} = \mathbf{p} \times \mathbf{E}. \tag{1.12}$$

Similarly, a loop of current or a permanent magnet placed in an external magnetic field (of magnetic induction **B**) experiences a torque given by

$$\boldsymbol{\tau} = \boldsymbol{\mu} \times \mathbf{B}, \tag{1.13}$$

where $\boldsymbol{\mu}$ is the magnetic dipole moment of the current loop or magnet, and a characteristic of the magnet or the current loop. For example, $\boldsymbol{\mu}$ for a planar loop of current i enclosing an area A has a magnitude of iA and the direction is along the perpendicular to A. Similar to the electric dipole moment, **p**, the magnetic dipole moment, $\boldsymbol{\mu}$, can be a permanent property of the object (independent of external fields) or it can be induced by the presence of an external magnetic field. When any medium is placed in a region with magnetic induction **B**, there will be an amount of induced magnetic dipole moment in the medium due to the presence of **B**. The magnetization vector **M** is defined to be the total induced magnetic moment per unit volume. From experiments it was found that in presence of a current density **J** (defined as the amount of charge crossing a unit area of surface per unit time) the vectors **B**, **M**, and **J** are related by the generalized Ampere's law

$$\nabla \times \left(\frac{1}{\mu_0} \mathbf{B} - \mathbf{M} \right) = \mathbf{J} \tag{1.14}$$

where μ_0 is called the permeability constant, with a value of $4\pi \times 10^{-7}$ kg m/coulomb^2 (μ_0 is unrelated to the magnetic dipole moment $\boldsymbol{\mu}$). Incorporating the continuity equation for charge flow in any region of space

$$\nabla \cdot \mathbf{J} + \frac{\partial \rho}{\partial t} = 0 \tag{1.15}$$

in the generalized Ampere's equation, Maxwell extended Ampere's law to the equation

$$\nabla \times \mathbf{H} = \mathbf{J} + \frac{\partial \mathbf{D}}{\partial t}, \tag{1.16}$$

where the *magnetic field strength* **H** is defined as

$$\mathbf{H} = \frac{1}{\mu_0} \mathbf{B} - \mathbf{M}. \tag{1.17}$$

For paramagnetic and diamagnetic materials, it is experimentally found that **B** is directly proportional to **H**, i.e.,

$$\mathbf{B} = \mu \mathbf{H} \tag{1.18}$$

where μ is known as the permeability of the magnetic medium (and again it should not be confused with the magnetic dipole moment $\boldsymbol{\mu}$). The magnetization **M** is related to the field strength H by

$$\mathbf{M} = \left(\frac{\mu}{\mu_0} - 1 \right) \mathbf{H}. \tag{1.19}$$

The relative permeability is defined by the symbol

$$\kappa_m = \frac{\mu}{\mu_0}. \tag{1.20}$$

The relative permeability κ_m is equal to 1 in vacuum and in nonmagnetic materials, κ_m is close to 1.

Although not used further in this book, the magnetic susceptibility χ_m is defined by the relation

$$\chi_m \equiv \kappa_m - 1. \tag{1.21}$$

Equations 1.8 and 1.18 are known as the constitutive relations. The four equations 1.5, 1.10, 1.11, and 1.16 constitute the Maxwell's equations for macroscopic media (which were expressed in this form by Oliver Heaviside) and are summarized here for convenience:

$$\nabla \cdot \mathbf{D} = \rho \tag{1.22}$$

$$\nabla \cdot \mathbf{B} = 0 \tag{1.23}$$

$$\nabla \times \mathbf{E} = -\frac{\partial \mathbf{B}}{\partial t} \tag{1.24}$$

$$\nabla \times \mathbf{H} = \mathbf{J} + \frac{\partial \mathbf{D}}{\partial t} \tag{1.25}$$

Maxwell's equations and the relationships between $\mathbf{E}$, $\mathbf{D}$, $\mathbf{B}$, and $\mathbf{H}$ are discussed in further detail in Chapter 2.

1.5 Vector and Scalar Potentials

Since $\nabla \cdot \mathbf{B} = 0$, $\mathbf{B}$ can be expressed as the curl of some vector, say denoted by $\mathbf{A}$, i.e.,

$$\mathbf{B} = \nabla \times \mathbf{A} \tag{1.26}$$

Substituting Eqs. 1.26 in Eqs. 1.24 we find that the curl of $\mathbf{E} + \dfrac{\partial \mathbf{A}}{\partial t}$ is equal to 0, so that

$$\mathbf{E} + \frac{\partial \mathbf{A}}{\partial t} = -\nabla \phi. \tag{1.27}$$

where ϕ is some scalar function. $\mathbf{A}$ and ϕ are known as the vector and scalar potentials, respectively. From the four values of the three components of $\mathbf{A}$ and of ϕ, all six components of the vectors $\mathbf{E}$ and $\mathbf{B}$ can be obtained using Eqs. 1.26 and 1.27.

To do that, the values of $\mathbf{A}$ and of ϕ must of course be obtained first. By substituting equations 1.26 and 1.27 in equation 1.16 we obtain the following equations for $\mathbf{A}$ and ϕ:

$$\nabla \times (\nabla \times \mathbf{A}) + \mu\epsilon\frac{\partial^2 \mathbf{A}}{\partial t^2} + \mu\epsilon\nabla\frac{\partial \phi}{\partial t} = \mu\mathbf{J} \tag{1.28}$$

and

$$\nabla^2\phi + \frac{\partial}{\partial t}\nabla \cdot \mathbf{A} = -\frac{\rho}{\epsilon}. \tag{1.29}$$

Without loss of generality and without affecting $\mathbf{E}$ or $\mathbf{B}$, *two conditions* can be chosen for $\mathbf{A}$ and ϕ. The first choice is known as the *Coulomb Gauge*, where the divergence of the vector potential is zero, i.e.,

$$\nabla \cdot \mathbf{A} = 0. \tag{1.30}$$

Using the Coulomb Gauge, the equation for the scalar potential reduces to

$$\nabla^2 \phi = -\frac{\rho}{\epsilon}. \tag{1.31}$$

For static charges, i.e., when all time derivatives of ρ are zero, Eqs. 1.31 has the solution

$$\phi(\mathbf{r}) = \frac{1}{4\pi\epsilon} \int \frac{\rho(\mathbf{r})}{|\mathbf{r}-\mathbf{r}'|} dv', \tag{1.32}$$

where the unprimed vector, $\mathbf{r}$, represents the three-dimensional position vector of the point of interest, and the primed vector, $\mathbf{r}'$, represents the three-dimensional position vector of a source point and the integration is over all space containing source points. We see that the scalar potential per unit volume is just the electric Coulomb potential of the particular charge distribution $\rho(\mathbf{r}')$, hence the origin of the name "Coulomb gauge."

A second condition known as the *Lorentz Gauge* [1] consists of the restriction that

$$\nabla \cdot \mathbf{A} + \mu\epsilon \frac{\partial \phi}{\partial t} = 0 \tag{1.33}$$

under which the equations 1.28 and 1.29 for the potentials reduce to

$$\nabla^2 \mathbf{A} - \mu\epsilon \frac{\partial^2 \mathbf{A}}{\partial t^2} = -\mu \mathbf{J}, \tag{1.34}$$

and

$$\nabla^2 \phi - \mu\epsilon \frac{\partial^2 \phi}{\partial t^2} = -\frac{\rho}{\epsilon}. \tag{1.35}$$

Solving for the vector potential of a given ("static") current density, $\mathbf{J}$, with primed source coordinates, yields the relation

$$\mathbf{A}(\mathbf{r}) = \frac{\mu}{4\pi} \int \frac{\mathbf{J}(\mathbf{r}')}{|\mathbf{r}-\mathbf{r}'|} dv'. \tag{1.36}$$

The method for obtaining the solutions are described in detail in [2]. It should also be noted here that the solution for the scalar potential, Eq. 1.32, is the same both for the Coulomb gauge and the Lorentz gauge.

For moving, i.e., time varying charges or currents, the scalar and vector potential solutions are *retarded* or *delayed* with respect to the changes in the sources, as shown by [3], and [2], and are given by

$$\phi(\mathbf{r},t) = \frac{1}{4\pi\epsilon} \int \frac{\rho(\mathbf{r}, t-(R/c))}{R} dv', \tag{1.37}$$

and

$$\mathbf{A}(\mathbf{r},t) = \frac{\mu}{4\pi} \int \frac{\mathbf{J}(\mathbf{r}', t-(R/c))}{R} dv'. \tag{1.38}$$

where $R = |\mathbf{r}-\mathbf{r}'|$.

For known distribution of static or moving charges, the scalar and vector potentials $\mathbf{A}$ and ϕ provide a convenient approach for determining electromagnetic fields.

1.6   Hertz Vector Potential

The four components of the electric potentials (the three dimensions of $\mathbf{A}$, and the scalar value of ϕ) do simplify problems in electrostatic and electrodynamics in homogeneous and isotropic media but even further simplification is possible through the use of a single vector known as the Hertz vector $\boldsymbol{\Pi}$, also known as the polarization vector. Similar to using the traditional vector electric potential, $\mathbf{A}$, and the scalar electric potential, ϕ, for solving problems in electrodynamics, the Hertz vector provides an alternative method for solving problems in electrodynamics using a single vector potential.

The method used to determine the Hertz vector will be described in Chapters 7, 8, and 10 for a variety of cases. Once the Hertz vector is known, the scalar and vector electric potentials can be determined from the relations

$$\mathbf{A} = \mu\epsilon \frac{\partial \boldsymbol{\Pi}}{\partial t} \quad \text{and} \quad \phi = -\boldsymbol{\nabla} \cdot \boldsymbol{\Pi}. \tag{1.39}$$

Due to the fact that the Hertz vector is already a solution to the wave equation, the potentials calculated by Eq. 1.39 automatically satisfy the Lorentz condition, Eq. 1.33.

The electric field and the magnetic induction can be determined from the Hertz vector through

$$\mathbf{E} = \boldsymbol{\nabla} \left(\boldsymbol{\nabla} \cdot \boldsymbol{\Pi} \right) - \mu\epsilon \frac{\partial^2 \boldsymbol{\Pi}}{\partial t^2}, \tag{1.40}$$

and

$$\mathbf{B} = \mu\epsilon \boldsymbol{\nabla} \times \frac{\partial \boldsymbol{\Pi}}{\partial t}. \tag{1.41}$$

Equations 1.40 and 1.41 are general and independent of which complex notation is used. (See Appendix A for a discussion of "engineering" versus "physicist" complex notation.) Using physicist complex notation (where the time dependence of the fields follows $e^{-i\omega t}$) and carrying out the time derivative of the Hertz vector, the electric and magnetic fields can be written as

$$\mathbf{E} = k^2 \boldsymbol{\Pi} + \boldsymbol{\nabla} \left(\boldsymbol{\nabla} \cdot \boldsymbol{\Pi} \right), \tag{1.42}$$

and

$$\mathbf{H} = -ik\sqrt{\frac{\epsilon}{\mu}} \boldsymbol{\nabla} \times \boldsymbol{\Pi}, \tag{1.43}$$

where k is known as the 'wave number' and can be defined by

$$k = \omega\sqrt{\epsilon\mu}. \tag{1.44}$$

The wave number is related to the angular frequency and propagation speed of the wave, v,

$$v = \frac{1}{\sqrt{\epsilon\mu}}, \tag{1.45}$$

by

$$k = \frac{\omega}{v}. \tag{1.46}$$

Using engineering complex notation (where the time dependence of the fields follows $e^{+i\omega t}$) the electric and magnetic fields are

$$\mathbf{E} = k^2 \mathbf{\Pi} + \mathbf{\nabla}\left(\mathbf{\nabla} \cdot \mathbf{\Pi}\right), \tag{1.47}$$

and

$$\mathbf{H} = ik\sqrt{\frac{\epsilon}{\mu}}\mathbf{\nabla} \times \mathbf{\Pi}. \tag{1.48}$$

The sign difference in the magnetic field expressions obtained using the two different complex notations is noteworthy.

The use of the Hertz vector potentials is very convenient in solving many problems that arise in electromagnetism, such as in propagation of radiation through a waveguide or from apertures. The generation of harmonic (or single frequency) radiation from periodic motion of charges as an inevitable consequence of the validity of the Maxwell's equations can be demonstrated using the vector potential formalism [3, 4, 5] as well as directly from the vector and scalar potential [2, 6, 7]. We will follow the treatment of Wangsness in the next section to describe the generation of radiation and will return to the use of Hertz vector potentials in a later chapters (Chapters 7 and 8) when we describe propagation of light transmitted through apertures of various shapes.

1.7   Radiation from an Orbiting Charge

Refs. [3] and [2] describe the generation of electromagnetic radiation from an oscillating linear dipole in which the distance d between the two equal and opposite charges is changing sinusoidally with time t as

$$d = d_0 e^{-i\omega t}. \tag{1.49}$$

It is implied that the real part of equation 1.49 is to be taken to calculate physical quantities like distance. The dipole axis is assumed to be oriented along one of the Cartesian axes (say the z axis) and the electric and magnetic fields at a distance R from the dipole center is found by evaluating the components of the vector potential, which is found by evaluating the differentials using the spherical coordinate system. Using this formalism, the case of an electric charge uniformly orbiting about a fixed charge of equal magnitude and opposite sign can of course be resolved into the motion of two perpendicular dipoles linearly oscillating 90 degrees out of phase with each other.

Another method, which we use here, is to treat such a circular motion explicitly and derive the field distributions generated by the orbiting charge

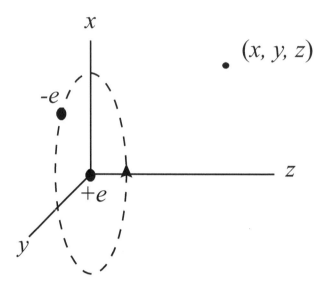

FIGURE 1.1
An electron orbiting in the x-y plane around a nucleus of charge $+e$ at the origin, and the observation point (x, y, z).

at a distance. Such a description of course pertains to the classical picture of an atom with an electron (charge $-e$) orbiting around a screened nucleus, having a charge e. Instead of using the spherical coordinate system we will use the Cartesian coordinates, with the benefit that the differential relations developed here will be useful later when describing propagation of light through apertures.

As shown in Figure 1.1, say an electron is orbiting in the x-y plane around a nucleus of charge e fixed at the origin of the coordinate system. Our goal is to find the field distribution at a point (x, y, z) located on a parallel plane at a distance z. We assume that other than the charges, the entire space under consideration is vacuum.

For the charge harmonically oscillating as given by Eqs. 1.49, Ref. [2] expands Eqs. 1.38 in terms of multipoles and shows that at a large distance from the charge, the electric dipole part of the vector potential **A** (designated $\mathbf{A}_{ed}$) is given by

$$\mathbf{A}_{ed} = -\frac{i\mu_0\omega\mathbf{P}_0}{4\pi r}e^{i(kr-\omega t)}, \tag{1.50}$$

where k is defined by the equation

$$k \equiv \omega\sqrt{\mu_0\epsilon_0} \tag{1.51}$$

and r, the distance from the origin to the observation point, is assumed to be much bigger than d_0. For the orbiting electron shown in Figure 1.1, the instantaneous position is given by

$$\mathbf{d} = \mathbf{d}_0(\cos\theta\hat{x} + \sin\theta\hat{y}) \tag{1.52}$$

where θ denotes the azimuthal angle. For the electron orbiting with a constant angular velocity ω, the angle $\theta = \omega t$ so that

$$\mathbf{d} = \mathbf{d}_0 \left(\frac{e^{i\omega t} + e^{-i\omega t}}{2}\hat{x} + \frac{e^{i\omega t} - e^{-i\omega t}}{2i}\hat{y} \right) \tag{1.53}$$

$$= \frac{\mathbf{d}_0}{2} \left(e^{i\omega t}(\hat{x} - i\hat{y}) + e^{-i\omega t}(\hat{x} + i\hat{y}) \right). \tag{1.54}$$

Following the logic of retarded potential that led to Eqs. 27.41 in Wangsness, we find that Eqs. 1.50 needs to be replaced by

$$\mathbf{A}_{ed} = -\frac{i\mu_0\omega\mathbf{p}_0}{8\pi r}e^{i(kr-\omega t)}(\hat{x} - i\hat{y}) + \text{cc}, \tag{1.55}$$

where cc stands for the complex conjugate. In Eqs. 1.55, the vector potential is explicitly real, instead of being only implicitly real as in 1.50. For convenience of calculations below, let's define

$$\mathbf{A}_{1(ed)} = -\frac{i\mu_0\omega\mathbf{p}_0}{8\pi r}e^{i(kr-\omega t)}(\hat{x} - i\hat{y}) \tag{1.56}$$

so that

$$\mathbf{A}_{ed} = \mathbf{A}_{1(ed)} + \text{cc}. \tag{1.57}$$

The magnetic induction $\mathbf{B}$ can be evaluated from the vector potential $\mathbf{A}$ using Eqs. 1.26. The electric field can then be evaluated from the magnetic field using Eqs. 1.16, noting that in free space the current density vector $\mathbf{J} = 0$, and that in vacuum, $\mathbf{D} = \epsilon_0 \mathbf{E}$ and $\mathbf{B} = \mu_0 \mathbf{H}$. Further, for fields and charges oscillating harmonically with angular frequency ω, the time derivative in the Ampere-Maxwell equation 1.16 can be replaced by $-i\omega$, so that

$$\mathbf{E} = \frac{i}{\mu_0\epsilon_0\omega}\nabla \times \mathbf{B}. \tag{1.58}$$

Once the magnetic induction $\mathbf{B}$ is known (from Eqs. 1.26), the electric field can be obtained by using Eqs. 1.58.

From Eqs. 1.56 and Eqs. 1.26 we obtain

$$\mathbf{B} = -\frac{i\mu_0\omega p_0}{8\pi} \begin{vmatrix} \hat{x} & \hat{y} & \hat{z} \\ \dfrac{\partial}{\partial x} & \dfrac{\partial}{\partial y} & \dfrac{\partial}{\partial z} \\ \dfrac{e^{ikr}}{r} & -i\dfrac{e^{ikr}}{r} & 0 \end{vmatrix} e^{-i\omega t} + \text{cc}. \tag{1.59}$$

Taking the observation point to be at (x, y, z) on the plane at a distance z from the plane of oscillation of the electron, as shown in Figure 1.1, we have

$$r = \sqrt{x^2 + y^2 + z^2}. \tag{1.60}$$

It can be easily shown that

$$\frac{\partial}{\partial x}\left(\frac{e^{ikr}}{r}\right) = xg(r), \quad \frac{\partial}{\partial y}\left(\frac{e^{ikr}}{r}\right) = yg(r), \quad \frac{\partial}{\partial z}\left(\frac{e^{ikr}}{r}\right) = zg(r) \tag{1.61}$$

where

$$g(r) = e^{ikr}\left(\frac{ik}{r^2} - \frac{1}{r^3}\right). \tag{1.62}$$

Using Eqs. 1.61 **B** from Eqs. 1.59, the following can be found

$$\mathbf{B} = -\frac{i\mu_0\omega p_0}{8\pi}g(r)[iz\hat{x} + z\hat{y} + (-ix - y)\hat{z}]e^{-i\omega t} + \text{cc}, \tag{1.63}$$

and using Eqs. 1.58, the electric field **E** is obtained to be

$$\mathbf{E} = \frac{p_0}{8\pi\epsilon_0}e^{-i\omega t}\left[-\hat{x}\{2g(r) + (ix + y)\frac{\partial g(r)}{\partial y} + z\frac{\partial g(r)}{\partial z}\}\right. \tag{1.64}$$

$$+\hat{y}\{2ig(r) + (ix + y)\frac{\partial g(r)}{\partial x} + iz\frac{\partial g(r)}{\partial z}\} \tag{1.65}$$

$$\left.+\hat{z}\{z\frac{\partial g(r)}{\partial x} - iz\frac{\partial g(r)}{\partial y}\}\right]. \tag{1.66}$$

Noting that

$$\frac{\partial g(r)}{\partial x} = \frac{x}{r}\frac{\partial g(r)}{\partial r}, \quad \frac{\partial g(r)}{\partial y} = \frac{y}{r}\frac{\partial g(r)}{\partial r}, \quad \frac{\partial g(r)}{\partial z} = \frac{z}{r}\frac{\partial g(r)}{\partial r} \tag{1.67}$$

and that

$$\frac{\partial g(r)}{\partial x} = e^{ikr}\left(-\frac{k^2}{r^2} - \frac{3ik}{r^3} + \frac{3}{r^4}\right) \tag{1.68}$$

and assuming that the transverse distances x and y are small compared to the longitudinal distance z and further that $z \approx r \gg \lambda$, the only terms in **E** and **B** that are non-negligible are those proportional to $1/r$, so that

$$\mathbf{B} = \frac{\mu_0\omega p_0 k}{8\pi}(i\hat{x} + \hat{y})\frac{e^{i(kr-\omega t)}}{r} + \text{cc} \tag{1.69}$$

and

$$\mathbf{E} = \frac{\mu_0\omega^2 p_0}{8\pi}(\hat{x} - i\hat{y})\frac{e^{i(kr-\omega t)}}{r} + \text{cc}. \tag{1.70}$$

The $e^{i(kr-\omega t)}$ term in the solutions for both the **E** and the **B** vectors show that the field distributions at a large distance from the dipole ($r \gg \lambda, d_0$) do not begin or end at the charges they originate from. Instead they both are periodically oscillating with distance r and time t. Since a wave traveling with uniform speed v can be defined to be a function (say f) of position (r) and time (t) such that $f(r,t) = f(r - vt)$, i.e., in the function f, r and t occur in the combination $r - vt$, we see that $\mathbf{E}$ and $\mathbf{B}$ are both waves, traveling with a speed $v = \omega/k$. This wave of electric and magnetic field was designated electro-magnetic radiation, which can propagate in space without any intervening material medium.

Recalling that $k \equiv \omega\sqrt{\mu_0\epsilon_0}$, the speed v with which this wave propagates through space is:

$$v = \frac{\omega}{k} = \frac{1}{\sqrt{\mu_0\epsilon_0}}. \tag{1.71}$$

Since the values of μ_0 and ϵ_0 are known from experiments (through Ampere's Law and Coulomb's law) to be $4\pi \times 10^{-7}$ kg/m/Coulomb^2 and 8.85×10^{-12} Coulomb^{2}s^2/(kg m^3) respectively, the value of v is calculated to be 3×10^8 m/s. This very closely matched the measured speed of light, and that fact led to the understanding that light is an *electro-magnetic radiation*. The speed of propagation of light (and all forms of electro-magnetic radiation) in vacuum is one of the most important fundamental constants, and instead of the symbol v, the symbol c, possibly from the Latin word *celeritas* for speed, is reserved for it.

Eqs. 1.69 and 1.70 show that the electromagnetic waves repeat themselves in every $2\pi/k$ units of distance. This distance is designated the *wavelength* and denoted by λ, or

$$\lambda = \frac{2\pi}{k} = \frac{2\pi}{\omega\sqrt{\mu_0\epsilon_0}} = \frac{c}{\nu}, \tag{1.72}$$

where $\nu = \omega/2\pi$ is as the frequency of oscillation of the dipole giving rise to the propagating wave. Both $\mathbf{B}$ and $\mathbf{E}$ vectors are transverse i.e., they lie in the $x - y$ plane perpendicular to the propagation direction z and oscillate 90 degrees out of phase with each other. The inverse dependence of the fields with r show that they are in the form of an expanding spherical wave.

1.8   Poynting Vector

For oscillating electric and magnetic fields $\mathbf{E}$ and $\mathbf{H}$, the rate of flow of electromagnetic energy across a closed surface S is given by

$$\oint_S (\mathbf{E} \times \mathbf{H}) \cdot d\mathbf{a} \tag{1.73}$$

where $d\mathbf{a}$ denotes a surface element on S [2]. The integrand is interpreted as the rate of flow of electromagnetic energy per unit area, i,e., the power density, and is called the *Poynting vector* $\mathbf{S}$ defined as

$$\mathbf{S} = \mathbf{E} \times \mathbf{H}. \tag{1.74}$$

with the direction of $\mathbf{S}$ the same as that of the instantaneous flow of energy.

The magnitude (or the time averaged amplitude) of the Poynting vector is the power density of light. In this book we will assume that the terms *irradiance* or *intensity* are synonymous and that both terms mean the power density of light.

1.9   Radiation from a Classical Atom

Using the expressions for E and B from Eqs. 1.70 and 1.69 in 1.74 we can easily find that the time average energy flow rate per unit area $<\mathbf{S}>$ for a rotating charge is given by

$$< \mathbf{S} > = \frac{\mu_0 \omega^4 p_0^2}{32 \pi^2 c r^2} \hat{z}, \tag{1.75}$$

which (apart from an angular factor) is the same expression as that for a linearly oscillating dipole, for example in Ref. [2].

In Bohr's classical model of an atom, one or more electrons orbit around a nucleus in an elliptical orbit under a Coulomb force, much as the planets orbit around the sun through gravitational interaction. Assuming that an electron having mass m and charge e is orbiting in a circular orbit of radius d, the angular frequency ω can be determined using Newton's laws of motion:

$$|\mathbf{F}_{\text{Coulomb}}| = |\mathbf{F}_{\text{centripetal}}| \tag{1.76}$$

or

$$\frac{e^2}{4 \pi \epsilon_0 d^2} = m \omega^2 d \tag{1.77}$$

i.e.,

$$\omega = \frac{e}{\sqrt{4 \pi \epsilon_0 m d^3}}. \tag{1.78}$$

Assuming a typical size of an atom ranging from, say, 50 pm to 500 pm, the frequency ω ranges from about 1.5×10^{15} rad/sec to about 1.5×10^{16} rad/sec, corresponding to wavelength range from about 0.10 μm to about 1.3 μm, i.e., spanning the X-ray to infrared spectral range. But if an orbiting electron emitted radiation as predicted by electromagnetic theory, it would lose energy at the rate given by Eqs. 1.75. For an atomic radius d the rotational kinetic energy is

$$\mathcal{E} = \frac{1}{2}m\omega^2 d^2, \tag{1.79}$$

whereas the amount of energy radiated per unit time through a surface at a distance r, obtained from Eqs. 1.75 is

$$\mathcal{P} = 4\pi r^2 < S > = \frac{\mu_0 \omega^4 e^2 d^2}{2\pi c}. \tag{1.80}$$

The amount of time taken to radiate out the entire rotational kinetic energy of the electron is therefore $\mathcal{E}/\mathcal{P}$, which is

$$\tau \equiv \frac{\mathcal{E}}{\mathcal{P}} = \frac{m\pi c}{\mu_0 \omega^2 e^2}. \tag{1.81}$$

For ω ranging from 1.5×10^{15} rad/sec to 1.5×10^{16} rad/sec, τ, the lifetime of the atom ranges from about 120 ps to about 12 ns, i.e., such atoms would not be very long lived. That is obviously not the case because we are all made of stable atoms with much longer lifetimes.

Another problem with the classical model of the atom is that as the electron is losing its energy through radiation, it would spiral in closer to the nucleus with its oscillation frequency changing continuously. It is therefore expected to emit a continuum of wavelengths. However, light emitted by atoms were found to consist of a multitude of sharp and discrete single frequencies. These contradictions are of course resolved by the fact that the laws of classical mechanics are invalid when applied to an electron orbiting around a nucleus at sub-nanometer distances.

1.10 A Quantum Mechanical Interlude

Classical electromagnetic theory well describes most of the observed phenomena related to light *propagation*, but as we just saw, it fails to explain the origin of light from atoms. Since motion of electrons, governed by the laws of quantum mechanics, give rise to atomic radiation, it is necessary to invoke principles of quantum mechanics to understand how atoms can emit light and still be stable. In this section, we provide descriptions of a few phenomena and quantum mechanical concepts pertinent to the generation of light by atoms.

1.10.1 Blackbody Radiation

Even before the proposal of atomic theory, phenomena were observed that could not be explained by classical electrodynamics. One such phenomenon

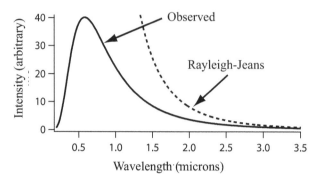

FIGURE 1.2

Intensity spectrum of a blackbody with a temperature of 5000 Kelvin compared to the emissions spectrum according to Rayleigh-Jeans' pure electromagnetic wave theory.

was blackbody radiation, a continuum of wavelengths of light emitted by all bodies, which had a spectral distribution that depended on the temperature of the body. During the end of the 1800s, and the beginning of the 1900s, many leading scientists studied and tried to model the radiation and absorption of light by thermal sources.

Among the available wave theory models for the production of light at the time, Rayleigh-Jeans theory had the best fit to the observed spectral emissions of blackbody source. This model used existing electromagnetic theory and was based upon the number of possible standing electromagnetic wave modes per unit volume within the object and attributed an average energy per mode. However, Rayleigh-Jeans theory only correctly fits observed blackbody spectra for long wavelengths. As observed in Figure 1.2, it does not fit very well for shorter wavelengths or predict the observed shape of the spectral distribution of a blackbody emitter. Additionally, the area under the curve is infinite (indicating an infinite amount of emitted energy) because of the asymptotic behavior for shorter wavelengths, resulting in what came to be known as the "ultraviolet catastrophe."

1.10.2   Planck's Theory of Light Quanta

At the turn of the century, Max Planck approached the problem from a different point of view. Instead of trying to fit current wave theory to the data, he decided to fit a mathematical function to the data and use the mathematical fit as a foundation to develop a new theory. Planck's idea was that all light frequencies are harmonics of some fundamental frequency or "quanta." Each of these harmonic frequencies has an associated energy (E) that is proportional to the frequency (ν)

$$E = h\nu. \tag{1.82}$$

The proportionality constant, h, is known as Planck's constant. The value of Planck's constant is

$$h = 6.626 \times 10^{-34} \text{ J-s.} \tag{1.83}$$

Planck's work evolved the understanding of light from being classical waves having arbitrary continuum of frequency values to waves having a set of quantized frequencies, but light was still considered to be purely a wave.

1.10.3 Photoelectric Effect

Another phenomenon that electromagnetic wave theory could not explain is the photoelectric effect, first discovered by Heinrich Hertz in 1887. In this effect, light incident upon a metal gives rise to electrons with kinetic energies within the metal. According to wave theory, the kinetic energy gained by the electrons would be a function of the intensity (irradiance) of the light. Electrons that are bound by a certain energy should be able to be freed and subsequently given kinetic energies by simply turning up the light irradiance to increase the amount of energy transferred to them. Thus, according to the wave theory of light, the kinetic energy gained by the electrons should be a function of the incident light irradiance.

Experimentally, however, it was observed that the electrons' kinetic energies were not a function of the irradiance, but rather a linear function of the *frequency* of the light as illustrated in Figure 1.3. In addition, a threshold frequency was observed where for light frequencies below the threshold no electron kinetic energies are observed even when the light irradiance is increased.

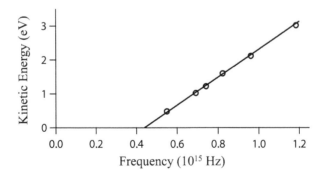

FIGURE 1.3
Demonstration of photoelectric effect in sodium by Millikan. (Reprinted figure with permission from R. A. Millikan, Phys. Rev., 7, 355–388, 1916. Copyright 1916 by the American Physical Society.)

1.10.4 Einstein's Theory of Photons

In 1905 Albert Einstein presented a revolutionary new theory that explained the photoelectric effect. The idea was that light is composed of quantized packets of energy that behave like particles (which are called photons). The energy of each photon scales linearly with frequency via Planck's relationship, Eq. 1.82. These quantized particles can transfer their momentum and kinetic energy to other particles through collisions. The particle-like collisions between the photons and the electrons in a metal explained the observations of photoelectric effect. Incident light of a particular frequency is a stream of particles, all of which have the same energy, and it is a collision between the photons and the electrons that can give rise to the kinetic energy of the electron. The energy threshold observed in the electron kinetic energy spectrum is the "work function," or the amount of energy required to free the electron. The difference between the photon's energy and the work function gives the kinetic energy of the electron, or

$$KE = h\nu - \phi \tag{1.84}$$

where ϕ is the work function of the material. The lack of kinetic electrons when light frequency was below a threshold value can be explained by Eq. 1.84. Since kinetic energy cannot be negative, ν must exceed ϕ/h for electrons to be emitted. Also, the kinetic energy of the electron is only a function of the energy of each of the incident particles (i.e., frequency of incident light) and independent of how many particles are in the beam of light (i.e., its irradiance).

The result of Einstein's work on the photoelectric effect is that light is *both* a particle *and* a wave, simultaneously. The fact that an object can be both a particle and a wave led to the term "particle-wave duality." Einstein's Nobel prize winning work on the photoelectric was a pivotal point in physics because it effectively merged the two previously mutually exclusive fields of particle physics and wave physics.

The particle nature of light was further supported by experiments conducted by Arthur H. Compton in 1923 on scattering of X-rays by a target, by Raman and Krishnan in 1928 on scattering of visible light by liquids and gases, and by Mandelstam and Landsberg, also in 1928, on scattering of light by crystals. (These discoveries in 1928 are collectively known as Raman scattering). Compton's experiments directly measured momentum transfer and demonstrated conservation of energy and momentum in a collision between X-ray photons and electrons. In Raman scattering, incident light either gives up some of its energy to excite vibrational states of molecules, so that the scattered light loses energy, which manifests itself as lower frequency of the scattered beam, or the scattered light gains energy from the vibrational states and then has higher frequency than the incident light.

1.10.5   Wave Particle Duality of Matter

In 1924 Louis De Broglie extended Einstein's particle-wave duality of photons to *all* matter. He postulated that if any particle or object has momentum then it simultaneously has an associated wavelength, λ, of

$$\lambda = \frac{h}{p} \tag{1.85}$$

where h is Planck's constant and p is the object's momentum. Soon after, also in 1924, Davisson and Germer published work on diffraction patterns (a key property of wave behavior) for high energy electrons scattering off of crystals of nickel, demonstrating particle-wave duality for electrons. Several experiments following these also demonstrated particle-wave duality for neutrons, hydrogen atoms, and helium atoms.

Thus in the domain of quantum mechanics, all objects have the properties of waves. One of these properties is the inability to say *exactly* where the object is located. Waves occupy a *region of space* and the edges of the object are impossible to clearly identify, whereas particles occupy *specific points* in space with clearly identifiable edges.

1.10.6   The Particle-Function of Classical Mechanics

According to classical mechanics, objects behave as particles where the position of the object can be known precisely, objects have well-defined edges, and objects collide and exchange momentum when they come into contact with each other. First, let us examine something that is typically the final result of a kinematics problem, the equations of motion, or

$$x(t),\ y(t),\ \text{and}\ z(t), \tag{1.86}$$

where these represent the time-dependent positions of the particle for each of the three Cartesian axes.

For example, a particle under the influence of only gravity would have equations of motion with the general form of:

$$
\begin{aligned}
x(t) &= x_o + v_{ox}t \\
y(t) &= y_o + v_{oy}t \\
z(t) &= z_o + v_{oz}t - \frac{1}{2}gt^2.
\end{aligned}
\tag{1.87}
$$

These equations, put together as a vector, can be written as

$$\mathbf{r}(x,y,z,t) = x(t)\,\hat{i} + y(t)\,\hat{j} + z(t)\,\hat{k}. \tag{1.88}$$

and this position and time dependent function of the particle is called the **classical particle-function**, which can serve as the source for desired kinematic information concerning the object. Table 1.1 illustrates some examples

Desired information	Mathematical operation on the particle-function, $\mathbf{r}$
Position	No operation
Velocity	$\dfrac{\partial \mathbf{r}}{\partial t}$
Momentum	$m\dfrac{\partial \mathbf{r}}{\partial t}$
Kinetic energy	$\dfrac{1}{2}m\left\lvert\dfrac{\partial \mathbf{r}}{\partial t}\right\rvert^{2}$
Acceleration	$\dfrac{\partial^{2}\mathbf{r}}{\partial t^{2}}$
Net force	$m\dfrac{\partial^{2}\mathbf{r}}{\partial t^{2}}$

TABLE 1.1
Some examples of extracting desired information from the particle-function using classical mechanics.

of using mathematical operations to extract desired information about the object. The idea of using mathematical operations on a function as a tool to extract information about an object is also used in the next section when we discuss an equivalent **quantum wave-function**.

1.10.7   The Wavefunction of Quantum Mechanics

A wavefunction is to a quantum mechanical object what our previously discussed particle-function is to a classical object. It is a mathematical function that contains all information about the object of interest. In order to extract information about the object, one must perform mathematical operations on the object's wavefunction.

Just like the particle-function of a classical object, the wavefunction of a quantum object is dependent upon all available dimensions of space and dependent upon time. Both the particle-function and the wavefunction are obtained by solving fundamental equations for either classical mechanics or quantum mechanics. For classical mechanics, the equations to solve for the particle-function are Newton's laws. For quantum mechanics, the equation to solve is Schrödinger's equation. Table 1.2 summarizes these similarities along with many of the differences between particle-functions and wavefunctions of objects.

Similar to the particle-function for classical objects, the wavefunction of a quantum object serves as the source of all possible information concerning the object. Classical mechanics is the study of classical particles using a toolbox of

	Particle-function	Wavefunction
Similarities	function of position and	function of position and time
	time obtained by solving Newton's equations	obtained by solving Schrödinger's equation
	used to extract information from classical objects	used to extract information from quantum objects
Differences	represents a particle	represents a wave
	vector	scalar
	real	complex
	cannot be written as a separation of variables	can be written as a separation of variables
	function itself gives a precise position	modulus square gives a *probability* density

TABLE 1.2
Some examples of the similarities and differences between a particle-function and a wavefunction.

mathematical methods. Quantum mechanics is the study of quantum objects using a different toolbox of mathematical methods. As described previously, one can perform mathematical operations on the particle-function in order to extract specific desired information about the object. For quantum objects one can also perform mathematical operations on the wavefunction in order to extract desired information. If one wants to know where a classical particle is located, one does not need to perform any mathematical operations on the object's particle function, as illustrated in Table 1.1, because the particle-function itself yields the vector position of the object. If one wants to know where a quantum particle is located, a precise location is not possible due to the wave nature of the object. However, extracting a *probability* of finding the object at a particular location is possible. Unlike classical mechanics and the particle-function where no mathematical operations are required to find the object, a mathematical operation must be performed on a quantum wavefunction in order to extract location information. The modulus square of the object's wavefunction reveals information concerning the location of the object, actually a probability distribution of finding the object as a function of position. In many cases, extracting information about quantum objects in-

Desired information	Mathematical operation on the wavefunction, ψ
exact position	not possible
probability density	$\lvert\psi\rvert^2$, or $\psi^*\psi$
probability of observing the object within a given volume	$\int \psi^*\psi dv$
average position in x	$\int \psi^* x\psi dx$
momentum in x	$\int \psi^* \left(\dfrac{\hbar}{i}\dfrac{d}{dx}\right)\psi dx$
kinetic energy	$\int \psi^* \left(-\dfrac{\hbar^2}{2m}\nabla^2\right)\psi dv$

TABLE 1.3
Some examples of extracting desired information from the wavefunction using quantum mechanics.

volves mathematical operations that have the form of operators. Table 1.3 is a brief summary of just a few of the different ways of extracting information from quantum objects using the object's wavefunction, ψ.

1.10.8 The Schrödinger Equation

The space-time dynamics of the wavefunction of an object is governed by the Schrödinger equation. The time-independent and the time-dependent forms of this equation are

$$\left[-\frac{\hbar^2}{2m}\nabla^2 + V(\mathbf{r})\right]\psi(\mathbf{r}) = E\psi(\mathbf{r}), \tag{1.89}$$

and

$$\left[-\frac{\hbar^2}{2m}\nabla^2 + V(\mathbf{r})\right]\Psi(\mathbf{r},\,t) = i\hbar\frac{\partial}{\partial t}\Psi(\mathbf{r},\,t), \tag{1.90}$$

where ψ and Ψ denote the time-independent and time-dependent wavefunctions, respectively. The other parameters of Schrödinger's equations are $\hbar$, m, V, and E, where $\hbar$ is a mathematically convenient function of Planck's constant,

$$\hbar = \frac{h}{2\pi}, \tag{1.91}$$

m is the mass of the object, $V(\mathbf{r})$ is the position-dependent potential energy of the object, and E is the total energy of the particle. One of the most common uses of Schrödinger's equations is to determine the wavefunction of the object. As discussed previously, once the wavefunction for an object is determined, *anything* can be determined about the object in that quantum state.

Finding analytical solutions to these multi-dimensional differential equations can be quite an overwhelming task. Fortunately, we can make an assumption (which after testing the results turns out to be valid) that the wavefunction for quantum objects can be expressed using **separation of variables**, i.e., the complete function is a product of functions that are each only dependent upon one of the variables. In spherical coordinates and separation of variables Ψ and ψ can be written as

$$\Psi(r, \theta, \phi, t) = \psi(r, \theta, \phi) f(t) = R(r)\,\Theta(\theta)\,\Phi(\phi)\, f(t). \qquad (1.92)$$

For atomic states of bound electrons the spatial variables of the wavefunction are frequently represented by the functions of θ and ϕ combined into one function, $Y(\theta, \phi)$, where solutions are known as *spherical harmonics*, or

$$\psi(r, \theta, \phi) = R(r)\, Y(\theta, \phi). \qquad (1.93)$$

1.10.9 Wavefunctions of Electrons in a Stable Atom

For a quantum mechanical object in a particular state, Ψ, one 'observable' is the probability of measuring the particle to be in that state, which is the modulus square of the wavefunction. If the probability of finding an electron anywhere in one particular atomic state does not change over time then we will define that atomic state to be stable. In addition, we can also limit the definition of a stable state to be only states where the total energy of the state does not decay over time.

Due to the fact that quantum objects have all of the properties of waves, we can also include wave mechanics when forming our definition of atomic stability. Waves bound to a particular region can be said to be stable when the wave is a standing wave with a particular frequency. Using these definitions of stability, it can be found from Schrödinger's equation that particular atomic states exist where the time dependent function of the separation of variables, $f(t)$ is

$$f(t) = e^{-i\omega t}, \qquad (1.94)$$

and ω represents the frequency of the standing wave.

The time-dependent wavefunction for stable states of an electron bound to a nucleus using spherical coordinates now has the form of

$$\Psi(r, \theta, \phi, t) = R(r)\, Y(\theta, \phi)\, e^{-i\omega t}. \qquad (1.95)$$

For a hydrogen-like ion, i.e., an atom in which an electron is bound to a nucleus of charge Z, the potential energy term would have the form

$$V(r, \theta, \phi) = \frac{1}{4\pi\epsilon_o} \frac{q_e q_n}{r}, \tag{1.96}$$

in spherical coordinates, where q_e is the electron charge, or just "e," and q_n is the charge of the nucleus, which can be represented as Ze, or

$$V(r, \theta, \phi) = \frac{1}{4\pi\epsilon_o} \frac{Ze^2}{r}. \tag{1.97}$$

Solutions of the partial differential equation of the Schrödinger equation involve three different integers: n, ℓ, and m. Incorporating these integers into their associated separation of variable functions of the general solutions to Schrödinger's equation yields

$$\Psi(\mathbf{r}, t) = R_{n,\ell}(r) Y_{\ell,m}(\theta, \phi) e^{-i\omega t} \tag{1.98}$$

where the position vector $\mathbf{r}$ is expressed in terms of the spherical coordinates (r, θ, ϕ), and the variables n, ℓ, and m can only have integer values. The integer set (n, ℓ, m) are known as the **quantum numbers**, which define a particular stable atomic wavefunction. The functional dependence of the radial part of the wavefunction, $R_{n,\ell}$, and the spherical harmonics, $Y_{\ell,m}$, on each of their dependent spatial coordinates are tabulated in many textbooks for various values of n, ℓ, and m.

The energy of each solution of Schrödinger's equation for hydrogen-like ions is dependent only upon the principle quantum number, n, and can be found to be

$$E_n = -\frac{mk^2 e^4 Z^2}{2\hbar^2 n^2}, \tag{1.99}$$

or in a more convenient form of

$$E_n = -Z^2 \frac{E_R}{n^2}, \tag{1.100}$$

where E_R is known as the Rydberg energy, or

$$E_R = -\frac{mk^2 e^4}{2\hbar^2} = -13.6 \text{eV}. \tag{1.101}$$

Since n takes on only non-zero positive integer values, the energy of the electron in orbit around a nucleus is restricted to only certain discrete values.

1.10.10   Atomic Radiation

According to quantum mechanics, the measurable value of a physical property (say denoted by A) of an electron having a wave function ψ is given by its expectation value, defined as

$$\langle A \rangle = \int \psi^* A \psi \, dv \tag{1.102}$$

where ψ^* denotes the complex conjugate of ψ and dv denotes the differential volume element. The dipole moment of an electron at a distance r from a screened nucleus of charge e is then given by the expectation value

$$\langle \mathbf{p} \rangle = e \int \psi^* \mathbf{r} \psi dv \qquad (1.103)$$

where $\mathbf{r}$ denotes the position vector. It can be shown in general that for any state of the electron (say with quantum numbers n and ℓ), the expectation value $\langle \mathbf{p} \rangle$ is equal to 0.

One way to show this is to calculate the expectation value of the transition dipole moment of an electronic transition from state (n_1, ℓ_1) to (n_2, ℓ_2) as

$$< \mathbf{p}_{12} >= e \int \psi^*_{n_2, \ell_2} \mathbf{r} \psi_{n_1, \ell_1} dv. \qquad (1.104)$$

Since in spherical coordinates, the volume element $dv = r^2 \sin\theta dr d\theta d\phi$ and

$$\mathbf{r} = r \sin\theta \cos\phi \hat{x} + r \sin\theta \sin\phi \hat{y} + r \cos\theta \hat{z} \qquad (1.105)$$

Equation 1.104 reduces to

$$\langle \mathbf{p}_{12} \rangle = e \int R^*_{n_2, \ell_2}(r) R_{n_1, \ell_1}(r) r^3 dr \{ J_x(\ell_1, \ell_2)\hat{x} + J_y(\ell_1, \ell_2)\hat{y} + J_z(\ell_1, \ell_2)\hat{z} \} \qquad (1.106)$$

where

$$J_x(\ell_1, \ell_2) = \int Y^*_{\ell_2} Y_{\ell_1} \sin^2\theta \cos\phi d\theta d\phi \qquad (1.107)$$

$$J_y(\ell_1, \ell_2) = \int Y^*_{\ell_2} Y_{\ell_1} \sin^2\theta \sin\phi d\theta d\phi \qquad (1.108)$$

$$J_z(\ell_1, \ell_2) = \int Y^*_{\ell_2} Y_{\ell_1} \sin\theta \cos\phi d\theta d\phi. \qquad (1.109)$$

The mathematical proof is a little involved, but it can be rigorously shown that the terms $J_x(\ell_1, \ell_2)$, $J_y(\ell_1, \ell_2)$, and $J_z(\ell_1, \ell_2)$ are all equal to zero unless ℓ_1 and ℓ_2 differ by 1. So for Eqs. 1.103, for which $\ell_1 = \ell_2$, $\langle \mathbf{p} \rangle$ must be zero.

Thus quantum mechanics shows why electrons bound in atoms do not radiate light when they are in a stationary state and can therefore have long lifetimes. Physically, this is made plausible by the understanding that for an electron orbiting under Coulomb force of the nucleus in one direction (say clockwise) there is an equal probability of it orbiting in the opposite direction (anticlockwise). The electric fields produced by the two cases cancel each other so the net radiating electric field from the stationary state is zero.

However, the expectation value of the "transition" dipole moment between two states (with quantum numbers n_1 and n_2) is not zero and it oscillates with an angular frequency $\omega_{12} = 2\pi(E_{n_1} - E_{n_2})/h$. The electric and magnetic fields corresponding to the transition dipole moment should therefore be oscillating with angular frequency ω_{12} and give rise to radiation at the frequency $\omega_{12}/2\pi$, which explains the sharp lines with single frequencies observed in atomic emission [11].

Type	Physical Quantity	Name	Symbol	Unit	Dimension
Fundamental	Mass	kilogram	m	kg	M
	Length	meter	λ	m	L
	Time	second	t	s	T
	Charge	Coulomb	q	C	Q
Composite	Force	Newton	N	$\dfrac{\text{kgm}}{\text{s}^2}$	MLT^{-2}
	Energy	Joule	J	Nm	ML^2T^{-2}
	Current	Ampere	A	$\dfrac{\text{C}}{\text{s}}$	$T^{-1}Q$
	Electric Potential	Volt	V	$\dfrac{\text{J}}{\text{C}}$	$ML^2T^{-2}Q^{-1}$
	Magnetic Flux	Weber	Wb	$\dfrac{\text{J}}{\text{A}}$	$ML^2T^{-3}Q$
	Magnetic Field	Tesla	T	$\dfrac{\text{Wb}}{\text{m}^2}$	$MT^{-3}Q$
	Inductance	Henry	H	$\dfrac{\text{Wb}}{\text{A}}$	ML^2T^{-2}
	Capacitance	Farad	F	$\dfrac{\text{C}}{\text{V}}$	$M^{-1}L^{-2}T^2Q^2$

TABLE 1.4
Symbols, units, and dimensionality for the fundamental units and some composite units.

1.11   Units and Dimensions

Throughout this book we use SI system of units (mass in kilograms, length in meters, time in seconds, and charge in Coulombs). However, as convenient, the distance units of centimeters instead of meters will often be used, we hope with not much confusion. (For example, irradiance values are commonly expressed as W/cm^2 instead of W/m^2). Tables 1.4 and 1.5 are a summary of the fundamental and some composite units used in this book.

Category	Physical Quantity	Symbol	Unit	Dimension
Potentials	Scalar Electric Potential	ϕ	V	$ML^2T^{-2}Q^{-1}$
	Vector Electric Potential	$\mathbf{A}$	$\frac{Vs}{m}$	$MLT^{-1}Q^{-1}$
	Hertz Vector Potential	$\boldsymbol{\Pi}$	V m	$ML^3T^{-2}Q^{-1}$
Vectors and Fields	Current Density	$\mathbf{J}$	$\frac{A}{m^2}$	$L^{-2}T^{-1}Q$
	Electric Field	$\mathbf{E}$	$\frac{V}{m}$	$MLT^{-2}Q^{-1}$
	Electric Displacement Vector	$\mathbf{D}$	$\frac{C}{m}$	$L^{-1}Q$
	Magnetic Induction	$\mathbf{B}$	T	$MLT^{-2}Q^{-1}$
	Magnetic Field	$\mathbf{H}$	$\frac{A}{m}$	$L^{-1}T^{-1}Q$
	Poynting Vector	$\mathbf{S}$	$\frac{W}{m^2}$	MT^{-3}

TABLE 1.5
Symbols, units, and dimensionality for some potentials, fields, and vectors used in this book.

Bibliography

[1] R. Nevels, C.-S. Shin, "Lorenz, Lorentz, and the gauge," IEEE Antennas Prop. Mag. 43, 3, pp. 701, 2001.

[2] R.K. Wangsness, *Electromagnetic Fields*, John Wiley & Sons, New York, 1979.

[3] J.A. Stratton, *Electromagnetic Theory*, McGraw Hill Book Co., New York, 1941.

[4] W.K.H. Panofsky and M. Phillips, *Classical Electricity and Magnetism*, Addison-Wesley Publishing, Boston, 1964.

[5] A. Ishimaru, *Electromagnetic Wave Propagation, Radiation, and Scattering*, Prentice Hall, Upper Saddle River, New Jersey, 1991.

[6] J. Griffiths, *Introduction to electrodynamics*, Prentice Hall, Upper Saddle River, New Jersey, 1999.

[7] J.M. Stone, *Radiation and Optics*, McGraw-Hill Book Company, Inc., New York, 1963.

[8] E. Hecht, *Optics*, Third Edition, Addison Wesley, Reading, MA 1998.

[9] M. Born and W. Wolf, *Principles of Optics*, Seventh (Expanded) Edition, Cambridge University Press, Cambridge, UK 1999.

[10] R. A. Millikan, "A direct photoelectric determination of Plank's 'h' constant," Phys. Rev. 7, 355–388 (1916).

[11] M. Born, *Atomic Physics (8th.ed.)*, Blackie & Son Ltd., Glasgow, 1969

2

Electromagnetic Waves in Linear Media

> In this chapter we present mathematical models describing the vector electric and magnetic fields for light propagating within a linear medium. Mathematical models presented cover:
>
> > 1. General forms for Maxwell's equations and the wave equation
> >
> > 2. Maxwell's Equations for source-free media
> >
> > 3. Maxwell's equations for vacuum
> >
> > 4. Polarized light fields
>
> We then present and explain a few of the common forms of light that are solutions to these equations.

To simplify the mathematics for this chapter, we place the following restrictions on the propagation of the light through the medium:

- the volume through which the light is propagating is composed of only a single medium

- the medium's electric permittivity, ϵ, and magnetic permeability, μ, are both constants

- there are no boundary conditions for the volume, nor diffractive or spatially limiting elements within the volume

2.1 Maxwell's Equations in Linear Media

As discussed in Sections 1.2–1.4, when electromagnetic waves propagate through a medium other than a vacuum the electrodynamic properties of the medium must be taken into consideration. Here, we will assume that the medium can have any one, two, or all three of the following properties:

1. the electric field of the EM wave, $\widetilde{\mathbf{E}}$, induces a dipole moment per unit volume, $\widetilde{\mathbf{P}}$, within the medium,

2. the magnetic field of the EM wave, $\widetilde{\mathbf{H}}$, induces a magnetic moment per unit volume, $\widetilde{\mathbf{M}}$, within the medium, or

3. the medium can have free charges per unit volume, or a free charge density, ρ.

The vectors here are written with the tilde sign to signify their time dependence.

The response of the medium to the incident electromagnetic fields can either be a linear response or a nonlinear response. In general, the magnitude of the induced polarization in the medium, P, can be expressed as a power series of the incident light field's electric component, or

$$P = \chi^{(1)} E + \chi^{(2)} E^2 + \chi^{(3)} E^3 + \cdots, \qquad (2.1)$$

where the variables represent the optical susceptibilities of the medium. The first-order term represents the linear response of the medium, and χ_2 and χ_3 represent the second-order and third-order nonlinear optical susceptibilities of the medium.

For linear optical media, the response of the medium is only a *linear* function of the incident electric and magnetic fields. Thus, only the linear response of the medium is non-zero, or

$$\chi^{(1)} \neq 0 \text{ , and } \chi^{(2)} = \chi^{(3)} = \cdots = 0, \qquad (2.2)$$

and the vector electric and magnetic polarization responses of the material can be expressed as

$$\widetilde{\mathbf{P}} = \epsilon_o \chi_e \widetilde{\mathbf{E}}, \qquad (2.3)$$

and

$$\widetilde{\mathbf{M}} = \chi_m \widetilde{\mathbf{H}}, \qquad (2.4)$$

where χ_e and χ_m are the electric and magnetic linear optical susceptibilities, respectively. For isotropic media, the linear susceptibilities are constants and effect all components of the induced polarization equally. For anisotropic media, the susceptibilities are tensors and relate the components of the induced polarization to the components of the incident electric field according to the properties of the medium, and are further discussed in Chapter 3.

The total effect of the incident electric field, $\widetilde{\mathbf{E}}$, and the induced polarization, $\widetilde{\mathbf{P}}$, within the medium can be represented by the *electric displacement* vector, $\widetilde{\mathbf{D}}$, where

$$\widetilde{\mathbf{D}} = \epsilon_o \widetilde{\mathbf{E}} + \widetilde{\mathbf{P}} = \epsilon_o \widetilde{\mathbf{E}} + \epsilon_o \chi_e \widetilde{\mathbf{E}}, \qquad (2.5)$$

or

$$\widetilde{\mathbf{D}} = \epsilon_o \left(1 + \chi_e\right) \widetilde{\mathbf{E}}. \qquad (2.6)$$

Equation 2.6 can be written in a more convenient form of

$$\widetilde{\mathbf{D}} = \epsilon \widetilde{\mathbf{E}} \qquad (2.7)$$

where the parameter, ϵ, has been defined as

$$\epsilon \equiv \epsilon_o \left(1 + \chi_e\right), \tag{2.8}$$

and represents the permittivity of the medium.

Similarly, the total effect of the magnetic field, $\widetilde{\mathbf{H}}$, and the magnetic polarization, $\widetilde{\mathbf{M}}$, within the medium can be represented by the *magnetic induction*, $\widetilde{\mathbf{B}}$, where

$$\widetilde{\mathbf{B}} = \mu_o \widetilde{\mathbf{H}} + \mu_o \widetilde{\mathbf{M}} = \mu_o \widetilde{\mathbf{H}} + \mu_o \chi_m \widetilde{\mathbf{H}}, \tag{2.9}$$

or

$$\widetilde{\mathbf{B}} = \mu_o \left(1 + \chi_m\right) \widetilde{\mathbf{H}}. \tag{2.10}$$

The more convenient form of Eq. 2.10 is

$$\widetilde{\mathbf{B}} = \mu \widetilde{\mathbf{H}}, \tag{2.11}$$

where the parameter, μ, is defined to be

$$\mu \equiv \mu_o \left(1 + \chi_m\right), \tag{2.12}$$

and represents the permeability of the medium. The relationships between $\widetilde{\mathbf{B}}$, $\widetilde{\mathbf{H}}$, and $\widetilde{\mathbf{M}}$ are also commonly expressed as

$$\widetilde{\mathbf{H}} = \frac{1}{\mu}\widetilde{\mathbf{B}} \ , \ \text{or} \ \widetilde{\mathbf{H}} = \frac{1}{\mu_o}\widetilde{\mathbf{B}} - \widetilde{\mathbf{M}}. \tag{2.13}$$

If the medium contains a free charge density, ρ, the electric and magnetic fields can give rise to a free current density, $\mathbf{J}$. As discussed in section 1.4 the various electromagnetic fields ($\widetilde{\mathbf{E}}$, $\widetilde{\mathbf{D}}$, $\widetilde{\mathbf{H}}$, and $\widetilde{\mathbf{B}}$) in regions with a free charge density can be shown to obey Maxwell's equations, or

$$\nabla \cdot \widetilde{\mathbf{D}} \ = \ \rho \tag{2.14}$$

$$\nabla \cdot \widetilde{\mathbf{B}} \ = \ 0 \tag{2.15}$$

$$\nabla \times \widetilde{\mathbf{E}} \ = \ -\frac{\partial \widetilde{\mathbf{B}}}{\partial t} \tag{2.16}$$

$$\nabla \times \widetilde{\mathbf{H}} \ = \ \mathbf{J} + \frac{\partial \widetilde{\mathbf{D}}}{\partial t}. \tag{2.17}$$

The four partial differential equations above comprising the Maxwell's equations, along with the constitutive relations (Eqns. 2.7 and 2.11) and a charge continuity equation (which leads to Ohm's law) explain an enormous number of phenomena. These range from electric circuits, classical electrodynamics, optics and electromagnetic waves spanning x-rays to radio waves. Maxwell's equations are justly celebrated as the culmination of classical physics. Modern day life is unthinkable without electricity and magnetism and Maxwell's equations provide the theoretical foundation of most of modern technology. The solutions obtained in laser beam propagation models need to be checked for consistency with Maxwell's equations to ensure their validity. Although Maxwell's equations, being of classical origin, are not exact laws of nature but classical approximations of the more fundamental theory of quantum electrodynamics, for propagation of light as electromagnetic waves they can be relied upon as being accurate enough for all practical purposes.

2.2 Electromagnetic Waves in Linear Source-Free Media

Throughout most of this book we consider light to be conceptually, and mathematically, represented as an oscillating electromagnetic field propagating through a region of space that is:

1. entirely free of electric charges or sources, and

2. only contains media that have linear responses to the incident electromagnetic fields.

Within a charge-free linear optical medium, Maxwell's equations simplify to

$$\nabla \cdot \widetilde{\mathbf{D}} = 0 \tag{2.18}$$

$$\nabla \times \widetilde{\mathbf{E}} = -\frac{\partial \widetilde{\mathbf{B}}}{\partial t} \tag{2.19}$$

$$\nabla \cdot \widetilde{\mathbf{B}} = 0 \tag{2.20}$$

$$\nabla \times \widetilde{\mathbf{H}} = \frac{\partial \widetilde{\mathbf{D}}}{\partial t}, \tag{2.21}$$

where the vectors $\mathbf{E}$, $\mathbf{D}$, $\mathbf{H}$, and $\mathbf{B}$ are related by the equations

$$\widetilde{\mathbf{D}} = \epsilon \widetilde{\mathbf{E}}, \tag{2.22}$$

and

$$\widetilde{\mathbf{B}} = \mu \widetilde{\mathbf{H}}. \tag{2.23}$$

Denoting the amplitudes of the vectors $\widetilde{\mathbf{D}}$, $\widetilde{\mathbf{B}}$, $\widetilde{\mathbf{H}}$, and $\widetilde{\mathbf{E}}$ by D_o, B_o, H_o, and E_o, respectively, we have

$$D_o = \epsilon E_o \tag{2.24}$$

$$B_o = \sqrt{\mu \epsilon} E_o \tag{2.25}$$

$$H_o = \sqrt{\frac{\epsilon}{\mu}} E_o. \tag{2.26}$$

Maxwell's equations couple various electromagnetic field vectors to one another. However, the fields can be decoupled to yield a pair of equations, one dependent only on $\widetilde{\mathbf{D}}$ and the other only on $\widetilde{\mathbf{H}}$. To do so, we begin by taking the curl of both sides of 2.19, and we get

$$\nabla \times \left(\nabla \times \widetilde{\mathbf{E}} \right) = \nabla \times \left(-\frac{\partial \widetilde{\mathbf{B}}}{\partial t} \right) = -\frac{\partial}{\partial t} \left(\nabla \times \widetilde{\mathbf{B}} \right). \tag{2.27}$$

Substitution of Eq.2.23 into Eq. 2.21, and Eq. 2.21 into the right hand side of Eq. 2.27, and replacing $\widetilde{\mathbf{E}}$ in the left hand side with $\widetilde{\mathbf{D}}/\epsilon$ yields

$$\nabla \times \left(\nabla \times \widetilde{\mathbf{D}} \right) = -\mu\epsilon \frac{\partial^2 \widetilde{\mathbf{D}}}{\partial t^2}. \tag{2.28}$$

Using the vector relationship

$$\nabla \times (\nabla \times \mathbf{A}) = \nabla (\nabla \cdot \mathbf{A}) - \nabla^2 \mathbf{A} \tag{2.29}$$

which is valid for any vector $\mathbf{A}$, and assuming

$$\nabla \cdot \widetilde{\mathbf{E}} = \frac{1}{\varepsilon} \nabla \cdot \widetilde{\mathbf{D}} = 0, \tag{2.30}$$

we obtain

$$\nabla^2 \widetilde{\mathbf{D}} - \mu\epsilon \frac{\partial^2 \widetilde{\mathbf{D}}}{\partial t^2} = 0. \tag{2.31}$$

Similar application of Eq. 2.29 to the magnetic field and use of Maxwell's equations in a source-free region yields

$$\nabla^2 \widetilde{\mathbf{H}} - \mu\epsilon \frac{\partial^2 \widetilde{\mathbf{H}}}{\partial t^2} = 0. \tag{2.32}$$

Equations 2.31 and 2.32 are known as the wave equations for electromagnetic waves.

Now it will be assumed that the vector fields composing the electromagnetic wave can be written using a separation of the spatial variables and of the time-dependence, or

$$\widetilde{\mathbf{D}} (\mathbf{r}, t) = \mathbf{D}_r (\mathbf{r}) T (t), \tag{2.33}$$

where $\mathbf{D}_r (\mathbf{r})$ is the function governing the spatial dependence, and $T (t)$ is the function of the temporal dependence of the electric field. As with many waves in physics, the temporal dependence, $T (t)$ can be assumed to have a simple sinusoidal dependence on t, and it can be represented by a complex exponential with an angular frequency, ω, or

$$T (t) = e^{-i\omega t}. \tag{2.34}$$

The electric displacement can now be expressed as

$$\widetilde{\mathbf{D}} (\mathbf{r}, t) = \mathbf{D}_r (\mathbf{r}) e^{-i\omega t}, \tag{2.35}$$

and taking the second derivative with respect to time, Eq. 2.31 becomes

$$\nabla^2 \mathbf{D} + \mu\epsilon\omega^2 \mathbf{D} = 0. \tag{2.36}$$

Similar treatment of the temporal dependence of the magnetic field and carrying out the second derivative with respect to time yields

$$\nabla^2 \mathbf{H} + \mu\epsilon\omega^2 \mathbf{H} = 0. \tag{2.37}$$

Ignoring the transverse (x and y) variation of the fields, Equations 2.31 and 2.32 have solutions in the form of functions of $z - vt$, i.e., of waves traveling in the z direction with speed v given by

$$v = \frac{1}{\sqrt{\mu\epsilon}}. \tag{2.38}$$

For electromagnetic waves in linear optical media, the angular frequency ω is independent of the medium, i.e., the frequency of oscillation (ν) of the electric and magnetic fields is the same whether the wave is traveling through vacuum or through a linear material. Using the relations for an electromagnetic wave traveling through vacuum

$$\omega = 2\pi\nu \quad , \quad \nu\lambda = c, \tag{2.39}$$

and the speed of light in a vacuum,

$$c = \frac{1}{\sqrt{\mu_o \epsilon_o}}, \tag{2.40}$$

the coefficient of the second term on Eqs. 2.36 and 2.37 can be written as

$$\mu\epsilon\omega^2 = \mu\epsilon \left(\frac{2\pi}{\lambda\sqrt{\mu_o\epsilon_o}} \right)^2. \tag{2.41}$$

Defining a variable k called the *wave number* as

$$k \equiv \sqrt{\mu\epsilon\omega^2} = \frac{2\pi}{\lambda} \sqrt{\frac{\mu\epsilon}{\mu_o\epsilon_o}}, \tag{2.42}$$

Eqs. 2.36 and 2.37 can be rewritten as

$$\nabla^2 \mathbf{D} + k^2 \mathbf{D} = 0, \quad \text{and} \quad \nabla^2 \mathbf{H} + k^2 \mathbf{H} = 0. \tag{2.43}$$

The wave number k can also be written as

$$k = \frac{2\pi n}{\lambda}, \tag{2.44}$$

where n is the *linear refractive index* of the material, or

$$n = \sqrt{\frac{\mu\epsilon}{\mu_o\epsilon_o}}. \tag{2.45}$$

Combining Eqs. 2.40 and 2.45 we come to a simple relationship for the speed of the wave in the medium to be

$$v = \frac{c}{n}. \tag{2.46}$$

2.3 Maxwell's Equations in Vacuum

In vacuum there are no free charges or currents, so

$$\rho = |\mathbf{J}| = 0. \tag{2.47}$$

Also there is no induced electric or magnetic polarization, leading to

$$\mathbf{P} = 0 \quad , \text{ and } \quad \mathbf{M} = 0, \tag{2.48}$$

so that

$$\epsilon \longrightarrow \epsilon_o \quad , \text{ and } \quad \mu \longrightarrow \mu_o, \tag{2.49}$$

which forces the refractive index and the wave speed to be

$$n = 1 \quad , \text{ and } \quad v = c = \frac{1}{\sqrt{\mu_o \epsilon_o}}. \tag{2.50}$$

Since the refractive index becomes unity in vacuum, the wave number simplifies to

$$k_0 = \frac{2\pi}{\lambda}. \tag{2.51}$$

For light propagating through vacuum, Maxwell's Equations for the electric field $\widetilde{\mathbf{E}}$ and the magnetic induction $\widetilde{\mathbf{B}}$ simplify to

$$\boldsymbol{\nabla} \cdot \widetilde{\mathbf{E}} = 0 \tag{2.52}$$

$$\boldsymbol{\nabla} \cdot \widetilde{\mathbf{B}} = 0 \tag{2.53}$$

$$\boldsymbol{\nabla} \times \widetilde{\mathbf{E}} = -\frac{\partial \widetilde{\mathbf{B}}}{\partial t} \tag{2.54}$$

$$\boldsymbol{\nabla} \times \widetilde{\mathbf{B}} = \mu_o \epsilon_o \frac{\partial \widetilde{\mathbf{E}}}{\partial t}. \tag{2.55}$$

The associated wave equations for $\widetilde{\mathbf{E}}$ and $\widetilde{\mathbf{B}}$ are

$$\nabla^2 \widetilde{\mathbf{E}} - \mu_o \epsilon_o \frac{\partial^2 \widetilde{\mathbf{E}}}{\partial t^2} = 0, \tag{2.56}$$

$$\nabla^2 \widetilde{\mathbf{B}} - \mu_o \epsilon_o \frac{\partial^2 \widetilde{\mathbf{B}}}{\partial t^2} = 0, \tag{2.57}$$

or

$$\left(\nabla^2 + k_0^2 \right) \mathbf{E} = 0, \tag{2.58}$$

$$\left(\nabla^2 + k_0^2 \right) \mathbf{B} = 0, \tag{2.59}$$

and the relationship between the amplitudes of the electric and magnetic fields obeys

$$B_o = \frac{E_o}{c}. \tag{2.60}$$

2.4 Plane Waves

Mathematical Plane Waves

To understand what exactly is a plane wave, let us first start with a simple assumption that some arbitrary vector, $\mathbf{s}$, exists. By the properties of vectors, we can translate this vector anywhere in space as long as we maintain its fixed direction (arrow orientation) and magnitude (arrow length). Now, for each location in three-dimensional space to which we move the vector $\mathbf{s}$ we establish a location vector, $\mathbf{r}$, which is from the origin to that particular point in space. If it is restricted that the dot product of these two vectors, $\mathbf{s} \cdot \mathbf{r}$, is equal to a particular constant, then the locations for which this restriction is true maps out a plane in three-dimensional space, as illustrated in Figure 2.1. Two of the properties of this plane are that it resides a distance away from the origin of $|\mathbf{s}|$, and the normal unit vector for the plane is $\hat{s} = \mathbf{s}/|\mathbf{s}|$.

Let us now assume that the position dependence of $\mathbf{E}$ is a function of the function $f(\mathbf{r})$ where it is restricted that

$$f(\mathbf{r}) = \mathbf{s} \cdot \mathbf{r} = \text{constant}. \tag{2.61}$$

Let us also assume that the exact dependence of $\widetilde{\mathbf{E}}$ on the function $f(\mathbf{r})$ is

$$\widetilde{\mathbf{E}}(\mathbf{r}, t) = \mathbf{E}_r [f(\mathbf{r})] e^{-i\omega t}, \tag{2.62}$$

Equation 2.58 can be expressed as

$$\nabla^2 \mathbf{E}_r [f(\mathbf{r})] + k^2 \mathbf{E}_r [f(\mathbf{r})] = 0. \tag{2.63}$$

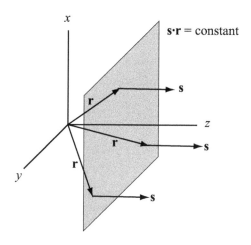

FIGURE 2.1

A vector, $\mathbf{s}$, the position vector, $\mathbf{r}$, and the plane normal to $\mathbf{s}$ on which $\mathbf{s} \cdot \mathbf{r}$ is constant.

The x-component of the first partial derivative of the Laplacian of $\mathbf{E}_r$ $[f(\mathbf{r})]$ becomes

$$\frac{\partial \mathbf{E}_r}{\partial x} = \frac{\partial \mathbf{E}_r}{\partial f(\mathbf{r})} \frac{\partial f(\mathbf{r})}{\partial x} = s_x \frac{\partial \mathbf{E}_r}{\partial f(\mathbf{r})}. \tag{2.64}$$

Differentiating again we get

$$\frac{\partial^2 \mathbf{E}_r}{\partial x^2} = s_x^2 \frac{\partial^2 \mathbf{E}_r}{\partial f(\mathbf{r})^2}. \tag{2.65}$$

Following the same routine for y and z, Eq. 2.58 becomes

$$s^2 \frac{\partial^2 \mathbf{E}_r}{\partial f(\mathbf{r})^2} + k^2 \mathbf{E}_r = 0, \tag{2.66}$$

whose magnitude has the solution

$$E_r(\mathbf{r}) = e^{i\frac{k}{s}f(\mathbf{r})}. \tag{2.67}$$

Identifying the wave vector $\mathbf{k}$ with our previously chosen arbitrary vector $\mathbf{s}$

$$E_r(\mathbf{r}) = e^{i\mathbf{k}\cdot\mathbf{r}}. \tag{2.68}$$

Including the time dependence, the electric field is be expressed as

$$\widetilde{\mathbf{E}}(\mathbf{r}, t) = \mathbf{E}e^{i(\mathbf{k}\cdot\mathbf{r}-\omega t)}, \tag{2.69}$$

where $\mathbf{E}$ is the amplitude of the electric field. A similar treatment of Eqs. 2.61 – 2.68 yields the magnetic field to be

$$\widetilde{\mathbf{H}}(\mathbf{r}, t) = \mathbf{H}e^{i(\mathbf{k}\cdot\mathbf{r}-\omega t)}, \tag{2.70}$$

where $\mathbf{H}$ is the amplitude of the magnetic field.

Note that the seemingly arbitrary restriction placed on the function $f(\mathbf{r})$ of $\mathbf{s} \cdot \mathbf{r} = \text{constant}$ forces the phase of the electric field to be the same throughout the plane perpendicular to $\hat{k}$. The function $e^{i\mathbf{k}\cdot\mathbf{r}}$ is the same for all values of $\mathbf{k} \cdot \mathbf{r} + m2\pi$ where m is an integer. At a given instant of time, a light beam will have the same amplitude and constant phase on a series of parallel planes, the spacing between which is $2\pi/k$, or the wavelength of the electromagnetic field, λ. On these planes the electric and the magnetic field vectors are uniform. Due to the unique planar properties of these types of electromagnetic waves they are commonly referred to as plane waves. Thus a plane wave

- propagates only in one direction, and

- has uniform amplitudes of $\mathbf{E}$ and $\mathbf{H}$ in a plane perpendicular to the propagation direction.

Relative Orientation of E, H, and k

The relationship between the directions of the vectors (E, H) and k can be determined from Maxwell's equations. Substitution of the plane wave solution for **E** into the left side of the Maxwell equation for the divergence of **E** yields

$$\nabla \cdot \mathbf{E} = ik_x E_x + ik_y E_y + i\,k_z E_z = i\mathbf{k} \cdot \mathbf{E} \qquad (2.71)$$

Setting this equal to the right side of the divergence of **E** shows that

$$\mathbf{k} \cdot \mathbf{E} = 0; \qquad (2.72)$$

i.e., the electric field vector must be perpendicular to the wave vector and must reside in the plane of uniform phase. Similarly, substitution of the magnetic field plane wave solution into the divergence of **H** shows that

$$\mathbf{k} \cdot \mathbf{H} = 0. \qquad (2.73)$$

The magnetic field must also be perpendicular to the wave vector and resides within the plane of uniform phase. To determine the relationship between E and H within the plane of uniform phase we should consider the two equations that couple the electric field to the magnetic field, Eqs. 2.19 and 2.21. Both of these equations are cross products with the electric and magnetic field directions on either side of the equal sign. Thus, by definition of a cross product, the directions of E and H have to be perpendicular to each other. The net result is that *all three* vectors, **E**, **H**, and **k**, must be *orthogonal* to each other at all points in space, as illustrated in Figure 2.2.

Experimental Plane Waves

Creating mathematical functions for electromagnetic waves that satisfy Maxwell's equations and the wave equation, and duplicating these waves in an experimental laboratory environment are very different tasks. The fact that a mathematical plane wave does not have any spatial limitations or boundary conditions means that a 'true' plane wave extends out to infinity in the plane perpendicular to the propagation direction. This fact alone proves that a 'true' experimental plane wave cannot exist (as it would need to be infinitely large). However, plane waves can be experimentally approximated within a localized region of space.

There are two key properties of mathematical plane waves that must be experimentally locally reproduced (within a spatially limited region of interest) in order to create an approximate experimental plane wave:

- an electromagnetic wave that propagates only in one direction, and

- uniform amplitudes of **E** and **H** within the region of interest.

Under experimental conditions, these two properties can be reproduced in a variety of ways for limited regions of space. For example, two common experimental methods for locally producing plane waves are through the use of either laser light or point sources.

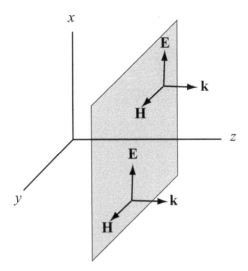

FIGURE 2.2
Orthogonality between the electric field, **E**, the magnetic field, **H**, and the wave vector, **k**, for a plane wave where **E** and **H** reside within the plane and **k** is normal to the plane.

Laser Light Plane Waves

If any laser beam is *collimated* then the first criteria is approximately true for all points within the laser beam. Satisfaction of the second criteria (uniform field amplitudes) can be obtained on a localized scale. One type of laser beam where an approximate plane wave exists for anywhere within the beam is that of a 'top hat' laser beam. A top hat laser beam is a beam that contains many higher order modes such that the superposition of all of the modes results in a fairly uniform intensity across the beam profile.[1]

If a laser beam is a pure (TEM$_{00}$) Gaussian beam (see Section 6.2) then a localized approximate plane wave exists at the very center of it. Figure 2.3(a) is an illustration of the intensity profile of a pure Gaussian beam. As a whole, the amplitude of a Gaussian beam is far from uniform. However, if only a small localized sub-section of the beam profile is considered, a relatively uniform amplitude can be observed on an absolute scale, as illustrated in Figure 2.3(b). To obtain a localized plane wave from a Gaussian beam the width of the area of interest must be much smaller than the width of the beam, or

$$\omega_{pw} \ll \omega_o, \tag{2.74}$$

---

[1] For a mathematical explanation and an example of superimposing multiple waves to create a uniform amplitude in a localized region see Section 3.3.1 and Figure 3.19 of D. J. Griffiths, *Introduction to Electrodynamics*, Third Edition, (Prentice Hall, Upper Saddle River, New Jersey, 1999.)

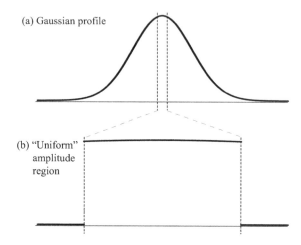

FIGURE 2.3

Creating an experimental plane wave from a Gaussian laser beam. On an absolute scale, the center portion of a Gaussian beam is approximately both uni-directional and uniform in amplitude.

where ω_{pw} is the width of the localized area of interest for creating a plane wave, and ω_o is the width of the Gaussian beam.

Point Source Plane Waves

Point sources of light can also experimentally produce approximate plane waves for localized regions of space. A point source of light can be anything where the dimensions of the source of the light are relatively small compared to the observational distance from the source. A point source can vary in size from a single atom or molecule to the size of a planet to the size of a star. In a laboratory setting, point sources are typically considered to be small light sources whose radiation is emitted into a large solid angle, such as: filaments of light bulbs, LEDs, highly divergent laser beams, etc. As long as the observation distance is large compared to the dimensions of the source, the light emitted from these objects can be considered to be that of a point source and the light field is approximately that of a spherical wave with the object at the center (see Section 2.6 for a discussion of spherical waves).

Figure 2.4 is an illustration of experimentally creating a localized plane wave from a point source. Suppose that the point source is located at the orgin, and the area of interest for creating an experimental plane wave is a small spherical surface area, dA, mapped out by the vector $\mathbf{r}$ over a small solid angle. Due to the fact that the light from experimental point sources is typically uniform in intensity over a large solid angle for a fixed distance from the source, the second criteria for experimentally creating plane waves

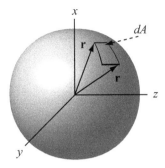

FIGURE 2.4
Creating an experimental plane wave from a point source. A spherical surface created by a uniform distance, $|\mathbf{r}|$, has a uniform amplitude. If the dimensions of the area dA are small compared to the distance from the point source, the propagation direction within dA is approximately uniform as well.

(uniform field amplitudes) is fulfilled for all points on the spherical surface mapped out by the vector $\mathbf{r}$. However, the first criteria (uniform propagation direction) is only fulfilled if the dimensions of the area of interest is very small compared to the observation distance from the light source, or

$$\omega_{pw} \ll r, \tag{2.75}$$

where ω_{pw} is once again the width of the localized area of interest for creating a plane wave, and r is the distance between the light source and the localized area of interest. In other words, if dA is small enough, or if the area of interest is far enough away from the point source then the direction of $\mathbf{r}$ is approximately the same for all points within the area of interest.

Figure 2.5 is another depiction of approximately creating a plane wave on a small planar surface due to a point source some distance away. Using a small angle approximation, the variation in the direction of $\mathbf{r}$ can be assumed to be negligible as long as the distance from the point source is much smaller

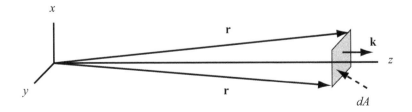

FIGURE 2.5
Creating an experimental plane wave from a point source. If the direction of the vector from the point source to the area of interest, $\mathbf{r}$ is approximately constant over the area of interest, dA, then the surface of interest approximately represents a plane wave traveling in the propagation direction, $\mathbf{k}$.

than the dimensions of the plane. Therefore, the direction of propagation of the plane wave is equal to that of the average direction of **r** over the area of interest.

2.5 Polarization States of Light

Without lack of generality, the Cartesian coordinate axes can be chosen so that the wave vector **k** is along the z direction. The electric and magnetic field vectors, **E** and **H**, then lie on the xy plane. The direction along which the electric field vector points is called the polarization direction of light.

Using Eq. 2.69, the electric field for a plane wave with the **k** vector along the z direction can be written as

$$\widetilde{\mathbf{E}}\,(z,t) = \mathbf{E}e^{i(kz-\omega t)}, \tag{2.76}$$

where **E** lies on the xy plane. Denoting the components of **E** along the x and y axes by E_x and E_y, we have

$$\mathbf{E} = E_x\hat{x} + E_y\hat{y}. \tag{2.77}$$

Since E_x and E_y are in general complex, they can be expressed as

$$E_x = A_x e^{i\phi_x} \quad \text{and} \quad E_y = A_y e^{i\phi_y} \tag{2.78}$$

where A_x, A_y, ϕ_x, and ϕ_y are all real quantities. The phases ϕ_x and ϕ_y for the x and y components of the fields can be different if the two components have propagated through different distances before reaching a common plane.

Electric fields, being physical quantities measurable in terms of force on a charge, are real. The complex notation used to denote fields has inherent in it the assumption that only the real part of the field expression will eventually be used for comparison with experiments. Here we revert to the physical field by taking the real part of Eq. 2.76, which becomes, using Eqs. 2.78

$$\begin{aligned}
\widetilde{\mathbf{E}}_{\mathbf{R}}\,(z,t) &\equiv Re\,\widetilde{\mathbf{E}}\,(z,t) \\
&\equiv \widetilde{E}_{Rx}\hat{x} + \widetilde{E}_{Ry}\hat{y} \\
&= A_x\cos(\phi_x + \psi)\hat{x} + A_y\cos(\phi_y + \psi)\hat{y}
\end{aligned} \tag{2.79}$$

where we have defined

$$\psi \equiv kz - \omega t \tag{2.80}$$

Further defining four variables ϕ_1, ϕ_2, a, and b by the relations

$$\phi_1 \equiv \phi_x + \psi, \quad \phi_2 \equiv \phi_y + \psi, \quad a \equiv \frac{\widetilde{E}_{Rx}}{A_x} \quad \text{and} \quad b \equiv \frac{\widetilde{E}_{Ry}}{A_y} \tag{2.81}$$

we have from Eq. 2.79

$$a = \cos\phi_1, \quad b = \cos\phi_2, \tag{2.82}$$

so that

$$
\begin{aligned}
\sin^2(\phi_1 - \phi_2) &= \sin^2\phi_1\cos^2\phi_2 + \cos^2\phi_1\sin^2\phi_2 - 2\sin\phi_1\cos\phi_1\sin\phi_2\cos\phi_2 \\
&= (1 - a^2)b^2 + (1 - b^2)a^2 - 2ab\sqrt{(1 - a^2)(1 - b^2)} \\
&= a^2 + b^2 - 2ab(ab + \sqrt{(1 - a^2)(1 - b^2)}) \\
&= a^2 + b^2 - 2ab\cos(\phi_1 - \phi_2) \tag{2.83}
\end{aligned}
$$

The x and y components of the electric field vector $\tilde{\mathbf{E}}_{\mathbf{R}}$ are then related by

$$\left(\frac{\tilde{E}_{Rx}}{A_x}\right)^2 + \left(\frac{\tilde{E}_{Ry}}{A_y}\right)^2 - 2\left(\frac{\tilde{E}_{Rx}}{A_x}\right)\left(\frac{\tilde{E}_{Ry}}{A_y}\right)\cos(\phi_x - \phi_y) = \sin^2(\phi_x - \phi_y) \tag{2.84}$$

The components $\tilde{E}_{Rx}$ and $\tilde{E}_{Ry}$ are functions of z and t, since $\tilde{E}_{Rx} = A_x\cos(\phi_x + kz - \omega t)$ and $\tilde{E}_{Ry} = A_y\cos(\phi_y + kz - \omega t)$. For constant z, i.e., on one plane wavefront, the field components $\tilde{E}_{Rx}$ and $\tilde{E}_{Ry}$ both change with time while maintaining the relationship shown in Eq. 2.84, which describes an ellipse with its axes tilted with respect to the coordinate axes.

Assuming the axes of the ellipse to be along a new set of coordinate axes denoted by x' and y', with an angle θ between the x and x' axes as shown in Figure 2.6, the components of the electric field vector in the new coordinate axes, $\tilde{E}_{Rx'}$ and $\tilde{E}_{Ry'}$, are related to $\tilde{E}_{Rx}$ and $\tilde{E}_{Ry}$ components

$$
\begin{aligned}
\tilde{E}_{Rx} &= \tilde{E}_{Rx'}\cos\theta - \tilde{E}_{Ry'}\sin\theta \\
\tilde{E}_{Ry} &= \tilde{E}_{Ry'}\sin\theta + \tilde{E}_{Ry'}\cos\theta \tag{2.85}
\end{aligned}
$$

Substituting Eq. 2.85 in Eq. 2.84 and requiring the coefficient of the $\tilde{E}_{Rx'}\tilde{E}_{Ry'}$ term in the resulting elliptical relationship between $\tilde{E}_{Rx'}$ and $\tilde{E}_{Ry'}$ to be 0 leads to the relation

$$\tan 2\theta = \cos(\phi_x - \phi_y)\tan 2\alpha \tag{2.86}$$

where

$$\alpha = \tan^{-1}\frac{A_y}{A_x} \tag{2.87}$$

Since for a given z, the components $\tilde{E}_{Rx'}$ and $\tilde{E}_{Ry'}$ oscillate with an angular frequency ω (in s^{-1}), the tip of the $\tilde{\mathbf{E}}_{\mathbf{R}}$ vector traces out an ellipse in every $1/\omega$ second and the beam of light is said to be *elliptically polarized*.

The state of elliptical polarization is obtained for a plane wave in the general case when the relationship between the phases ϕ_x and ϕ_y is not specified. For certain relationships between the phases, the ellipse takes the special forms of a line or a circle. These cases, known as linear and circular polarization cases, respectively, are described next.

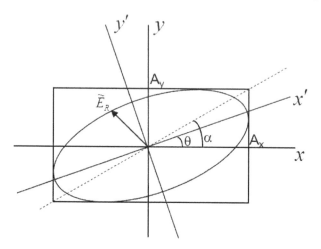

FIGURE 2.6

The tip of the resultant electric field vector $\widetilde{\mathbf{E}}_{\mathbf{R}}$ maps out an ellipse. In general the axes of this ellipse (along x' and y') are inclined with respect to the coordinate axes x and y, with an angle θ between x and x'. The ellipse is contained within a rectangle with sides $2A_x$ and $2A_y$. The diagonal of the rectangle is at an angle α with the x axis.

2.5.1   Special Case 1: Linear Polarization

If $\phi_x - \phi_y$ is 0 or an integer multiple of π, the quadratic relationship between the components $\widetilde{E}_{Rx}$ and $\widetilde{E}_{Ry}$ shown in Eq. 2.84 degenerates into a linear relationship,

$$\frac{\widetilde{E}_{Rx}}{A_x} = \frac{\widetilde{E}_{Ry}}{A_y}. \tag{2.88}$$

In this case, the magnitude of the vector $\mathbf{E}_{\mathbf{R}}$ oscillates in time, but its direction with respect to the coordinate axes remains the same. The angle between $\widetilde{\mathbf{E}}_{\mathbf{R}}$ and the x axis is given by $\tan^{-1}(A_x/A_y)$. The angle is independent of time and remains constant. The light beam in such a case is said to be *linearly polarized*.

With $\phi_x = \phi_y$, Eq. 2.86 also shows that $\theta = \alpha$ in Figure 2.6. Moreover, the component of $\widetilde{\mathbf{E}}_{\mathbf{R}}$ along the y' axis is

$$
\begin{aligned}
\widetilde{E}_{Ry'} &= \widetilde{E}_{Ry} \cos\theta - \widetilde{E}_{Rx} \sin\theta \\
&= A_y \cos\phi_2 \cos\theta - A_x \cos\phi_1 \sin\theta \\
&= \frac{A_x}{\cos\alpha} (\sin\alpha \cos\phi_2 \cos\theta - \cos\alpha \cos\phi_1 \sin\theta) \\
&= 0 \quad \text{for} \quad \phi_1 = \phi_2 \quad \text{and} \quad \theta = \alpha.
\end{aligned}
\tag{2.89}
$$

The ellipse in Figure 2.6 thus collapses to a line along the x' axis for this

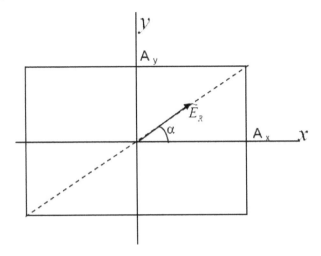

FIGURE 2.7
When ϕ_x and ϕ_y are equal or differ by a multiple of π, the vector $\mathbf{E_R}$ oscillates in time along a straight line at an angle α with the x axis and the beam of light is said to be linearly polarized.

case and the electric field vector for a linear polarized plane wave is along the diagonal of the rectangle with sides $2\,A_x$ and $2\,A_y$, as also illustrated in Figure 2.7.

Figures 2.8 and 2.9 illustrate the spatial and temporal behavior of linearly polarized light. Figure 2.8 is a series of plots of the net electric field vector from the viewpoint along the z-axis looking towards the origin, and in the $z = 0$ plane for a variety of time values.

In order to illustrate the spatial oscillation of $\mathbf{E}$ and $\mathbf{H}$, Figure 2.9 first assumes that $\alpha = 0$, forcing the electric field to be linearly polarized along the x-axis. It is also assumed that both $\mathbf{E}$ and $\mathbf{H}$ are frozen in time, $t = 0$. Note that the electric field oscillates along x and the magnetic field oscillates along y.

2.5.2 Special Case 2: $|\phi_x - \phi_y| = \dfrac{\pi}{2}$

If the phases ϕ_x and ϕ_y are related by

$$\phi_x = \phi_y \pm \frac{\pi}{2}, \tag{2.90}$$

Eq. 2.84 simplifies to

$$\left(\frac{\widetilde{E}_{Rx}}{A_x}\right)^2 + \left(\frac{\widetilde{E}_{Ry}}{A_y}\right)^2 = 1. \tag{2.91}$$

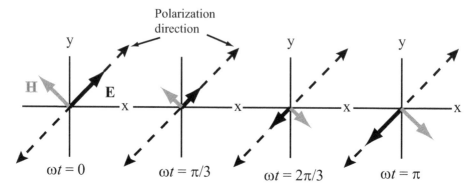

FIGURE 2.8
Oscillatory behavior of the electric field, **E**, and the magnetic field, **H**, for a series of times in the $z = 0$ plane for *linearly* polarized light traveling in the z-direction.

Eq. 2.86 shows that in this case $\theta = 0$, so that in Figure 2.6 the axes x' and x coincide as do the axes y' and y and the ellipse traced out by the tip of vector $\mathbf{E_R}$ has its major and minor axes along the coordinate axes x and y. This special case is shown in Figures 2.10 and 2.11. Figure 2.11 illustrates the time-dependent rotational behavior of the electric and magnetic fields for $\phi_y = \phi_x + \frac{\pi}{2}$ and $A_x < A_y$.

Whether the sign is plus or minus in the phase difference of Eq. 2.90 determines what "handedness" the light beam has. If

$$\phi_y = \phi_x + \frac{\pi}{2}, \tag{2.92}$$

the light beam is **right hand** elliptically polarized, and if

$$\phi_y = \phi_x - \frac{\pi}{2} \tag{2.93}$$

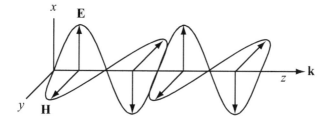

FIGURE 2.9
Spatial sinusoidal behavior of the electric field, **E**, and the magnetic field, **H**, for a linearly polarized plane wave traveling in the z-direction. The z-dependence is shown at a given time t, taken here as $t = 0$.

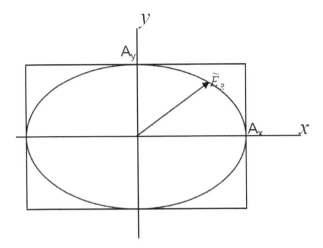

FIGURE 2.10
When the phases ϕ_x and ϕ_y differ by $\pi/2$ the tip of vector $\mathbf{E_R}$ traces out an ellipse, which has its axes along the coordinate axes x and y.

the light beam is **left hand** elliptically polarized.

When the phases ϕ_x and ϕ_y differ by integral multiples of $\pi/2$ and in addition the amplitudes A_x and A_y are equal, the ellipse becomes a circle and the plane wave is said to be *circularly* polarized. As with elliptically polarized light, the circularly polarized light is said to be **left hand** or **right hand** circularly polarized depending on the plus or minus sign in Eq. 2.90.

As an example, Figure 2.12 is an illustration of right-handed circularly polarized light, which shows the electric and magnetic field vectors with a phase relationship of Eq. 2.92 in the $z = 0$ plane for a sequence of time values.

Figure 2.13 is an illustration of a left-hand circularly polarized field traveling in the z-direction. Here the phase difference between the x and y components follows that of Eq. 2.93.

2.5.3 Special Case 3: $A_x = A_y$

One special case of elliptically polarized light is when the amplitude components are equal, i.e.,

$$A_x = A_y = A_0 \tag{2.94}$$

and the phase constants are arbitrary.

In this case, Eqs. 2.87 and 2.86 show that the angles α and θ are both equal to 45° and that Eq. 2.84 in the primed coordinates given in Eqs. 2.85 reduces to

$$(1 - \cos\phi)E_{Rx'}^2 + (1 + \cos\phi)E_{Ry'}^2 = A_0^2 \sin^2\phi \tag{2.95}$$

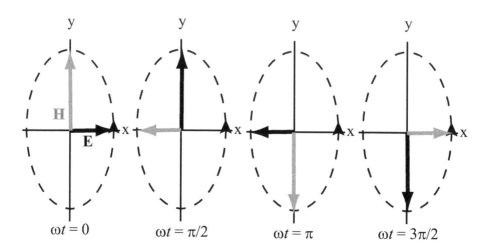

FIGURE 2.11
Rotational behavior of the electric field, **E**, and the magnetic field, **H**, for a series of times in the $z = 0$ plane for right-hand elliptically polarized light with $A_x < A_y, \phi_y = \phi_x + \pi/2$, and traveling in the z-direction.

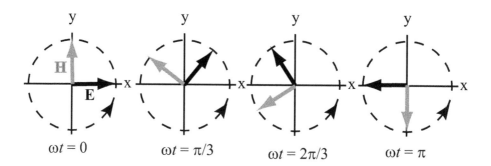

FIGURE 2.12
Rotational behavior of the electric field, **E**, and the magnetic field, **H**, for a series of times in the $z = 0$ plane for *right-hand* circularly polarized light traveling in the z-direction.

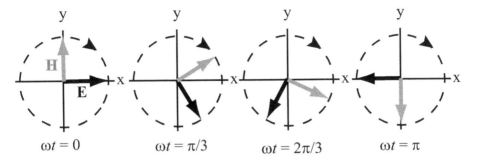

FIGURE 2.13
Rotational behavior of the electric field, **E**, and the magnetic field, **H**, for a series of times in the $z = 0$ plane for *left-hand* circularly polarized light traveling in the z-direction.

where

$$\phi = \phi_x - \phi_y \qquad (2.96)$$

In terms of the field components along the x- and y-coordinates, Eq. 2.84 can also be written in this case as

$$\frac{E_{Rx}}{A_0} = \cos\phi \frac{E_{Ry}}{A_0} \pm \sin\phi \sqrt{1 - \left(\frac{E_{Ry}}{A_0}\right)^2} \qquad (2.97)$$

The plots of E_{Ry}/A_0 against E_{Rx}/A_0 as given in Eq. 2.97 are shown in Figure 2.14 for different values of ϕ. For $\phi = 0$, the relationship is linear, with $E_{Rx} = E_{Ry}$. When $\phi = m\pi/2$, with m being a nonzero integer, the circular polarization case is obtained.

2.6 Spherical Waves

Spherical electromagnetic waves are physically created by oscillating point charges. An oscillating point charge will create electromagnetic fields that propagate away from the charge. If the amplitude of the oscillation, or the maximum displacement of the charge from the equilibrial position, is small compared to the distance away from the point charge, then the electromagnetic waves will be spherical waves with the center of the wave located at the equilibrial position.

Using the spherical coordinates depicted in Figure 2.15, the electromagnetic wave propagates away from the charge uniformly in the r-direction. Due to the spherical symmetry of the problem the phases of the waves will

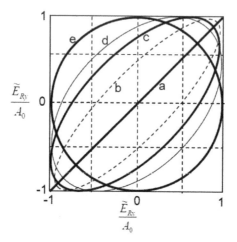

FIGURE 2.14
The plots of E_{Ry}/A_0 against E_{Rx}/A_0 when $A_x = A_y \equiv A_0$, for different values
of ϕ: (a) 0, (b) 30°, (c) 45°,(d) 60°, and (e) 90°.

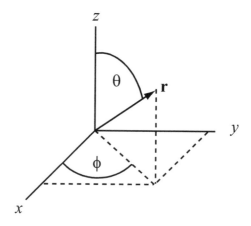

FIGURE 2.15
Spherical coordinates for a point (r, θ, ϕ), and its equivalent Cartesian coordinates (x, y, z).

also be spherically symmetric. Along the surface of each sphere of radius r the phase will be constant; i.e., the phase function $f(r)$ of Sec. 2.4 has the form

$$f(r) = \mathbf{k} \cdot \mathbf{r} = kr. \tag{2.98}$$

In other words, the wave vector $\mathbf{k}$ has the form

$$\mathbf{k} = k\hat{r}, \tag{2.99}$$

representing a wave propagating radially outward from the location of the oscillating dipole.

Unpolarized Spherical Waves

Unpolarized, or randomly polarized, spherical waves can be mathematically represented as scalar fields. The scalar wave equation for a spherical wave with a radial frequency ω, and the phase restriction of Eq. 2.98 has the form of

$$\frac{\partial^2}{\partial r^2}(rE) - \frac{\omega^2}{k^2}\frac{\partial^2}{\partial t^2}(rE) = 0, \tag{2.100}$$

which has the general scalar solution

$$E(r,t) = \frac{E_o}{r}e^{i(kr-\omega t)} + \frac{E_o}{r}e^{i(kr+\omega t)}. \tag{2.101}$$

The first term of Eq. 2.101 represents a diverging spherical wave propagating outward from the origin in the $+\hat{r}$ direction, and the second term represents a spherical wave propagating towards the origin. In this section we will only consider diverging spherical waves. However, the inclusion of a second wave propagating in the $-\hat{r}$ direction is straightforward.

Polarized spherical waves

Similar to our vector wave equation treatment of plane waves in Sec. 2.4, a vector form of the scalar spherical wave equation (Eq. 2.100 can be found to be

$$\frac{\partial^2}{\partial r^2}(r\mathbf{E}) - \frac{\omega^2}{k^2}\frac{\partial^2}{\partial t^2}(r\mathbf{E}) = 0. \tag{2.102}$$

One solution to the wave equation of the form of Eq. 2.102 for the out-goinging wave from an electric charge oscillating along the z-axis is the vector electric field

$$\mathbf{E}(r,t) = \frac{E_o}{r}e^{i(kr-\omega t)}\hat{\theta}, \tag{2.103}$$

with the corresponding magnetic field

$$\mathbf{H}(r,t) = \frac{H_o}{r}e^{i(kr-\omega t)}\hat{\phi}. \tag{2.104}$$

Figure 2.16 is an illustration of the constant spherical wave front of a spherical wave, and the orintation of the electric, magnetic, and wave vectors for a single point located on the wave front.

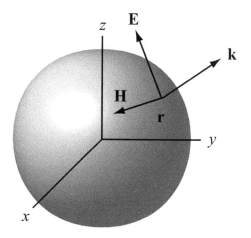

FIGURE 2.16
Orientation of **E**, **H**, and **k** at a single point on the spherical wave front for a spherical wave propagating in the $\hat{r}$-direction.

3

Light Propagation in Anisotropic Crystals

This chapter introduces the *principal axes* and *principal refractive indices* of a crystal, and addresses these questions: For light propagating in an anisotropic crystal, if the direction of the **k** vector is known with respect to the principal axes,

1. What are the directions along which the electric field vector **E** and the displacement vector **D** oscillate?

2. What is the angle between the **k** vector and the Poynting vector **S**?

3. What is the propagation equation describing the spatial change of the electric field amplitude for the propagation of an electromagnetic wave through an anisotropic medium?

Propagation in a biaxial crystal in an arbitrary direction will be considered first. Then, special directions of propagation in a biaxial crystal (along the principal dielectric axes and then along the principal planes) will be considered. Finally, propagation in an arbitrary direction in a uniaxial crystal will be described.

3.1 Introduction

Light propagates in vacuum with a constant speed c, which is given by $(\epsilon_0 \mu_0)^{-1/2}$, with ϵ_0, the vacuum permittivity, equal to $8.85 \times 10^{-12} \mathrm{CV}^{-1}\mathrm{m}^{-1}$ and μ_0, the vacuum permeability, equal to $4\pi \times 10^{-7} \mathrm{Vs}^2\mathrm{C}^{-1}\mathrm{m}^{-1}$. In material media, the speed of light (v) depends on the properties of the medium, with v equal to c/n, where n, the refractive index of the medium, depends on the properties of the medium. Gases, liquids, most glasses, and some crystalline solids are *optically isotropic*, and in these media the value of n is independent of the propagation direction of light. However, many other optically transparent solids are *optically anisotropic* because of their anisotropic crystalline structure, and for light propagation in them, the value of n is not the same for all directions of propagation. Moreover, the allowed directions of oscillation of the electric and the magnetic fields in anisotropic media also depend on the propagation direction.

The main aim of this chapter is to provide the explicit expressions for the components of the electric field for a general propagation direction of light in a general optical medium in terms of the propagation angles and the principal refractive indices. These components are used in the calculation of effective nonlinear optical coefficients in the field of nonlinear optics. There are numerous textbooks describing light propagation in anisotropic media, particularly Ref. [1]. Ref. [2] has an extensive collection of the needed equations and literature references relevant to nonlinear optics. But to our knowledge the electric field components are not available (or at least not easily available) in past literature, and usually only the components of the displacement vector are given. Explicit expressions for the walk-off angles between the propagation vector and the Poynting vector for different crystal classes are also presented here.

After the presentation of the general formalism, some special cases (such as, light traveling along principal axes and principal planes of biaxial crystals, and in uniaxial media) are discussed here. Since results for the special cases can be directly derived from the general equations, this discussion (which is lengthy because there are several cases to consider) is admittedly redundant. But this detailed discussion of the special cases is included here in the hope that it may add to the understanding of the topic and that the results, summarized in tables, may be useful as a reference. A detailed derivation of the propagation equation for light traveling in an anisotropic medium is also provided.

3.2 Vectors Associated with Light Propagation

In a charge and current free, non-magnetic medium (i.e., with $\mu = \mu_0$), Maxwell's equations are written as

$$\nabla \cdot \widetilde{\mathbf{D}} = 0 \tag{3.1}$$

$$\nabla \cdot \widetilde{\mathbf{H}} = 0 \tag{3.2}$$

$$\nabla \times \widetilde{\mathbf{E}} = -\mu_0 \frac{\partial \widetilde{\mathbf{H}}}{\partial t} \tag{3.3}$$

$$\nabla \times \widetilde{\mathbf{H}} = \frac{\partial \widetilde{\mathbf{D}}}{\partial t}. \tag{3.4}$$

where $\widetilde{\mathbf{D}}$, $\widetilde{\mathbf{E}}$, and $\widetilde{\mathbf{H}}$ denote the electric displacement vector and the electric and magnetic fields, respectively. The tilde symbol is used to indicate that these fields are oscillating in time and space.

For a propagating electromagnetic wave, the direction of energy flow is given by the Poynting vector $\widetilde{\mathbf{S}}$, where

$$\widetilde{\mathbf{S}} = \widetilde{\mathbf{E}} \times \widetilde{\mathbf{H}}. \tag{3.5}$$

Equations 3.3, 3.4, and 3.5 show that the vectors $\widetilde{\mathbf{E}}$, $\widetilde{\mathbf{D}}$, and $\widetilde{\mathbf{S}}$ are all perpen-

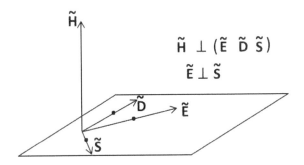

FIGURE 3.1
The vectors $\widetilde{\mathbf{E}}$, $\widetilde{\mathbf{D}}$, and $\widetilde{\mathbf{S}}$ are all perpendicular to $\widetilde{\mathbf{H}}$ and are coplanar. The vectors $\widetilde{\mathbf{E}}$ and $\widetilde{\mathbf{S}}$ are also perpendicular to each other. Vectors lying on the same plane are indicated by the dots.

dicular to $\widetilde{\mathbf{H}}$, which implies $\widetilde{\mathbf{E}}$, $\widetilde{\mathbf{D}}$, and $\widetilde{\mathbf{S}}$ must be co-planar, and also that $\widetilde{\mathbf{E}}$ and $\widetilde{\mathbf{S}}$ are perpendicular to each other. This is shown in Figure 3.1.

3.2.1 Plane Waves

Assuming monochromatic plane wave solutions for the fields $\widetilde{\mathbf{E}}$, $\widetilde{\mathbf{D}}$, and $\widetilde{\mathbf{H}}$, i.e., with

$$\widetilde{\mathbf{E}} = \mathbf{E_0} e^{i(\mathbf{k}\cdot\mathbf{r}-\omega t)} \tag{3.6}$$

$$\widetilde{\mathbf{D}} = \mathbf{D_0} e^{i(\mathbf{k}\cdot\mathbf{r}-\omega t)} \tag{3.7}$$

$$\widetilde{\mathbf{H}} = \mathbf{H_0} e^{i(\mathbf{k}\cdot\mathbf{r}-\omega t)} \tag{3.8}$$

it can be easily shown that

$$\nabla \times \widetilde{\mathbf{E}} = i e^{i(\mathbf{k}\cdot\mathbf{r}-\omega t)}(\mathbf{k} \times \mathbf{E_0}). \tag{3.9}$$

Using Eqns. 3.3, 3.6, 3.8, and 3.9, we obtain

$$(\mathbf{k} \times \mathbf{E_0}) = \mu_0 \omega \mathbf{H_0} \tag{3.10}$$

and also, similarly

$$(\mathbf{k} \times \mathbf{H_0}) = -\omega \mathbf{D_0}. \tag{3.11}$$

Thus $\mathbf{k}$ is also perpendicular to $\widetilde{\mathbf{H}}$ and must lie in the plane with $\widetilde{\mathbf{E}}$, $\widetilde{\mathbf{D}}$, and $\widetilde{\mathbf{S}}$. In addition, $\mathbf{k}$ is perpendicular to $\widetilde{\mathbf{D}}$. The vectors $\widetilde{\mathbf{E}}$, $\widetilde{\mathbf{D}}$, $\mathbf{k}$, and $\widetilde{\mathbf{S}}$ are all coplanar. Since $\widetilde{\mathbf{E}} \perp \widetilde{\mathbf{S}}$ and $\widetilde{\mathbf{D}} \perp \mathbf{k}$, the angle between $\widetilde{\mathbf{E}}$ and $\widetilde{\mathbf{D}}$ must be the same as the angle between $\mathbf{k}$ and $\widetilde{\mathbf{S}}$. This angle of deviation of the energy

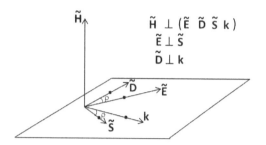

FIGURE 3.2

For plane waves and paraxial beams, the propagation vector $\mathbf{k}$ is coplanar with the vectors $\widetilde{\mathbf{E}}$, $\widetilde{\mathbf{D}}$, and $\widetilde{\mathbf{S}}$. Moreover, the vectors $\widetilde{\mathbf{D}}$ and $\mathbf{k}$ are perpendicular to each other as are the vectors $\widetilde{\mathbf{E}}$ and $\mathbf{S}$. ρ denotes the walk-off angle.

propagation direction ($\widetilde{\mathbf{S}}$) from the phase propagation direction ($\mathbf{k}$) is denoted by ρ and is called the walk-off angle. In isotropic media, $\widetilde{\mathbf{E}}$ and $\widetilde{\mathbf{D}}$ are parallel, and ρ is zero. In anisotropic media, ρ can be zero only for certain special directions of $\mathbf{k}$, but in general $\rho \neq 0$ for $\mathbf{k}$ in other directions. These results are shown in Figure 3.2.

3.2.2 Non-Plane Waves

Laser beams of interest propagating in anisotropic media are often cylindrical (collimated) in shape, or they may be converging or diverging, i.e., in general they are not truly plane waves. It is important to know if $\widetilde{\mathbf{D}}$ remains perpendicular to $\mathbf{k}$ for such beams because the expression for the refractive index of an anisotropic medium is usually deduced assuming $\mathbf{k} \cdot \widetilde{\mathbf{D}} = 0$, i.e., under the plane wave assumption.

   Say we consider a light beam which is *not* a plane wave, i.e., for which

$$\widetilde{\mathbf{E}} = \mathbf{A}e^{i(\mathbf{k}\cdot\mathbf{r}-\omega t)} \tag{3.12}$$

$$\widetilde{\mathbf{D}} = \mathfrak{D}e^{i(\mathbf{k}\cdot\mathbf{r}-\omega t)} \tag{3.13}$$

$$\widetilde{\mathbf{H}} = \mathcal{H}e^{i(\mathbf{k}\cdot\mathbf{r}-\omega t)} \tag{3.14}$$

where the amplitudes $\mathbf{A}$, $\mathfrak{D}$, and $\mathcal{H}$ are not necessarily constants but can be slowly varying functions of $\mathbf{r}$. Defining

$$\psi \equiv e^{i\mathbf{k}\cdot\mathbf{r}}$$

so that $\nabla\psi = i\mathbf{k}\psi$, and using the vector identity

$$\nabla \cdot (\psi\mathfrak{D}) = \mathfrak{D} \cdot \nabla\psi + \psi\nabla \cdot \mathfrak{D} \tag{3.15}$$

in the Maxwell's equation $\nabla \cdot \widetilde{\mathbf{D}} = 0$, we obtain

$$\nabla \cdot \widetilde{\mathbf{D}} = e^{-i\omega t}\nabla \cdot (\psi\mathfrak{D}) = e^{-i\omega t}\psi(i\mathbf{k} \cdot \mathfrak{D} + \nabla \cdot \mathfrak{D}) = 0. \tag{3.16}$$

If θ denotes the angle between the $\mathbf{k}$ and $\widetilde{\mathbf{D}}$, Eqs. 3.16 shows that

$$|\cos\theta| = \left|\frac{\nabla \cdot \mathfrak{D}}{k\mathfrak{D}}\right| \tag{3.17}$$

For a beam of light confined to a spatial region of the order of r_0 (the beam spot size), $\nabla \cdot \mathfrak{D}/\mathfrak{D}$ is approximately equal to $1/r_0$ so that

$$|\cos\theta| \approx \frac{1}{kr_0} = \frac{\lambda}{2\pi r_0}. \tag{3.18}$$

For most cases of interest here, $r_0 \geq \lambda$ so that $\cos\theta$ is small, implying $\theta \approx 90°$, i.e., for most practical purposes, $\mathbf{k}$ is perpendicular to $\widetilde{\mathbf{D}}$. However, this is only an approximate relationship, not as ironclad as the orthogonality of $\widetilde{\mathbf{E}}$ and $\widetilde{\mathbf{H}}$ and of $\widetilde{\mathbf{E}}$ and $\widetilde{\mathbf{S}}$, which follow directly from Maxwell's equations and the definition of $\widetilde{\mathbf{S}}$.

3.3 Anisotropic Media

The electric displacement vector $\widetilde{\mathbf{D}}$ is related to the electric field $\widetilde{\mathbf{E}}$ by the relation

$$\widetilde{\mathbf{D}} = \epsilon_0\widetilde{\mathbf{E}} + \widetilde{\mathbf{P}} \tag{3.19}$$

where $\widetilde{\mathbf{P}}$ represents the macroscopically averaged electric dipole density of the material medium in the presence of the applied field [3]. If the medium is isotropic, the polarization $\widetilde{\mathbf{P}}$ induced by the electric field $\widetilde{\mathbf{E}}$ is parallel to $\widetilde{\mathbf{E}}$, That is, for isotropic media, we can write

$$\widetilde{\mathbf{P}} = \varepsilon_0\chi\widetilde{\mathbf{E}}. \tag{3.20}$$

The parameter χ relating $\widetilde{\mathbf{P}}$ and $\widetilde{\mathbf{E}}$ is called the *electric susceptibility*. For isotropic media it is a scalar quantity independent of the relative direction between $\widetilde{\mathbf{P}}$ and $\widetilde{\mathbf{E}}$.

In an *anisotropic medium*, the electric field imposes a force on the microscopic charges in the direction of the field, but the net displacement of the charges can be in a different direction, imposed by the crystal structure. The macroscopically averaged electric dipole density, while being linearly proportional to the field, is in general in a different direction, i.e., $\widetilde{\mathbf{P}}$ is *not* parallel to $\widetilde{\mathbf{E}}$. In general, the Cartesian components of $\widetilde{\mathbf{P}}$ and $\widetilde{\mathbf{E}}$ can be related as

$$\widetilde{P}_x = \kappa_{xx}\widetilde{E}_x + \kappa_{xy}\widetilde{E}_y + \kappa_{xz}\widetilde{E}_z \tag{3.21}$$

$$\widetilde{P}_y = \kappa_{yx}\widetilde{E}_x + \kappa_{yy}\widetilde{E}_y + \kappa_{yz}\widetilde{E}_z \tag{3.22}$$

$$\widetilde{P}_z = \kappa_{zx}\widetilde{E}_x + \kappa_{zy}\widetilde{E}_z + \kappa_{zz}\widetilde{E}_z. \tag{3.23}$$

This can be written in tensor notation as

$$\widetilde{P}_i = \kappa_{ij}\widetilde{E}_j \tag{3.24}$$

where κ_{ij} is a tensor, the indices i and j of which run over the Cartesian coordinates x, y, and z, and the Einstein summation convention has been assumed. The vector Eqs. 3.19 can then be rewritten in tensor form as

$$\widetilde{D}_i = \varepsilon_{ij}\widetilde{E}_j \tag{3.25}$$

where

$$\varepsilon_{ij} = \varepsilon_0(\delta_{ij} + \chi_{ij}). \tag{3.26}$$

ε_{ij} is called the *dielectric permittivity tensor*, δ_{ij} is the Kronecker delta symbol, and

$$\chi_{ij} = \kappa_{ij}/\varepsilon_0. \tag{3.27}$$

Using the simple Lorentz model for electron oscillator motion under an applied electric field [4] or from energy flow consideration [5], it can be shown that the dielectric permittivity tensor is symmetric, i.e.,

$$\varepsilon_{ij} = \varepsilon_{ji}. \tag{3.28}$$

3.3.1   The Principal Coordinate Axes

The coordinate system used to define the susceptibility and dielectric permittivity tensors has been left arbitrary so far. For any anisotropic crystal, a coordinate system (X, Y, Z) can always be chosen that diagonalizes the dielectric permittivity tensor, i.e., in which only the diagonal components of the tensor are non-zero. In such a coordinate system, the dielectric permittivity tensor takes the form

$$\varepsilon = \begin{pmatrix} \varepsilon_X & 0 & 0 \\ 0 & \varepsilon_Y & 0 \\ 0 & 0 & \varepsilon_Z \end{pmatrix} \tag{3.29}$$

X, Y, and Z are called the *principal coordinate axes* of the crystal. The electric displacement and the field are related in the principal coordinate axes system by the relations

$$\widetilde{D}_X = \varepsilon_X\widetilde{E}_X \qquad \widetilde{D}_Y = \varepsilon_Y\widetilde{E}_Y \qquad \widetilde{D}_Z = \varepsilon_Z\widetilde{E}_Z. \tag{3.30}$$

ε_X, ε_Y, and ε_Z are called the *principal dielectric permittivities* of the anisotropic crystal.

By convention, the principal coordinate axes are always chosen (in optics) such that the values of ε_X, ε_Y, and ε_Z are either monotonically increasing or decreasing, i.e., either $\varepsilon_X > \varepsilon_Y > \varepsilon_Z$ or $\varepsilon_X < \varepsilon_Y < \varepsilon_Z$. We will also assume here that at the wavelengths of interest, the dielectric permittivities are dominantly real numbers. The principal coordinate axes system may not always coincide with the crystallographic coordinate axes (denoted by a, b, c in [6]).

The three planes XY, YZ, and ZX are called the principal planes.

3.3.2 Three Crystal Classes

All crystals can be classified into three groups, isotropic, uniaxial, and biaxial, depending on the relations between ε_X, ε_Y, and ε_Z. In *isotropic* crystals all the principal dielectric permittivities are equal, i.e.,

$$\varepsilon_X = \varepsilon_Y = \varepsilon_Z.$$

In *biaxial* crystals all three principal dielectric permittivities are unequal, i.e.,

$$\varepsilon_X \neq \varepsilon_Y \neq \varepsilon_Z.$$

In *uniaxial* crystals only two principal dielectric permittivities are equal to each other. By convention, in a uniaxial crystal,

$$\varepsilon_X = \varepsilon_Y \neq \varepsilon_Z.$$

The reason behind choosing the names *uniaxial* and *biaxial* will be discussed in the next section.

3.3.3 The Principal Refractive Indices

We *define* here three *principal refractive indices* n_X, n_Y, and n_Z

$$n_X \equiv \sqrt{\varepsilon_X/\varepsilon_0} \qquad n_Y \equiv \sqrt{\varepsilon_Y/\varepsilon_0} \qquad n_Z \equiv \sqrt{\varepsilon_Z/\varepsilon_0} \qquad (3.31)$$

using which the Eqns. 3.30 take the forms

$$\tilde{D}_X = \varepsilon_0 n_X^2 \tilde{E}_X \qquad \tilde{D}_Y = \varepsilon_0 n_Y^2 \tilde{E}_Y \qquad \tilde{D}_Z = \varepsilon_0 n_Z^2 \tilde{E}_Z. \qquad (3.32)$$

In a biaxial crystal, n_X, n_Y, and n_Z are all unequal. Some examples of biaxial crystals of importance in nonlinear optics are Lithium Triborate (LiB$_3$O$_5$ or LBO), Potassium Niobate (KNbO$_3$), Potassium Titanyl Phosphate (KTiOPO$_4$ or KTP), Potassium Titanyl Arsenate (KTiOAsO$_4$ or KTA), alpha Iodic acid (α HIO$_3$), etc.

In uniaxial crystals, two of the principal refractive indices are equal. By convention, the two equal indices are chosen to be the n_X and n_Y values and the indices are renamed as

$$n_o = n_X = n_Y \qquad \text{for uniaxial crystals} \qquad (3.33)$$

If $n_Z > n_o$, the crystal is defined to be *positive uniaxial* and if $n_o > n_Z$, the crystal is defined to be *negative uniaxial*. Some examples of positive uniaxial crystals are Zinc Germanium Phosphide ($ZnGeP_2$ or ZGP), Cadmium Selenide (CdSe), Cadmium Germanium Arsenide ($CdGeAs_2$ or CGA), Cinnabar (HgS), Selenium (Se), Tellurium (Te), and Quartz (SiO_2). Some examples of negative uniaxial crystals are Potassium Dihydrogen Phosphate (KH_2PO_4 or KDP), Lithium Iodate ($LiIO_3$), Lithium Niobate ($LiNbO_3$), Silver Gallium Sulfide ($AgGaS_2$), Silver Gallium Selenide ($AgGaSe_2$), Gallium Selenide (GaSe), Rubidium Dihydrogen Phosphate (RbH_2PO_4 or RDP), Mercury Thiogallate ($HgGa_2S_4$), Pyragyrite (Ag_3SbS_3), and Thallium Arsenide Sulfide (Tl_3AsS_3 or TAS).

3.4   Light Propagation in an Anisotropic Crystal

As shown in the last section, for a plane wave (or a paraxial beam of light) propagating in any medium, the direction of oscillation of the electric displacement vector $\widetilde{\mathbf{D}}$ is perpendicular to the propagation direction, i.e., $\mathbf{k} \cdot \widetilde{\mathbf{D}} = 0$, where $\mathbf{k}$ is the propagation vector. In an isotropic medium, the electric field $\widetilde{\mathbf{E}}$ is parallel to $\widetilde{\mathbf{D}}$, and for a given direction of $\mathbf{k}$, the $\widetilde{\mathbf{D}}$ and $\widetilde{\mathbf{E}}$ vectors can be in any direction as long as they lie on a plane perpendicular to $\mathbf{k}$.

For light propagation in anisotropic media, for a *general* direction of $\mathbf{k}$ (with respect to the principal axes), $\widetilde{\mathbf{D}}$ and $\widetilde{\mathbf{E}}$ are *not* parallel. Moreover, the direction of $\widetilde{\mathbf{D}}$ is not only restricted to a plane perpendicular to $\mathbf{k}$, it is also further specified to be only in two allowed directions in that plane, as will be shown below. Similarly, for the given $\mathbf{k}$, the electric field $\widetilde{\mathbf{E}}$ is also restricted to two fixed directions, which are in general not parallel to the allowed directions $\widetilde{\mathbf{D}}$.

In *biaxial* crystals (defined by the condition $\varepsilon_X \neq \varepsilon_Y \neq \varepsilon_Z$), there are two special directions called the directions of the *optic axes* such that if $\mathbf{k}$ is parallel to an optic axis, the $\widetilde{\mathbf{D}}$ and $\widetilde{\mathbf{E}}$ vectors are parallel and they can lie in *any* direction as long as they are restricted to the plane perpendicular to $\mathbf{k}$. Thus, for light propagation with $\mathbf{k}$ along the optic axes, the anisotropic medium behaves as an isotropic medium. It will be shown later that in biaxial crystals the two optic axes lie on the XZ principal plane, with the Z axis the bisector of the angle between the optic axes and the tangent of the angle between the two axes proportional to $\sqrt{n_X^2 - n_Y^2}$.

In uniaxial crystals, $n_X = n_Y$, so the angle between the two axes goes to zero and the two axes coalesce into one (pointed along the Z direction), thereby justifying the name of this crystal class.

3.4.1 Allowed Directions of $\widetilde{\mathbf{D}}$ and $\widetilde{\mathbf{E}}$ in an Anisotropic Medium

Here we determine the allowed oscillation directions of the propagating plane waves $\widetilde{\mathbf{D}}$ and $\widetilde{\mathbf{E}}$, for a general direction of $\mathbf{k}$ with respect to the principal axes. Say the unit vectors along the $\mathbf{k}$, $\widetilde{\mathbf{D}}$, and $\widetilde{\mathbf{E}}$ vectors are denoted by $\hat{m}$, $\hat{d}$, and $\hat{e}$, i.e.,

$$\mathbf{k} = \hat{m}k \quad \widetilde{\mathbf{D}} = \hat{d}\widetilde{D} \quad \widetilde{\mathbf{E}} = \hat{e}\widetilde{E}. \tag{3.34}$$

Maxwell's equations in a nonmagnetic medium given in Eqns. 3.1 to 3.4 can be rearranged in the form

$$\nabla \times (\nabla \times \widetilde{\mathbf{E}}) = -\mu_0 \frac{\partial^2 \widetilde{\mathbf{D}}}{\partial t^2}. \tag{3.35}$$

Rewriting Eqns. 3.12 and 3.13 as

$$\widetilde{\mathbf{E}} = \mathbf{E}e^{-i\omega t} \tag{3.36}$$

$$\widetilde{\mathbf{D}} = \mathbf{D}e^{-i\omega t} \tag{3.37}$$

where

$$\mathbf{E} = \mathbf{A}e^{i\mathbf{k}\cdot\mathbf{r}}, \quad \text{and} \quad \mathbf{D} = \mathfrak{D}e^{i\mathbf{k}\cdot\mathbf{r}} \tag{3.38}$$

Equation 3.35 is thus rewritten as

$$\nabla \times (\nabla \times \mathbf{E}) = \mu_0\omega^2\mathbf{D}. \tag{3.39}$$

For monochromatic plane waves, i.e., for $\mathbf{E}$ in Eqs. 3.38 constant, the ∇ operator can be replaced by $i\mathbf{k}$, i.e., by $ik\hat{m}$, so that Eqs. 3.39 becomes

$$-k^2(\hat{m} \times \{\hat{m} \times \mathbf{E}\}) = \mu_0\omega^2\mathbf{D}. \tag{3.40}$$

We will assume here that Eqs. 3.40 is valid even for non-planar waves, and that $\mathbf{k}$ and $\mathbf{D}$ are approximately perpendicular, as discussed before.

The waves propagate in the medium with speed ω/k. The refractive index n of the medium is the ratio of the wave speed to c, the speed of light in vacuum, i.e.,

$$n = \frac{c\,k}{\omega}. \tag{3.41}$$

From Eqns. 3.40 and 3.41 we obtain

$$-\hat{m} \times (\{\hat{m} \times \mathbf{E}\}) = \frac{\mathbf{D}}{\epsilon_0 n^2} \tag{3.42}$$

and using the vector identity $\mathbf{A} \times (\mathbf{B} \times \mathbf{C}) = \mathbf{B}(\mathbf{A} \cdot \mathbf{C}) - \mathbf{C}(\mathbf{A} \cdot \mathbf{B})$ we get

$$\mathbf{D} = \epsilon_0 n^2 \{\mathbf{E} - \hat{m}(\hat{m} \cdot \mathbf{E}).\} \tag{3.43}$$

For a given medium (i.e., one with known values of n_X, n_Y, and n_Z), the allowed oscillation directions of **D** and **E** fields, for a specified direction of $\hat{m}$, can be determined using Eqs. 3.30 and Eqs. 3.43. The speeds of these waves propagating with the **k** vector in the $\hat{m}$ direction are also determined by Eqs. 3.43 through the solution for n.

Rewriting Eqs. 3.43 in terms of the vector components along the three principal coordinate axes, and using Eqns. 3.30 and 3.31, we obtain

$$
\begin{aligned}
D_X &= \varepsilon_0 n^2 \{ E_X - m_X (m_X E_X + m_Y E_Y + m_Z E_Z) \} &= \varepsilon_0 n_X^2 E_X \\
D_Y &= \varepsilon_0 n^2 \{ E_Y - m_Y (m_X E_X + m_Y E_Y + m_Z E_Z) \} &= \varepsilon_0 n_Y^2 E_Y \\
D_Z &= \varepsilon_0 n^2 \{ E_Z - m_Z (m_X E_X + m_Y E_Y + m_Z E_Z) \} &= \varepsilon_0 n_Z^2 E_Z
\end{aligned}
\tag{3.44}
$$

which can be solved to obtain the values of the components of **E**:

$$
E_X = \frac{n^2 m_X}{n^2 - n_X^2} (\hat{m} \cdot \mathbf{E})
$$

$$
E_Y = \frac{n^2 m_Y}{n^2 - n_Y^2} (\hat{m} \cdot \mathbf{E})
\tag{3.45}
$$

$$
E_Z = \frac{n^2 m_Z}{n^2 - n_Z^2} (\hat{m} \cdot \mathbf{E}).
\tag{3.46}
$$

Figure 3.2 shows that if ρ is the angle between **D** and **E**, the angle between **E** and $\hat{m}$ is $90° - \rho$, i.e., $\hat{m} \cdot \mathbf{E} = E \sin \rho$, where E is the magnitude of the vector **E**. The Cartesian components of $\hat{e}$, the unit vector in the direction of **E**, are then given by

$$
e_X = \frac{n^2 m_X}{n^2 - n_X^2} \sin \rho
$$

$$
e_Y = \frac{n^2 m_Y}{n^2 - n_Y^2} \sin \rho
$$

$$
e_Z = \frac{n^2 m_Z}{n^2 - n_Z^2} \sin \rho.
\tag{3.47}
$$

Since $\hat{e}_X^2 + \hat{e}_Y^2 + \hat{e}_Z^2 = 1$, the walk-off angle ρ is given by

$$
\sin \rho = \frac{1}{n^2 \left[\left(\dfrac{m_X}{n^2 - n_X^2} \right)^2 + \left(\dfrac{m_Y}{n^2 - n_Y^2} \right)^2 + \left(\dfrac{m_Z}{n^2 - n_Z^2} \right)^2 \right]^{1/2}}.
\tag{3.48}
$$

The components of the unit vector $\hat{d}$ along the direction of **D** can similarly be determined. Taking the dot product of both sides of Eqs. 3.43 with themselves, we obtain

$$
\begin{aligned}
\mathbf{D} \cdot \mathbf{D} = D^2 &= \epsilon_0^2 n^4 [\mathbf{E} - \hat{m}(\hat{m} \cdot \mathbf{E})] \cdot [\mathbf{E} - \hat{m}(\hat{m} \cdot \mathbf{E})] \\
&= \epsilon_0^2 n^4 [\mathbf{E} \cdot \mathbf{E} - 2(\hat{m} \cdot \mathbf{E})(\hat{m} \cdot \mathbf{E}) + (\hat{m} \cdot \hat{m})(\hat{m} \cdot \mathbf{E})^2] \\
&= \epsilon_0^2 n^4 [\mathbf{E} \cdot \mathbf{E} - (\hat{m} \cdot \mathbf{E})^2] \\
&= \epsilon_0^2 n^4 [E^2 - E^2 \sin^2 \rho]
\end{aligned}
\tag{3.49}
$$

so that

$$
D = \epsilon_0 n^2 E \cos \rho.
\tag{3.50}
$$

Since $D_X = d_X D$, $E_X = e_X E$, and also $D_X = \varepsilon_0 n_X^2 E_X$, we obtain, using Eqns. 3.50 and 3.47

$$
\begin{aligned}
d_X = \frac{D_X}{D} &= \frac{\epsilon_0 n_X^2 E e_X}{D} \\
&= \frac{\epsilon_0 n_X^2 E}{\epsilon_0 n^2 E \cos \rho} e_X \\
&= \frac{n_X^2 m_X}{n^2 - n_X^2} \tan \rho.
\end{aligned}
\tag{3.51}
$$

Similarly,

$$
d_Y = \frac{n_Y^2 m_Y}{n^2 - n_Y^2} \tan \rho \quad \text{and} \quad d_Z = \frac{n_Z^2 m_Z}{n^2 - n_Z^2} \tan \rho.
\tag{3.52}
$$

3.4.2 Values of n for a Given Propagation Direction

Since $\mathbf{D}$ and $\hat{m}$ are perpendicular to each other, i.e., $\hat{m} \cdot \hat{d} = 0$, we obtain from Eqns. 3.51 and 3.52

$$
\frac{n_X^2 m_X^2}{n^2 - n_X^2} + \frac{n_Y^2 m_Y^2}{n^2 - n_Y^2} + \frac{n_Z^2 m_Z^2}{n^2 - n_Z^2} = 0.
\tag{3.53}
$$

Equation 3.53 easily reduces to a quadratic equation in n^2, which can be expressed as

$$
\mathcal{A} n^4 - \mathcal{B} n^2 + \mathcal{C} = 0
\tag{3.54}
$$

where

$$
\begin{aligned}
\mathcal{A} &= n_X^2 \, m_X^2 + n_Y^2 \, m_Y^2 + n_Z^2 \, m_Z^2 \\
\mathcal{B} &= n_X^2 \, m_X^2 (n_Y^2 + n_Z^2) + n_Y^2 \, m_Y^2 (n_Z^2 + n_X^2) + n_Z^2 \, m_Z^2 (n_X^2 + n_Y^2) \\
\mathcal{C} &= n_X^2 \, n_Y^2 \, n_Z^2.
\end{aligned}
\tag{3.55}
$$

The discriminant of the Eqs. 3.54 is $\mathcal{D}^2$, where

$$\mathcal{D} \equiv (\mathcal{B}^2 - 4\mathcal{A}\mathcal{C})^{1/2}. \tag{3.56}$$

If the values of n_X, n_Y, n_Z, and m_X, m_Y, m_Z are such that $\mathcal{B}^2$ is greater than $4\mathcal{A}\mathcal{C}$, then $\mathcal{D}$ is real and it is positive by definition. Equation 3.54 then has two roots for n^2, from which we get two possible values of n. Denoting these two values of n by n_s and n_f we have

$$n_s = \left(\frac{\mathcal{B} + \mathcal{D}}{2\mathcal{A}} \right)^{1/2} \quad \text{and} \quad n_f = \left(\frac{\mathcal{B} - \mathcal{D}}{2\mathcal{A}} \right)^{1/2}. \tag{3.57}$$

Since $\mathcal{B}$ is positive by its definition and $\mathcal{D}$ is less than or equal to $\mathcal{B}$ (from Eqs. 3.56), $n_s \geq n_f$. A wave propagating in the direction $\hat{m}$ in an anisotropic medium can have these two values for refractive index, corresponding to which there can be two waves traveling with speeds c/n_s and c/n_f for the given direction of propagation. These two waves are called the "slow" and "fast" waves, respectively, since the speed c/n_s is less than the speed c/n_f. Equations 3.47, 3.51, and 3.52 show that the directions of the unit vectors $\hat{e}$ and $\hat{d}$ depend on n, so for each direction of $\hat{m}$ there are two unit vectors $\hat{e}$, say denoted by $\hat{e}_s$ and $\hat{e}_f$ for $n = n_s$ and n_f, respectively, and correspondingly two unit vectors $\hat{d}_s$ and $\hat{d}_f$.

3.4.3   Directions of D and E for the Slow and Fast Waves

The components of $\hat{e}_s$, $\hat{e}_f$, $\hat{d}_s$, and $\hat{d}_f$ can be directly obtained from Eqns. 3.47, 3.51, and 3.52 by substituting n_s and n_f in place of n. From these equations we obtain

$$
\begin{aligned}
e_{sX} &= \frac{n_s^2 m_X}{n_s^2 - n_X^2} \sin \rho_s & e_{fX} &= \frac{n_f^2 m_X}{n_f^2 - n_X^2} \sin \rho_f \\[2mm]
e_{sY} &= \frac{n_s^2 m_Y}{n_s^2 - n_Y^2} \sin \rho_s & e_{fY} &= \frac{n_f^2 m_Y}{n_f^2 - n_Y^2} \sin \rho_f \\[2mm]
e_{sZ} &= \frac{n_s^2 m_Z}{n_s^2 - n_Z^2} \sin \rho_s & e_{fZ} &= \frac{n_f^2 m_Z}{n_f^2 - n_Z^2} \sin \rho_f
\end{aligned} \tag{3.58}
$$

and

$$
\begin{aligned}
d_{sX} &= \frac{n_X^2 m_X}{n_s^2 - n_X^2} \tan \rho_s & d_{fX} &= \frac{n_X^2 m_X}{n_f^2 - n_X^2} \tan \rho_f \\[2mm]
d_{sY} &= \frac{n_Y^2 m_Y}{n_s^2 - n_Y^2} \tan \rho_s & d_{fY} &= \frac{n_Y^2 m_Y}{n_f^2 - n_Y^2} \tan \rho_f \\[2mm]
d_{sZ} &= \frac{n_Z^2 m_Z}{n_s^2 - n_Z^2} \tan \rho_s & d_{fZ} &= \frac{n_Z^2 m_Z}{n_f^2 - n_Z^2} \tan \rho_f
\end{aligned} \tag{3.59}
$$

and from Eqs. 3.48 the angles ρ_s and ρ_f are given by

$$\sin \rho_s = \frac{1}{n_s^2 \left[\left(\frac{m_X}{n_s^2 - n_X^2} \right)^2 + \left(\frac{m_Y}{n_s^2 - n_Y^2} \right)^2 + \left(\frac{m_Z}{n_s^2 - n_Z^2} \right)^2 \right]^{1/2}}$$

$$\sin \rho_f = \frac{1}{n_f^2 \left[\left(\frac{m_X}{n_f^2 - n_X^2} \right)^2 + \left(\frac{m_Y}{n_f^2 - n_Y^2} \right)^2 + \left(\frac{m_Z}{n_f^2 - n_Z^2} \right)^2 \right]^{1/2}}.$$

$$(3.60)$$

Since Eqs. 3.48 was derived by taking a square root, the numerator on its right hand side could be -1 as well, so the angles ρ_s and ρ_f given in Eqs. 3.60 can be either positive or negative. We define here the angles ρ_s and ρ_f to both be positive. Because of this choice made here, the direction (but not the magnitudes of the components) of $\hat{d}$ determined here contradict that *chosen* in earlier work [7], [8]. We choose to maintain this definition of positive values for ρ_s and ρ_f for the sake of simplicity; the choice of negative values would of course be equally valid.

Rewriting Eqs. 3.53 as

$$\frac{m_X^2}{\frac{1}{n_X^2} - \frac{1}{n^2}} + \frac{m_Y^2}{\frac{1}{n_Y^2} - \frac{1}{n^2}} + \frac{m_Z^2}{\frac{1}{n_Z^2} - \frac{1}{n^2}} = 0, \qquad (3.61)$$

substituting n_s and n_f for n in succession, and subtracting we obtain

$$\frac{m_X^2}{\left(\frac{1}{n_X^2} - \frac{1}{n_s^2} \right) \left(\frac{1}{n_X^2} - \frac{1}{n_f^2} \right)} + \frac{m_Y^2}{\left(\frac{1}{n_Y^2} - \frac{1}{n_s^2} \right) \left(\frac{1}{n_Y^2} - \frac{1}{n_f^2} \right)}$$

$$+ \frac{m_Z^2}{\left(\frac{1}{n_Z^2} - \frac{1}{n_s^2} \right) \left(\frac{1}{n_Z^2} - \frac{1}{n_f^2} \right)}.$$

$$= 0 \qquad (3.62)$$

From Eqs. 3.59 the dot product of the unit vectors $\hat{d}_s$ and $\hat{d}_f$ is then equal to zero, since

$$
\hat{d}_s \cdot \hat{d}_f = \tan\rho_s \tan\rho_f
$$

$$
\times \left[\frac{m_X^2}{\left(\frac{1}{n_X^2} - \frac{1}{n_s^2}\right)\left(\frac{1}{n_X^2} - \frac{1}{n_f^2}\right)} + \frac{m_Y^2}{\left(\frac{1}{n_Y^2} - \frac{1}{n_s^2}\right)\left(\frac{1}{n_Y^2} - \frac{1}{n_f^2}\right)} \right.
$$

$$
\left. + \frac{m_Z^2}{\left(\frac{1}{n_Z^2} - \frac{1}{n_s^2}\right)\left(\frac{1}{n_Z^2} - \frac{1}{n_f^2}\right)} \right] = 0. \tag{3.63}
$$

Thus, displacement vectors $\mathbf{D}_s$ and $\mathbf{D}_f$ of the two plane waves propagating in a given direction in an anisotropic crystal are always perpendicular to each other. For a general propagation direction, the electric field vectors $\mathbf{E}_s$ and $\mathbf{E}_f$ are not necessarily perpendicular to each other.

To summarize, in a given anisotropic crystal, if the directions of the dielectric principal coordinate axes (X, Y, Z) are known (say through x ray diffraction measurements), and the values of the principal refractive indices n_X, n_Y and n_Z are known (through independent optical measurements), then for a beam of light with the propagation vector $\mathbf{k}$ having direction cosines m_X, m_Y, and m_Z with respect to the XYZ axes, there are two possible waves. These two waves, named "slow" and "fast," travel with speeds c/n_s and c/n_f where the values of n_s and n_f are given by Eqns. 3.57. The directions of the displacement vectors $\mathbf{D}_s$ and $\mathbf{D}_f$ of these two waves are given by Eqns. 3.59 and the directions of the electric fields $\mathbf{E}_s$ and $\mathbf{E}_f$ are given by Eqns. 3.58. The angles between the displacement vectors and the electric fields are denoted by ρ_s and ρ_f for the slow and the fast waves, respectively, and are determined by Eqns. 3.60.

These sets of equations provide almost all the information needed to describe light propagation in anisotropic materials for the purposes of this book. We use these equations to determine the characteristics of the polarization components of the slow and fast waves.

Since it is hard to find in the literature a detailed description of the *field components* for light propagation in an arbitrary direction in a biaxial crystal, we provide some calculations in the next section for the cases of interest here, assuming some hypothetical values for n_X, n_Y, and n_Z.

3.5 Characteristics of the Slow and Fast Waves in a Biaxial Crystal

The principal dielectric axes X, Y, and Z are chosen in such a way that the value of n_Y is in between the values of n_X and n_Z. In some crystals (such as cesium triborate, CBO, or lithium triborate, LBO), n_X has the smallest value, and in some other crystals (such as potassium niobate, KNbO$_3$, strontium formate dihydrate, Sr(COOH)$_2$· 2H$_2$O or barium sodium niobate, Ba$_2$NaNb$_5$O$_{15}$), n_X has the largest value of the three principal refractive indices. The five crystals mentioned here all "negative" biaxial crystals, and were chosen as examples to point out that for biaxial crystals the designations of "positive" or "negative" are independent of the ordering of the principal refractive indices in terms of their magnitudes. The definitions of "positive" and "negative" biaxial crystals will be provided later in terms of angles between the optic axes. The designation of "positive" or "negative" to uniaxial crystals does depend on the relative magnitudes of the principal refractive indices.

For biaxial crystals, we name the two cases of n_X the smallest and n_X the largest among the three principal refractive indices as Case 1 and Case 2, respectively, and consider these two cases in detail next. Since these are not "standard" definitions, we will try to define the two cases each time they occur in this book.

The propagation vector **k** is assumed to be oriented with polar and azimuthal angles θ, ϕ with respect to the principal axes X, Y, Z, as shown in Figure 3.3. The unit vector in the direction of **k**, denoted by $\hat{m}$, has components

$$m_X = \sin\theta \, \cos\phi \quad m_Y = \sin\theta \, \sin\phi \quad m_Z = \cos\theta. \quad (3.64)$$

The octants are numbered as shown in Figures 3.4 and 3.5, i.e., the four octants with positive Z (θ between 0 and $\pi/2$) with the angle ϕ going from 0 to $\pi/2$, $\pi/2$ to π, π to $3\pi/2$, and $3\pi/2$ to 2π, are numbered 1, 2, 3, and 4, respectively. Similarly, the four octants with negative Z (θ between $\pi/2$ and π), with the angle ϕ going over the same quadrants in order, are numbered 5, 6, 7 and 8, respectively.

We will assume here that $\hat{m}$ lies in the first octant, i.e., θ and ϕ both range from 0 to 90°, so that m_X, m_Y, and m_Z are all positive.

To illustrate the dependence of n_s, n_f, ρ_s, ρ_f, $\hat{d}_s$, $\hat{d}_f$, and $\hat{e}_s$, $\hat{e}_f$ on the propagation direction (characterized by the angles θ and ϕ), we assume a hypothetical crystal having the values $n_X = 1.65$, $n_Y = 1.75$, and $n_Z = 1.95$ for Case 1, and $n_X = 1.95$, $n_Y = 1.75$, and $n_Z = 1.65$ for Case 2.

3.5.1 n_s and n_f

For Case 1, ($n_X < n_Y < n_Z$), substituting Eqns. 3.64 in Eqns. 3.55 and 3.56, it can be shown that n_s ranges from n_Y to n_Z and that n_f ranges from n_X to

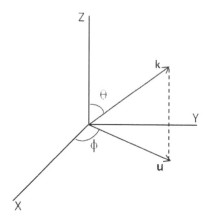

FIGURE 3.3

The orientation of **k** and **û** with respect to the principal axes X, Y, and Z.

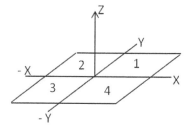 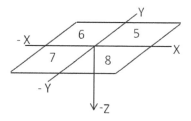

FIGURE 3.4

The four octants with positive Z, i.e., in the upper half of the XY plane. For a vector pointing in the octants 1, 2, 3, and 4, the angle θ is less than 90°, and the angle ϕ is between 0 and 90° for octant 1, between 90° and 180° for octant 2, between 180° and 270° for octant 3, and between 270 and 360° for octant 4.

FIGURE 3.5

The four octants with negative Z, i.e., in the lower half of the XY plane. For a vector pointing in the octants 5, 6, 7, and 8, the angle θ is greater than 90°, and the angle ϕ is between 0 and 90° for octant 5, between 90° and 180° for octant 6, between 180° and 270° for octant 7, and between 270 and 360° for octant 8.

n_Y. Similarly for Case 2, $(n_X > n_Y > n_Z)$, it can be shown that n_s ranges from n_Y to n_X and that n_f ranges from n_Z to n_Y.

The dependence of the values of n_s and n_f on θ for the hypothetical crystals described above (one with $n_X < n_Y < n_Z$ and the other with $n_X > n_Y > n_Z$) are shown in Figures 3.6 and 3.7 for a few values of ϕ.

Figure 3.6 shows that at $\phi = 0°$, as θ increases from 0 to 38°, the value of

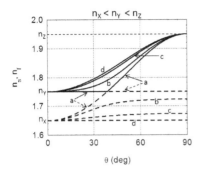

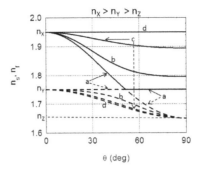

FIGURE 3.6

Dependence of n_s (shown by solid lines) and n_f (shown by dashed lines) on θ for different values of ϕ for a hypothetical crystal with $n_X < n_Y < n_Z$: (a) $\phi = 0°$, (b) $\phi = 30°$, (c) $\phi = 60°$, and (d) $\phi = 90°$.

FIGURE 3.7

Dependence of n_s (shown by solid lines) and n_f (shown by dashed lines) on θ for different values of ϕ for a hypothetical crystal with $n_X > n_Y > n_Z$: (a) $\phi = 0°$, (b) $\phi = 30°$, (c) $\phi = 60°$, and (d) $\phi = 90°$.

n_s is constant at n_Y, and n_f rises from n_X to n_Y. For θ increasing from 38° to 90°, n_s rises from n_Y to n_Z and n_f stays constant at n_Y. The values of θ and ϕ at which n_s and n_f are equal to each other provide the direction of the optic axis of the crystal ($\theta = 38°$, $\phi = 0°$). At values of ϕ other than 0°, n_s and n_f are not equal to each other for any value of θ.

Figure 3.7 shows that at $\phi = 0°$, as θ increases from 0 to 52°, the value of n_f is constant at n_Y, and n_s falls from n_X to n_Y. For θ increasing from 52° to 90°, n_f falls from n_Y to n_Z and n_s stays constant at n_Y. The values of θ, ϕ at which n_s and n_f are equal to each other provide the direction of the optic axis of the crystal for this case ($\theta = 52°$, $\phi = 0°$).

3.5.2 ρ_s and ρ_f

Figures 3.8 and 3.9 show the walk-off angles ρ_s and ρ_f for the two cases, $n_X < n_Y < n_Z$ and $n_X > n_Y > n_Z$, respectively, for the hypothetical crystal. For both cases, the walk-off angles are largest for $\phi = 0°$, i.e., on the YZ plane.

3.5.3 The Components of $\hat{d}_s$ and $\hat{d}_f$

The values of the components of the unit vectors $\hat{d}_s$ and $\hat{d}_f$ for the two cases, $n_X < n_Y < n_Z$ and $n_X > n_Y > n_Z$ are shown in Figures 3.10 through 3.15. Since $\tan \rho_s$ and $\tan \rho_f$ are defined to be positive and $n_s > n_x, n_s > n_y, n_s < n_z, n_f > n_x, n_f < n_y, n_f < n_z$, Eqns. 3.59 show that d_{sX} is positive,

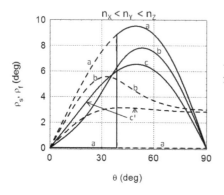

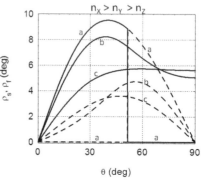

FIGURE 3.8
The walk-off angles ρ_s (solid lines) of the slow wave and ρ_f (dashed lines) of the fast wave as functions of the polar angle θ for a hypothetical crystal with $n_X < n_Y < n_Z$, for three values of ϕ: (a) $\phi = 0$, (b) $\phi = 30°$, and (c) $\phi = 60°$.

FIGURE 3.9
The walk-off angles ρ_s (solid lines) of the slow wave and ρ_f (dashed lines) of the fast wave as functions of the polar angle θ for a hypothetical crystal with $n_X > n_Y > n_Z$, for three values of ϕ: (a) $\phi = 0$, (b) $\phi = 30°$, and (c) $\phi = 60°$.

d_{sY} is positive, and d_{sZ} is negative, while d_{fX} is positive, d_{fY} is negative, and d_{fZ} is negative in Case 1 $(n_X < n_Y < n_Z)$. Thus for $\hat{m}$ in the first octant, the unit vectors $\hat{d}_s$ and $\hat{d}_f$ point in the fifth and eighth octants, respectively.

Similarly, for Case 2, $(n_X > n_Y > n_Z)$, it can be shown that n_s ranges from n_Y to n_X and that n_f ranges from n_Z to n_Y. Again, with $\tan\rho_s$ and $\tan\rho_f$ *defined* to be positive, Eqns. 3.59 show that when m is in the first octant, d_{sX} is negative, d_{sY} is positive, and d_{sZ} is positive, while d_{fX} is negative, d_{fY} is negative, and d_{fZ} is positive in this case, i.e., the unit vectors $\hat{d}_s$ and $\hat{d}_f$ point in the second and the third octant, respectively.

The signs of the components of the unit vectors $\hat{d}_s$ and $\hat{d}_f$ for the two cases are summarized in Table 3.1.

	d_{sX}	d_{sY}	d_{sZ}	d_{fX}	d_{fY}	d_{fZ}
Case 1 $(n_X < n_Y < n_Z)$	pos.	pos.	neg.	pos.	neg.	neg.
Case 2 $(n_X > n_Y > n_Z)$	neg.	pos.	pos.	neg.	neg.	pos.

TABLE 3.1
Signs of the components of the unit vectors $\hat{d}_s$ and $\hat{d}_f$ for Case 1 $(n_X < n_Y < n_Z)$ and case 2 $(n_X > n_Y > n_Z)$

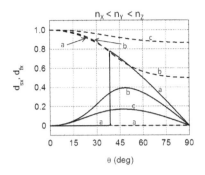

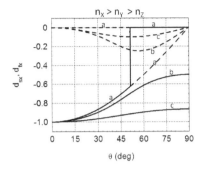

FIGURE 3.10

Dependence of d_{sX} (shown by solid lines) and d_{fX} (shown by dashed lines) on θ for different values of ϕ for a hypothetical crystal with $n_X < n_Y < n_Z$: (a) $\phi = 0$, (b) $\phi = 30°$, and (c) $\phi = 60°$.

FIGURE 3.11

Dependence of d_{sX} (shown by solid lines) and d_{fX} (shown by dashed lines) on θ for different values of ϕ for a hypothetical crystal with $n_X > n_Y > n_Z$: (a) $\phi = 0$, (b) $\phi = 30°$, and (c) $\phi = 60°$.

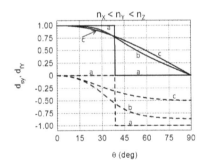

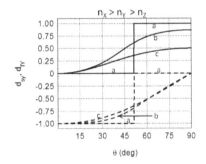

FIGURE 3.12

Dependence of d_{sY} (shown by solid lines) and d_{fY} (shown by dashed lines) on θ for different values of ϕ for a hypothetical crystal with $n_X < n_Y < n_Z$: (a) $\phi = 0°$, (b) $\phi = 30°$, and (c) $\phi = 60°$.

FIGURE 3.13

Dependence of d_{sY} (shown by solid lines) and d_{fY} (shown by dashed lines) on θ for different values of ϕ for a hypothetical crystal with $n_X > n_Y > n_Z$: (a) $\phi = 0$, (b) $\phi = 30°$, and (c) $\phi = 60°$.

3.5.4 The Components of $\hat{e}_s$ and $\hat{e}_f$

When the propagation vector k does not lie along a principal axis or on a principal plane, the components of the unit vectors $\hat{e}_s$ and $\hat{e}_f$ are different from those of the unit vectors $\hat{d}_s$ and $\hat{d}_f$. However, when k does lie on a

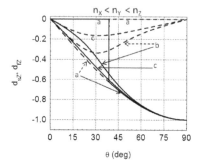

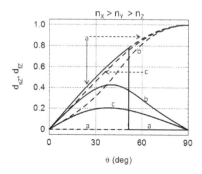

FIGURE 3.14

Dependence of d_{sZ} (shown by solid lines) and d_{fZ} (shown by dashed lines) on θ for different values of ϕ for a hypothetical crystal with $n_X < n_Y < n_Z$: 65a) $\phi = 0$, (b) $\phi = 30°$, and (c) $\phi = 60°$.

FIGURE 3.15

Dependence of d_{sZ} (shown by solid lines) and d_{fZ} (shown by dashed lines) on θ for different values of ϕ for a hypothetical crystal with $n_X > n_Y > n_Z$: 65a) $\phi = 0$, (b) $\phi = 30°$, and (c) $\phi = 60°$.

principal plane, the components of $\hat{e}_s$ and $\hat{e}_f$ are the same as those of $\hat{d}_s$ and $\hat{d}_f$ for certain ranges of θ or ϕ.

In Figures 3.16 and 3.17 the components of d_s and e_s are shown as functions of θ for an azimuthal angle $\phi = 45°$. Similarly, in Figures 3.18 and 3.19 the components of d_f and e_f are shown as functions of θ for an azimuthal angle $\phi = 45°$.

The discussion above shows how the directions of the **D** and **E** vectors can be determined for an arbitrary direction of the **k** vector in a biaxial crystal. For general values of m_X and m_Y, the expression for $\mathcal{D}$ given in Eqs. 3.56 does not reduce to a simple form so that the solutions for n_s and n_f in Eqns. 3.57 are algebraically complex, although computationally quite simple. For special cases, such as propagation along the principal axes or along principal planes, the expression for $\mathcal{D}$ and the expressions for n_s and n_f derived from it are much simpler. Also, in the special case of uniaxial crystals in which two of the principal refractive indices are equal, expressions for $\mathcal{D}$, n_s, and n_f are very similar to those obtained for propagation along principal planes of biaxial crystals. These special cases will be discussed later in this chapter. Before doing that, the discussion of the biaxial crystal is continued with a description of the optic axes. The values of the components of the **D** vector have been derived in the literature in terms of the angle between the optic axes. We will re-derive those expressions and will also derive the directions of the components of the **E** field in the next three sections.

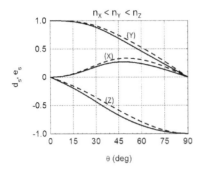

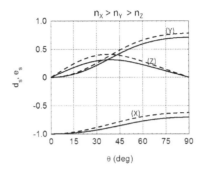

FIGURE 3.16

The components of $\hat{d}_s$ (solid lines) and $\hat{e}_s$ (dashed lines) as functions of the angle θ for $\phi = 45°$ for a hypothetical crystal with $n_X < n_Y < n_Z$: (X) indicates components $\hat{d}_{sX}, \hat{e}_{sX}$; (Y) indicates components $\hat{d}_{sY}, \hat{e}_{sY}$; and (Z) indicates the components $\hat{d}_{sZ}, \hat{e}_{sZ}$.

FIGURE 3.17

The components of $\hat{d}_s$ (solid lines) and $\hat{e}_s$ (dashed lines) as functions of the angle θ for $\phi = 45°$ for a hypothetical crystal with $n_X > n_Y > n_Z$: (X) indicates components d_{sX}, e_{sX}; (Y) indicates components d_{sY}, e_{sY}; and (Z) indicates the components d_{sZ}, e_{sZ}.

3.6 Double Refraction and Optic Axes

A plane wave traveling in an anisotropic medium with its wave vector **k** pointing in an arbitrary direction can have two speeds c/n_s and c/n_f, with n_s and n_f given by Eqs. 3.57. The directions of the unit vectors $\hat{e}_s$ and $\hat{e}_f$ along the electric field vectors associated with those waves are given by Eqns. 3.58 and 3.60. $\hat{s}_s$ and $\hat{s}_f$, the unit vectors in the directions of the corresponding Poynting vectors, can be obtained from $\hat{e}_s$, $\hat{e}_f$ and $\hat{m}$ using the relations

$$\hat{s}_s = \hat{e}_s \times (\hat{m} \times \hat{e}_s) = \hat{m} - \hat{e}_s(\hat{e}_s \cdot \hat{m})$$
$$\hat{s}_f = \hat{e}_f \times (\hat{m} \times \hat{e}_f) = \hat{m} - \hat{e}_f(\hat{e}_f \cdot \hat{m}), \qquad (3.65)$$

which can be derived from Eqns. 3.5 and 3.10.

Since $\hat{e}_s$ and $\hat{e}_f$ point in different directions, the directions of $\hat{s}_s$ and $\hat{s}_f$ are also different, i.e., the slow and fast waves having the same wave vector **k** have different directions for energy propagation, causing the two waves to be spatially separated after traveling a certain distance through the medium. This phenomenon is given the name *double refraction* and is also called *birefringence*.

It will be shown later that when $\hat{\mathbf{k}}$ is along the principal axes X, Y, or Z, the Poynting vector **S** is along $\hat{\mathbf{k}}$ for both the slow and the fast waves,

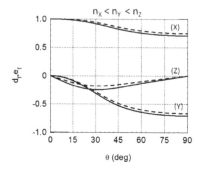

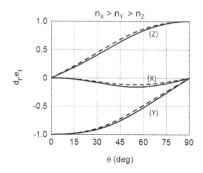

FIGURE 3.18
The components of $\hat{d}_f$ (solid lines) and $\hat{e}_f$ (dashed lines) as functions of the angle θ for $\phi = 45°$ for a hypothetical crystal with $n_X < n_Y < n_Z$: (X) indicates components $\hat{d}_{fX}$, $\hat{e}_{fX}$; (Y) indicates components $\hat{d}_{fY}$, $\hat{e}_{fY}$; and (Z) indicates the components $\hat{d}_{fZ}$, $\hat{e}_{fZ}$.

FIGURE 3.19
The components of d_f (solid lines) and e_f (dashed lines) as functions of the angle θ for $\phi = 45°$ for a hypothetical crystal with $n_X > n_Y > n_Z$: (X) indicates components d_{fX}, e_{fX}; (Y) indicates components d_{fY}, e_{fY}; and (Z) indicates the components d_{fZ}, e_{fZ}.

and there is no double refraction. In addition there are two other directions in biaxial crystals such that light propagating with the $\hat{\mathbf{k}}$ vectors along these directions undergoes no double refraction, i.e., along these directions, $n_s = n_f$ and the speeds of the fast and the slow waves are equal. These additional directions in biaxial crystals are defined to be the directions of the *optic axes*.

The directions of the optic axes can be determined from Section 3.4.2. Equation 3.57 shows that for $n_s = n_f$, $\mathcal{D}$ must be equal to 0, i.e., from Eqs. 3.56, $\mathcal{B}^2 = 4\mathcal{A}\mathcal{C}$. For simplicity of notation, let us define here (locally) three constants a, b, and c, and three variables x, y, and z as

$$a \equiv n_X^2, \quad b \equiv n_Y^2, \quad c \equiv n_Z^2$$
$$x \equiv m_X^2, \quad y \equiv m_Y^2, \quad z \equiv m_Z^2. \tag{3.66}$$

Using the relation $m_X^2 + m_Y^2 + m_Z^2 = 1$, i.e., $x + y + z = 1$, we write Eqs. 3.55 as

$$\mathcal{A} = x(a - c) + d, \quad \mathcal{B} = xbd + e, \quad \mathcal{C} = abc \tag{3.67}$$

where

$$d \equiv c + y(b - c), \text{and} \quad e \equiv ya(b - c) + c(a + b). \tag{3.68}$$

The condition $\mathcal{B}^2 = 4\mathcal{A}\mathcal{C}$ leads to a quadratic equation in x

$$A_1 x^2 + B_1 x + C_1 = 0 \tag{3.69}$$

where

$$A_1 = b^2(a-c)^2, \quad B_1 = 2b(a-c)(e-2ac) \quad \text{and} \quad C_1 = e^2 - 4abcd. \quad (3.70)$$

Equation 3.69 will have real solution(s) for x if the discriminant $D_1^2 = B_1^2 - 4A_1C_1$ is nonnegative. From Eqs. 3.70 we find

$$D_1^2 = -[16b^2ac(a-c)^2(c-b)(b-a)]y^2. \quad (3.71)$$

With the values of n_X, n_Y, and n_Z ordered from low to high or high to low (i.e., with n_Y in the middle), the quantity inside the square brackets in Eqs. 3.71 is always positive. The only way x can have a real solution is therefore with $y = 0$, i.e., with $m_Y = 0$.

With $m_Y = 0$, x and $z(= 1-x)$ are obtained from the solution of Eqs. 3.69:

$$x = m_X^2 = \frac{n_Z^2}{n_Y^2}\left(\frac{n_X^2 - n_Y^2}{n_X^2 - n_Z^2}\right)$$

$$z = m_Z^2 = \frac{n_X^2}{n_Y^2}\left(\frac{n_Y^2 - n_Z^2}{n_X^2 - n_Z^2}\right) \quad (3.72)$$

For these values of m_X and m_Z, the directions of the two optic axes, OA_1 and OA_2 are shown in Figure 3.20, with the angle Ω between the optic axes and the Z direction can be obtained from the ratio of m_X to m_Z:

$$\tan\Omega = \frac{n_Z}{n_X}\left(\frac{n_X^2 - n_Y^2}{n_Y^2 - n_Z^2}\right)^{1/2}. \quad (3.73)$$

Since for anisotropic crystals with $n_X \neq n_Y \neq n_Z$, there are two optical axes,

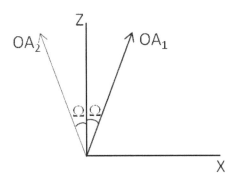

FIGURE 3.20
OA_1 and OA_2 denote the two *optic axes*. When the propagation vector $\mathbf{k}$ is along OA_1 or OA_2, the two waves associated with $\mathbf{k}$ travel with the same speed and there is no birefringence.

both lying in the XZ plane as shown above, these crystals are called *biaxial*. Biaxial crystals are designated **positive** or **negative** depending on whether the angle 2Ω between the two optic axes is less than or greater than $90°$, respectively.

Using the values of 1.65, 1.75, and 1.95 for n_X, n_Y, and n_Z, respectively, we find that Ω is $38°$. For n_X, n_Y, and n_Z, respectively, equal to 1.95, 1.75, and 1.65, Ω is equal to $52°$, confirming the results shown in Figures 3.6 and 3.7.

3.6.1   Expressions for Components of $\hat{d}$ in Terms of the Angles θ, ϕ, and Ω

To obtain expressions for components of $\hat{d}$ in terms of the angles θ, ϕ, and Ω, several new unit vectors pointing at different directions will be introduced next. In an attempt to make it a little easier to follow the logic of the selection of these vectors, we summarize the procedure here: we will consider only **k** vectors lying in the first octant, i.e., with ϕ between 0 and $90°$. Results for other values of ϕ can be determined using the same procedure. Since **D** is perpendicular to **k** (from Sections 3.2.1 and 3.2.2), both the slow and fast components of **D**, i.e, both $\hat{d}_s$ and $\hat{d}_f$, are perpendicular to **k** and therefore they both must lie on a plane perpendicular to **k**. The intersection of this plane (named, say, "the $k_\perp$ plane") with the k-Z plane containing **k** vector and the Z axis is a line, and the unit vector along this line is named $\hat{u}_1$. Thus $\hat{u}_1$ lies on the $k - Z$ plane and is perpendicular to **k**. Another unit vector, perpendicular to $\hat{u}_1$ and lying in the $k_\perp$ plane, is named $\hat{u}_2$. $\hat{u}_1$ and $\hat{u}_2$ are completely determined by the angles θ and ϕ of the **k** vector. Since $\hat{d}_s$ and $\hat{d}_f$ lie on the $\hat{u}_1$–$\hat{u}_2$ plane, we *define* two mutually perpendicular unit vectors named $\hat{d}_1$ and $\hat{d}_2$, with $\hat{d}_1$ at an angle δ with $\hat{u}_1$. Cartesian components of $\hat{d}_1$ and $\hat{d}_2$ are then determined in terms of the angles θ, ϕ, and δ. While the angle δ is still undetermined, it can be shown that for the two cases under consideration (Case 1, with $n_X < n_Y < n_Z$ and Case 2, with $n_X > n_Y > n_Z$), δ will be positive and acute if we assign $\hat{d}_s = \hat{d}_1$ and $\hat{d}_s = \hat{d}_2$ respectively. Next, the components of the unit vectors $\hat{e}_s$ and $\hat{e}_f$ of the slow and fast components of the electric fields are also determined in terms of the angles θ, ϕ, δ and the walk-off angles ρ_s and ρ_f.

The propagation vector **k** is oriented with polar and azimuthal angles θ, ϕ with respect to the principal axes X, Y, Z, as shown in Figure 3.3, and the components of the unit vector $\hat{m}$ are given by Eqs. 3.64.

As shown in Figure 3.3, the unit vector along the projection of **k** on the XY plane is denoted by $\hat{u}$, so that

$$\hat{u} = \hat{X} \, \cos\phi + \hat{Y} \, \sin\phi. \tag{3.74}$$

Figure 3.21 shows that the unit vector $\hat{u}_1$, lying on the $k - Z$ plane and

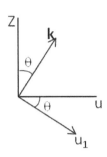

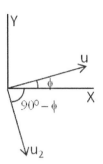

FIGURE 3.21
The orientation of **k** and **u₁** with respect to Z and **u**.

FIGURE 3.22
The orientation of the vectors $\hat{u}$ and $\hat{u}_2$ with respect to X and Y.

perpendicular to **k**, can be written as

$$
\begin{aligned}
\hat{u}_1 &= \hat{u} \, \cos\theta - \hat{Z} \, \sin\theta \\
&= \hat{X} \, \cos\theta \cos\phi + \hat{Y} \, \cos\theta \sin\phi - \hat{Z} \, \sin\theta \qquad (3.75)
\end{aligned}
$$

using Eqs. 3.74.

$\hat{u}_2$ denotes the unit vector perpendicular to both $\hat{u}_1$ and **k**. $\hat{u}_2$ must then also be perpendicular to $\hat{u}$ and Z and can be obtained by taking the cross product of $\hat{u}_1$ and **k** or any two of the vectors $\hat{u}_1$, $\hat{u}$, **k**, and $\hat{Z}$ in the Zu plane. Taking the cross product of the unit vectors $\hat{u}$ and $\hat{Z}$ we find

$$
\begin{aligned}
\hat{u}_2 &= \hat{u} \times \hat{Z} \\
&= \begin{vmatrix} \hat{X} & \hat{Y} & \hat{Z} \\ \cos\phi & \sin\phi & 0 \\ 0 & 0 & 1 \end{vmatrix} \\
&= \hat{X} \, \sin\phi - \hat{Y} \, \cos\phi \qquad (3.76)
\end{aligned}
$$

showing that $\hat{u}_2$ lies on the XY plane, along with $\hat{u}$ (Figure 3.22).

In the plane u_1–u_2 formed by the vectors $\hat{u}_1$ and $\hat{u}_2$, another pair of mutually perpendicular unit vectors $\hat{d}_1$ and $\hat{d}_2$ are drawn, with δ denoting the angle between $\hat{d}_1$ and $\hat{u}_1$ as shown in Figure 3.23.

Figure 3.23 shows that the unit vectors $\hat{d}_1$ and $\hat{d}_2$ can be expressed in the X, Y, Z coordinate system as

$$
\begin{aligned}
\hat{d}_1 &= \hat{u}_1 \, \cos\delta - \hat{u}_2 \, \sin\delta \\
&= \hat{X} \, (\cos\theta \, \cos\phi \, \cos\delta - \sin\phi \, \sin\delta) \\
&\quad + \hat{Y} \, (\cos\theta \, \sin\phi \, \cos\delta + \cos\phi \, \sin\delta) - \hat{Z} \, \sin\theta \cos\delta \qquad (3.77)
\end{aligned}
$$

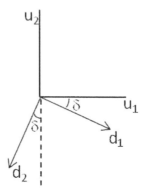

FIGURE 3.23
The orientation of $\mathbf{d_1}$ and $\mathbf{d_2}$ with respect to $\mathbf{u_1}$ and $\mathbf{u_2}$.

and

$$
\begin{aligned}
\hat{d}_2 &= -\hat{u}_1 \, \sin\delta - \hat{u}_2 \, \cos\delta \\
&= -\hat{X} \, (\cos\theta \, \cos\phi \, \sin\delta + \sin\phi \, \cos\delta) \\
&\quad -\hat{Y} \, (\cos\theta \, \sin\phi \, \sin\delta - \cos\phi \, \cos\delta) + \hat{Z} \, \sin\theta \sin\delta \quad (3.78)
\end{aligned}
$$

using Eqns. 3.75 and 3.76. The components of $\hat{d}_1$ and $\hat{d}_2$ are then given by

$$
\begin{aligned}
d_{1X} &= \cos\theta \, \cos\phi \, \cos\delta - \sin\phi \, \sin\delta \\
d_{1Y} &= \cos\theta \, \sin\phi \, \cos\delta + \cos\phi \, \sin\delta \\
d_{1Z} &= -\sin\theta \, \cos\delta
\end{aligned} \quad (3.79)
$$

and

$$
\begin{aligned}
d_{2X} &= -\cos\theta \, \cos\phi \, \sin\delta - \sin\phi \, \cos\delta \\
d_{2Y} &= -\cos\theta \, \sin\phi \, \sin\delta + \cos\phi \, \cos\delta \\
d_{2Z} &= \sin\theta \sin\delta.
\end{aligned} \quad (3.80)
$$

3.6.2   Relating the Angle δ to Ω, θ, and ϕ

To relate the angle δ to Ω, θ, and ϕ we start from the Eqns. 3.59 and express the refractive indices n_X, n_Y, and n_Z in terms of the Cartesian components of the unit vectors $\hat{d}_s$ as

$$
n_X^2 = \frac{n_s^2 d_{sX}}{d_{sX} + m_X \tan\rho_s} \quad , \quad n_Y^2 = \frac{n_s^2 d_{sY}}{d_{sY} + m_Y \tan\rho_s} \quad \text{and}
$$

$$
n_Z^2 = \frac{n_s^2 d_{sZ}}{d_{sZ} + m_Z \tan\rho_s}. \quad (3.81)
$$

Inserting these values of n_X, n_Y, and n_Z in Eqs. 3.73 we obtain (after a little algebra)

$$\cot^2 \Omega = \frac{d_{sX}}{d_{sZ}} \frac{m_Z d_{sY} - m_Y d_{sZ}}{m_Y d_{sX} - m_X d_{sY}}. \tag{3.82}$$

The angle δ has not yet been defined except for stipulating that it is the angle between some vector $\hat{d}_1$ and the vector $\hat{u}_1$. δ gets defined when we identify the vector $\hat{d}_1$ with the unit vector $\hat{d}_s$, i.e., when we require

$$\hat{d}_s = \hat{d}_1, \tag{3.83}$$

Then, using Eqns. 3.79 and 3.64, we obtain

$$
\begin{aligned}
m_Z d_{sY} - m_Y d_{sZ} &= m_Z d_{1Y} - m_Y d_{1Z} \\
&= \cos\theta(\cos\theta \sin\phi \cos\delta + \cos\phi \sin\delta) \\
&\quad + \sin\theta \sin\phi(\sin\theta \cos\delta) \\
&= \sin\phi \cos\delta + \cos\theta \cos\phi \sin\delta, \tag{3.84}
\end{aligned}
$$

and

$$
\begin{aligned}
m_Y d_{sX} - m_X d_{sy} &= m_Y d_{1X} - m_X d_{1X} \\
&= \sin\theta \sin\phi(\cos\theta \cos\phi \cos\delta - \sin\phi \sin\delta) \tag{3.85} \\
&\quad - \sin\theta \cos\phi(\cos\theta \sin\phi \cos\delta + \cos\phi \sin\delta) \\
&= -\sin\theta \sin\delta \tag{3.86}
\end{aligned}
$$

and

$$
\begin{aligned}
\frac{d_{sX}}{d_{sZ}} &= \frac{d_{1X}}{d_{1Z}} \\
&= -\frac{\cos\theta \cos\phi \cos\delta - \sin\phi \sin\delta}{\sin\theta \cos\delta} \tag{3.87}
\end{aligned}
$$

inserting which in Eqs. 3.82 we get

$$\cot^2 \Omega = \frac{(\cos\theta \, \cos\phi \, \cos\delta - \sin\phi \, \sin\delta)(\cos\theta \, \cos\phi \, \sin\delta + \sin\phi \, \cos\delta)}{\sin^2\theta \, \sin\delta \, \cos\delta} \tag{3.88}$$

which can be solved for the angle δ

$$\cot 2\delta = \frac{\cot^2 \Omega \, \sin^2\theta + \sin^2\phi - \cos^2\theta \cos^2\phi}{\cos\theta \sin 2\phi}. \tag{3.89}$$

If instead of the choice made in Eqs. 3.83, we make the alternate choice

$$\hat{d}_s = \hat{d}_2 \tag{3.90}$$

and go through the same algebra as above, we get the same expression for δ as given in Eqs. 3.89.

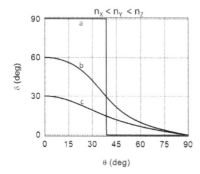

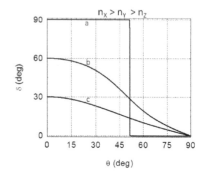

FIGURE 3.24

The angle δ is plotted against θ for three values of the angle ϕ for case 1, with $n_X = 1.65$, $n_Y = 1.75$, and $n_Z = 1.95$. (a) $\phi = 0°$, (b) $\phi = 30°$, and (c) $\phi = 60°$. At $\phi = 90°$, $\delta = 0$ (not shown here).

FIGURE 3.25

The angle δ is plotted against θ for three values of the angle ϕ for case 2, with $n_X = 1.95$, $n_Y = 1.75$, and $n_Z = 1.65$. (a) $\phi = 0°$, (b) $\phi = 30°$, and (c) $\phi = 60°$. At $\phi = 90°$, $\delta = 0$ (not shown here).

The dependence of δ on θ is shown for the two cases of $n_X < n_Y < n_Z$ and $n_X > n_Y > n_Z$ in Figures 3.24 and 3.25 for three values of the angle ϕ.

For a given crystal with Ω known from Eqs. 3.73, δ can be calculated from Eqs. 3.89 for any values of θ and ϕ. A series of values of δ, both positive and negative, separated by 90° can satisfy Eqs. 3.89. We adopt the convention here of choosing for δ only the lowest positive value among the series of values. For example, say, $\delta = 40°$ is a solution of Eqs. 3.89. Then the angles of 130°, 220°, 310°, etc. as well as -50°, - 140°, - 230° etc. are also solutions. From all these solutions, *only* $\delta = 40°$ is chosen for the calculation of the components of $\hat{d}$. This is different from the convention adopted in Ref. [8] and cited in Ref. [2] where negative values of δ that satisfy Eqs. 3.89 (and lie between 0 and $-90°$) are chosen to be associated with Case 2 ($n_X > n_Y > n_Z$).

With δ chosen to be positive, we see from Eqns. 3.79 and 3.80 that at $\phi = 0$, the components (d_{1X}, d_{1Y}, d_{1Z}) and (d_{2X}, d_{2Y}, d_{2Z}) are (positive, positive, negative) and (negative, positive, positive), respectively. Thus, comparing with the signs given in Table 3.1 we see that with $\hat{d}_s$ identified with $\hat{d}_1$, $\hat{d}_f$ needs to be along $-\hat{d}_2$ for Case 1, ($n_X < n_Y < n_Z$). Similarly, if $\hat{d}_s$ identified with $\hat{d}_2$, $\hat{d}_f$ needs to be along $-\hat{d}_1$ for Case 2, ($n_X > n_Y > n_Z$).

Writing out the expressions for the components of $\hat{d}_s$ and $\hat{d}_f$ explicitly, we have the following sets of equations for the two cases:

Case 1: $n_X < n_Y < n_Z$

$$\hat{d}_s = \hat{d}_1, \qquad \hat{d}_f = -\hat{d}_2$$

$$
\begin{aligned}
d_{sX} &= \cos\theta \, \cos\phi \, \cos\delta - \sin\phi \, \sin\delta \\
d_{sY} &= \cos\theta \, \sin\phi \, \cos\delta + \cos\phi \, \sin\delta \\
d_{sZ} &= -\sin\theta \, \cos\delta
\end{aligned}
\tag{3.91}
$$

and

$$
\begin{aligned}
d_{fX} &= \cos\theta \, \cos\phi \, \sin\delta + \sin\phi \, \cos\delta \\
d_{fY} &= \cos\theta \, \sin\phi \, \sin\delta - \cos\phi \, \cos\delta \\
d_{fZ} &= -\sin\theta \sin\delta
\end{aligned}
\tag{3.92}
$$

Case 2: $n_X > n_Y > n_Z$

$$\hat{d}_s = \hat{d}_2, \qquad \hat{d}_f = -\hat{d}_1$$

$$
\begin{aligned}
d_{sX} &= -\cos\theta \, \cos\phi \, \sin\delta - \sin\phi \, \cos\delta \\
d_{sY} &= -\cos\theta \, \sin\phi \, \sin\delta + \cos\phi \, \cos\delta \\
d_{sZ} &= \sin\theta \sin\delta
\end{aligned}
\tag{3.93}
$$

and

$$
\begin{aligned}
d_{fX} &= -\cos\theta \, \cos\phi \, \cos\delta + \sin\phi \, \sin\delta \\
d_{fY} &= -\cos\theta \, \sin\phi \, \cos\delta - \cos\phi \, \sin\delta \\
d_{fZ} &= \sin\theta \, \cos\delta
\end{aligned}
\tag{3.94}
$$

3.6.3 Directions of E and S

As discussed earlier, for light propagating in a biaxial crystal with propagation vector **k** in a general direction there are two oscillation directions of the electric fields, with unit vectors $\hat{e}_s$ and $\hat{e}_f$, which are different from the unit vectors $\hat{d}_s$ and $\hat{d}_f$, along the electric displacements. Since the Poynting vector (**S**) is perpendicular to the electric field, for a given **k** there are also two directions of **S** along which the light energy travels. Suppose the unit vectors along the two possible directions of **S** are denoted by $\hat{s}_s$ and $\hat{s}_f$. It was shown in Secs. 3.2.1 and 3.2.2 that the vectors $\hat{e}_s$, $\hat{d}_s$, $\hat{s}_s$, and **k** lie in one plane (named say, P_s) and so do the vectors $\hat{e}_f$, $\hat{d}_f$, $\hat{s}_f$, and **k**, (named say, P_f). Since **k**, $\hat{d}_s$ and $\hat{d}_f$ are mutually perpendicular, the planes P_s and P_f are also perpendicular to each other.

The directions of the vectors $\hat{e}_s$ and $\hat{s}_s$ in the plane P_s and the vectors $\hat{e}_f$ and $\hat{s}_f$ in the plane P_f are shown in Figures 3.26 and 3.27.

As shown earlier in Sec. 3.4.3, $\hat{d}_s$ and $\hat{d}_f$ are always perpendicular to each other, but in general $\hat{e}_s$ is not perpendicular to $\hat{e}_f$. The directions of $\hat{e}_s$ and

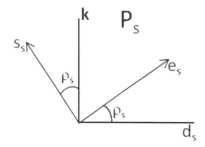

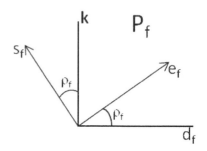

FIGURE 3.26

The unit vectors along the electric field, the displacement vector and the Poynting vector, for the slow wave.

FIGURE 3.27

The unit vectors along the electric field, the displacement vector and the Poynting vector, for the fast wave.

$\hat{e}_f$ can be completely determined from the directions of $\hat{d}_s$ and $\hat{d}_f$, as shown in Figures 3.26 and 3.27.

If ρ_s and ρ_f denote the angles between $\hat{e}_s$ and $\hat{d}_s$ and between $\hat{e}_f$ and $\hat{d}_f$, respectively, Figures 3.26 and 3.27 show that

$$\begin{aligned}
\hat{e}_s &= \hat{d}_s \, \cos\rho_s + \hat{m} \, \sin\rho_s \\
\hat{s}_s &= -\hat{d}_s \, \sin\rho_s + \hat{m} \, \cos\rho_s \\
\hat{e}_f &= \hat{d}_f \, \cos\rho_f + \hat{m} \, \sin\rho_f \\
\hat{s}_f &= -\hat{d}_f \, \sin\rho_f + \hat{m} \, \cos\rho_f
\end{aligned} \qquad (3.95)$$

where as before, $\hat{m}$ is the unit vector along **k**. Since the Cartesian components of $\hat{d}_s$, $\hat{d}_f$, and $\hat{m}$ are known (from Eqns. 3.91 to 3.94 and 3.64), the components of $\hat{e}_s$, $\hat{s}_s$, $\hat{e}_f$, and $\hat{s}_f$ can be obtained from Eqns. 3.95. Thus, for Case 1, $(n_X < n_Y < n_Z)$

$$\begin{aligned}
e_{sX} &= \cos\theta \, \cos\phi \, \cos\delta \, \cos\rho_s - \sin\phi \, \sin\delta \, \cos\rho_s + \sin\theta \, \cos\phi \, \sin\rho_s \\
e_{sY} &= \cos\theta \, \sin\phi \, \cos\delta \, \cos\rho_s + \cos\phi \, \sin\delta \, \cos\rho_s + \sin\theta \, \sin\phi \, \sin\rho_s \\
e_{sZ} &= -\sin\theta \, \cos\delta \, \cos\rho_s + \cos\theta \, \sin\rho_s
\end{aligned} \qquad (3.96)$$

$$\begin{aligned}
s_{sX} &= -\cos\theta \, \cos\phi \, \cos\delta \, \sin\rho_s + \sin\phi \, \sin\delta \, \sin\rho_s + \sin\theta \, \cos\phi \, \cos\rho_s \\
s_{sY} &= -\cos\theta \, \sin\phi \, \cos\delta \, \sin\rho_s - \cos\phi \, \sin\delta \, \sin\rho_s + \sin\theta \, \sin\phi \, \cos\rho_s \\
s_{sZ} &= \sin\theta \, \cos\delta \, \sin\rho_s + \cos\theta \, \cos\rho_s
\end{aligned} \qquad (3.97)$$

$$\begin{aligned}
e_{fX} &= \cos\theta \, \cos\phi \, \sin\delta \, \cos\rho_f + \sin\phi \, \cos\delta \, \cos\rho_f + \sin\theta \, \cos\phi \, \sin\rho_f \\
e_{fY} &= \cos\theta \, \sin\phi \, \sin\delta \, \cos\rho_f - \cos\phi \, \cos\delta \, \cos\rho_f + \sin\theta \, \sin\phi \, \sin\rho_f \\
e_{fZ} &= -\sin\theta \, \sin\delta \, \cos\rho_f + \cos\theta \, \sin\rho_f
\end{aligned} \qquad (3.98)$$

$$\begin{aligned}
s_{fX} &= -\cos\theta\ \cos\phi\ \sin\delta\ \sin\rho_f - \sin\phi\ \cos\delta\ \sin\rho_f + \sin\theta\ \cos\phi\ \cos\rho_f \\
s_{fY} &= -\cos\theta\ \sin\phi\ \sin\delta\ \sin\rho_f + \cos\phi\ \cos\delta\ \sin\rho_f + \sin\theta\ \sin\phi\ \cos\rho_f \\
s_{fZ} &= +\sin\theta\ \sin\delta\ \sin\rho_f + \cos\theta\ \cos\rho_f
\end{aligned} \tag{3.99}$$

and for Case 2, $(n_X > n_Y > n_Z)$

$$\begin{aligned}
e_{sX} &= -\cos\theta\ \cos\phi\ \sin\delta\ \cos\rho_s - \sin\phi\ \cos\delta\ \cos\rho_s + \sin\theta\ \cos\phi\ \sin\rho_s \\
e_{sY} &= -\cos\theta\ \sin\phi\ \sin\delta\ \cos\rho_s + \cos\phi\ \cos\delta\ \cos\rho_s + \sin\theta\ \sin\phi\ \sin\rho_s \\
e_{sZ} &= \sin\theta\ \sin\delta\ \cos\rho_s + \cos\theta\ \sin\rho_s
\end{aligned} \tag{3.100}$$

$$\begin{aligned}
s_{sX} &= \cos\theta\ \cos\phi\ \sin\delta\ \sin\rho_s + \sin\phi\ \cos\delta\ \sin\rho_s + \sin\theta\ \cos\phi\ \cos\rho_s \\
s_{sY} &= \cos\theta\ \sin\phi\ \sin\delta\ \sin\rho_s - \cos\phi\ \cos\delta\ \sin\rho_s + \sin\theta\ \sin\phi\ \cos\rho_s \\
s_{sZ} &= -\sin\theta\ \sin\delta\ \sin\rho_s + \cos\theta\ \cos\rho_s
\end{aligned} \tag{3.101}$$

$$\begin{aligned}
e_{fX} &= -\cos\theta\ \cos\phi\ \cos\delta\ \cos\rho_f + \sin\phi\ \sin\delta\ \cos\rho_f + \sin\theta\ \cos\phi\ \sin\rho_f \\
e_{fY} &= -\cos\theta\ \sin\phi\ \cos\delta\ \cos\rho_f - \cos\phi\ \sin\delta\ \cos\rho_f + \sin\theta\ \sin\phi\ \sin\rho_f \\
e_{fZ} &= \sin\theta\ \cos\delta\ \cos\rho_f + \cos\theta\ \sin\rho_f
\end{aligned} \tag{3.102}$$

$$\begin{aligned}
s_{fX} &= \cos\theta\ \cos\phi\ \cos\delta\ \sin\rho_f - \sin\phi\ \sin\delta\ \sin\rho_f + \sin\theta\ \cos\phi\ \cos\rho_f \\
s_{fY} &= \cos\theta\ \sin\phi\ \cos\delta\ \sin\rho_f + \cos\phi\ \sin\delta\ \sin\rho_f + \sin\theta\ \sin\phi\ \cos\rho_f \\
s_{fZ} &= -\sin\theta\ \cos\delta\ \sin\rho_f + \cos\theta\ \cos\rho_f.
\end{aligned} \tag{3.103}$$

3.6.4 The Walk-Off Angles ρ_s and ρ_f

The walk-off angles ρ_s and ρ_f expressed in Eqs. 3.60 in terms of n_s and n_f can be used in the last section to find the components of $\hat{e}$ and $\hat{s}$. Here we provide alternate expressions for ρ_s and ρ_f in terms of the components of $\hat{d}$. From Eqs. 3.32 we find

$$\begin{aligned}
\frac{e_{sZ}}{e_{sY}} &= \frac{n_Y^2}{n_Z^2}\frac{d_{sZ}}{d_{sY}} \\
\frac{e_{fZ}}{e_{fY}} &= \frac{n_Y^2}{n_Z^2}\frac{d_{fZ}}{d_{fY}}.
\end{aligned} \tag{3.104}$$

Using Eqns. 3.95 we obtain

$$\begin{aligned}
\frac{d_{sZ}\ \cos\rho_s + m_Z\ \sin\rho_s}{d_{sY}\ \cos\rho_s + m_Y\ \sin\rho_s} &= \frac{n_Y^2}{n_Z^2}\frac{d_{sZ}}{d_{sY}} \\
\frac{d_{fZ}\ \cos\rho_f + m_Z\ \sin\rho_f}{d_{fY}\ \cos\rho_f + m_Y\ \sin\rho_f} &= \frac{n_Y^2}{n_Z^2}\frac{d_{fZ}}{d_{fY}}
\end{aligned} \tag{3.105}$$

To derive Eqns. 3.105 we started from the ratio of the Z and Y components of the $\hat{e}$ and $\hat{d}$ fields in Eqs. 3.104. However, any other ratios (such as that of the Y to the X components or the X to the Z components) could equally well be taken to obtain the walk-off angles ρ_s and ρ_f. Using Eqns. 3.105 and similar equations for the other ratios, we obtain

$$
\begin{aligned}
\tan \rho_s &= \frac{\left(n_Z^2 - n_Y^2\right) d_{sZ} d_{sY}}{m_Y n_Y^2 d_{sZ} - m_Z n_Z^2 d_{sY}} = \frac{\left(n_Z^2 - n_X^2\right) d_{sZ} d_{sX}}{m_X n_X^2 d_{sZ} - m_Z n_Z^2 d_{sX}} \\
&= \frac{\left(n_X^2 - n_Y^2\right) d_{sX} d_{sY}}{m_Y n_Y^2 d_{sX} - m_X n_X^2 d_{sY}} \\
\tan \rho_f &= \frac{\left(n_Z^2 - n_Y^2\right) d_{fZ} d_{fY}}{m_Y n_Y^2 d_{fZ} - m_Z n_Z^2 d_{fY}} = \frac{\left(n_Z^2 - n_X^2\right) d_{fZ} d_{fX}}{m_X n_X^2 d_{fZ} - m_Z n_Z^2 d_{fX}} \\
&= \frac{\left(n_X^2 - n_Y^2\right) d_{fX} d_{fY}}{m_Y n_Y^2 d_{fX} - m_X n_X^2 d_{fY}}
\end{aligned}
$$

$$(3.106)$$

which can be expressed in terms of n_X, n_Y, n_Z and the angles θ, ϕ and δ using Eqns. 3.91 and 3.94.

3.6.5   An Interim Summary

Before moving to the next section, we summarize here what has been described thus far and present a preview of what will come next in this chapter:

1. For light traveling in an anisotropic crystal, the directions of the displacement vector **D** and the electric field vector **E** are in general different. The angle between the **D** and **E** vectors is the "walk-off" angle ρ, given in Eqs. 3.48, which is also the angle between the directions of the Poynting vector **S** and the propagation vector **k**.

2. Light traveling in a crystal with known values of the principal refractive indices n_X, n_Y, n_Z will have two waves (a *slow* wave and a *fast* wave) propagating with two speeds (c/n_s and c/n_f, respectively) associated with each direction of the propagation vector **k**. The values of n_s and n_f are given in Eqs. 3.57.

3. The displacement vectors of the slow wave and the fast wave are perpendicular to each other. The unit vectors $\hat{d}_s$ and $\hat{d}_f$ of the two waves along these perpendicular directions are given in Eqs. 3.59.

4. In general, the electric field directions of the *slow* and the *fast* waves are not perpendicular to each other. The unit vectors $\hat{e}_s$ and $\hat{e}_f$ of the two waves along the directions of the two electric field vectors are given in Eqs. 3.58.

5. The values of the walk-off angles of the *slow* and the *fast* waves are different in general, and are obtained from Eqns. 3.60.

6. When the propagation vector **k** lies along a particular direction on the $Z-X$ plane, the n_s and n_f values are equal, and light propagates as in an isotropic medium with no beam walk-off. This direction is called the direction of the optic axis. The value of the angle between the optic axis and the Z axis is denoted by Ω and given in Eqs. 3.73 in terms of the n_X, n_Y and n_Z values.

7. The unit vectors $\hat{d}_s$ and $\hat{d}_f$ are given in terms of angles in Eqns. 3.91, 3.92, 3.93, 3.94.

8. The unit vectors $\hat{e}_s$ and $\hat{e}_f$ as well as $\hat{s}_s$ and $\hat{s}_f$, which are the unit vectors along the directions of the Poynting vectors of the *slow* and the *fast* waves, respectively, are given in terms of angles in Eqns. 3.96 through 3.103.

Next, the special cases of propagation with the k vector along the principal axes and along the principal planes of a biaxial crystal are described. The case of uniaxial crystals is treated after that and finally the propagation equation with beam walk-off is derived.

3.7 Propagation along the Principal Axes and along the Principal Planes

3.7.1 Introduction

From the expressions for the components of the $\hat{d}$, $\hat{e}$, and $\hat{s}$ vectors for a general direction of the propagation vector **k**, the results for the special cases of propagation along the principal axes and along the principal planes can be obtained. However, the unit vectors $\hat{e}_s$, $\hat{e}_f$, $\hat{d}_s$, and $\hat{d}_f$ are determined here directly from the expressions for n_s and n_f so as to have a check on the results obtained from the general case, and also to have a direct method of determining the walk-off angles and the relative orientations of the $\hat{d}$, $\hat{e}$, and $\hat{s}$ vectors. Thus results will be obtained in this section for the cases of $n_X < n_Z$ and $n_X > n_Z$, for propagation directions along the X, Y, and Z axes and along the YZ, ZX, and XY planes. For propagation along the ZX plane ($\phi = 0$), the two cases of $\theta < \Omega$ and $\theta > \Omega$ are distinct and need to be considered separately. The results for these special cases of propagation are tabulated at the end of this section.

3.7.2 Propagation along the Principal Axes X, Y, and Z

Say light is propagating along the Z axis, i.e., **k** is parallel to Z. Then $\theta = 0°$ and ϕ is undefined. The equations of the last section can still be used, assuming *small* but non zero value of θ and finding the limits of the component values as θ goes to 0. However, instead of undertaking that algebraic complication, we find the directions of the **D** and **E** vectors in a different way.

When **k** is parallel to Z, $m_X = m_Y = 0$ and $m_Z = 1$. Since $\hat{m} \cdot \hat{d} = 0$ (from Sec. 3.2.1 and 3.2.2), $\hat{d}$ lies in the $X - Y$ plane, i.e., $d_Z = 0$, implying $e_Z = 0$ (from Eqs. 3.30).

For Case 1, $(n_X < n_Y < n_Z)$, Eqns. 3.55 and 3.56 give

$$\begin{aligned}
\mathcal{A} &= n_Z^2 \\
\mathcal{B} &= n_Z^2(n_X^2 + n_Y^2) \\
\mathcal{C} &= n_X^2\, n_Y^2\, n_Z^2 \\
\mathcal{D} &= n_Z^2(n_Y^2 - n_X^2).
\end{aligned}$$

(3.107)

Using Eqs. 3.57 the possible values of n are obtained as

$$n_s = n_Y \qquad n_f = n_X$$

(3.108)

With $n_s = n_Y$, and $m_X = 0$, we obtain $d_{sX} = 0$ and $e_{sX} = 0$ from Eqns. 3.59 and 3.58. Since d_{sZ} and e_{sZ} are also equal to zero, d_{sY} and e_{sY} are the only nonzero components of $\hat{d}_s$ and $\hat{e}_s$. Similarly, since $n_f = n_X$, and $m_Y = 0$, we obtain $d_{fY} = 0$ and $e_{fY} = 0$ from Eqns. 3.59 and 3.58. With d_{fZ} and e_{fZ} also equal to zero, d_{fX} and e_{fX} are the only nonzero components of $\hat{d}_f$ and $\hat{e}_f$. Thus, the electric field direction of the slow wave is along Y and that of the fast wave is along X, as shown in Figure 3.28.

Following the same arguments for case 2, i.e., for $n_X > n_Y > n_Z$, the electric field direction of the slow wave is along X and that of the fast wave is along Y, as shown in Figure 3.29.

When **k** is along the X or Y axes, the same arguments as those presented

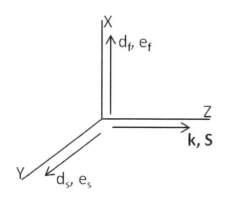

FIGURE 3.28
Propagation with **k** along the principal axis Z for Case 1, $n_X < n_Y < n_Z$.

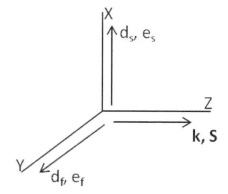

FIGURE 3.29
Propagation with **k** along the principal axis Z for Case 2, $n_X > n_Y > n_Z$.

Direction of **k**	Case 1, $n_X < n_Y < n_Z$		Case 2, $n_X > n_Y > n_Z$	
	$\hat{d}_s, \hat{e}_s$	$\hat{d}_f, \hat{e}_f$	$\hat{d}_s, \hat{e}_s$	$\hat{d}_f, \hat{e}_f$
X	Z	Y	Y	Z
Y	Z	X	X	Z
Z	Y	X	X	Y

TABLE 3.2
The directions of the $\hat{d}_s$, $\hat{e}_s$, $\hat{d}_f$, and $\hat{e}_f$ for **k** along the X, Y, or Z axes.

here can be used with appropriate substitution of variables to obtain the directions of the $\hat{d}_s$, $\hat{e}_s$, $\hat{d}_f$, and $\hat{e}_f$ as summarized in Table 3.2:

3.7.3 Propagation along the Principal Plane YZ

Another special case of light propagation in an anisotropic crystal is that of propagation along a principal plane, such as the XY, YZ, or ZX planes. We start with the propagation vector **k** lying in the YZ plane, so that $m_X = 0$ and $m_Y^2 + m_Z^2 = 1$. With $m_X = 0$, Eqns. 3.55 give

$$\mathcal{A} = n_Y^2 m_Y^2 + n_Z^2 m_Z^2, \quad \mathcal{B} = n_Y^2 n_Z^2 + n_X^2 \mathcal{A}, \quad \text{and} \quad \mathcal{C} = n_X^2 n_Y^2 n_Z^2. \ (3.109)$$

The two cases of $n_X < n_Y < n_Z$ and $n_X > n_Y > n_Z$ are considered separately in the next two subsections.

3.7.4 k along YZ Plane, Case 1: $n_X < n_Y < n_Z$

In this case, from 3.56 $\mathcal{D} = n_Y^2 n_Z^2 - n_X^2 \mathcal{A}$, since $\mathcal{D}$ is defined to be positive and $\mathcal{A}$ takes the values from n_Y^2 to n_Z^2 as m_Y goes from 1 to 0. Thus, the possible values of n (from Eqs. 3.57) are

$$n_s = \frac{n_Y\, n_Z}{(n_Y^2\, m_Y^2 + n_Z^2\, m_Z^2)^{1/2}} \quad \text{and} \quad n_f = n_X. \ (3.110)$$

As m_Y goes from 0 to 1, n_s increases in value from n_Y to n_Z, and is always greater than n_X. Since $m_X = 0$ and $n_s \neq n_X$, from the first lines of Eqns. 3.58 and 3.59 we obtain $e_{sX} = d_{sX} = 0$, i.e., the unit vectors $\hat{e}_s$ and $\hat{d}_s$ lie on the YZ plane, along with the propagation vector **k**. The unit vector $\hat{d}_f$, perpendicular to both **k** and to $\hat{d}_s$, must therefore lie along the X axis. Thus $d_{fY} = d_{fZ} = 0$, requiring $\rho_f = 0$ (from Eqs. 3.59), which in turn requires $e_{fY} = e_{fZ} = 0$. Thus, for **k** along the YZ plane and $n_X < n_Y < n_Z$, the **D** and **E** vectors of the fast wave are along the X axis, and those of the slow wave lie on the YZ plane.

Since n_s goes from n_Y to n_Z, i.e., for this case n_s is less than n_Z and greater than n_Y, Eqs. 3.58 and 3.59 show that e_{sY} and d_{sY} are positive and

$n_X < n_Y < n_Z$

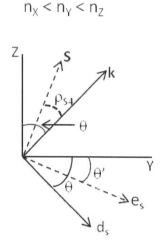

$n_X > n_Y > n_Z$

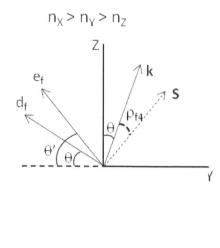

FIGURE 3.30

Directions of $\hat{d}_s$ and $\hat{e}_s$ for Case 1: $n_X < n_Y < n_Z$.

FIGURE 3.31

Directions of $\hat{d}_s$ and $\hat{e}_s$ for Case 2: $n_X > n_Y > n_Z$.

e_{sZ} and d_{sZ} are negative, i.e., the vectors **D** and **E** lie in the fourth quadrant of the YZ plane as shown in Figure 3.30.

With the angle between **k** and the Z axis denoted by θ, $\hat{d}_s$ makes an angle θ with the Y axis, so that $\tan\theta = |d_{sZ}|/|d_{sY}|$. If θ' is the angle between $\hat{e}_s$ and the Y direction, then $\tan\theta' = |e_{sZ}|/|e_{sY}|$.

From Eqs. 3.32 we have

$$\frac{E_Z}{E_Y} = \frac{n_Y^2}{n_Z^2}\frac{D_Z}{D_Y} \tag{3.111}$$

so that

$$\tan\theta' = \frac{n_Y^2}{n_Z^2}\tan\theta \tag{3.112}$$

For n_Y smaller than n_Z, θ' is smaller than θ, and $\hat{e}_s$ points between the directions of $\hat{d}_s$ and Y, as shown in Figure 3.30.

Denoting by ρ_{s4} the angle between $\hat{e}_s$ and $\hat{d}_s$, we have $\theta' = \theta - \rho_{s4}$. Using Eqs. 3.112

$$\tan(\theta - \rho_{s4}) = \left(\frac{n_Y}{n_Z}\right)^2 \tan\theta \tag{3.113}$$

which can be solved for ρ_{s4}:

$$\tan\rho_{s4} = \frac{(1 - p_4)\tan\theta}{1 + p_4\tan^2\theta} \tag{3.114}$$

where $p_4 \equiv (n_Y/n_Z)^2$ is less than 1.

The unit vectors $\hat{d}_s$ and $\hat{e}_s$ are given (from Figure 3.30) by

$$
\begin{aligned}
\hat{d}_s &= \cos\theta \hat{Y} - \sin\theta \hat{Z} \\
\hat{e}_s &= \cos(\theta - \rho_{s4})\hat{Y} - \sin(\theta - \rho_{s4})\hat{Z}
\end{aligned}
$$

$$(3.115)$$

Here and in the next few sub-sections below we have anticipated the notation that will be introduced in the next chapter with the numbers 4, 5, and 6 assigned to the combinations of coordinates YZ, ZX, and XY, respectively.

3.7.5   k along YZ Plane, Case 2: $n_X > n_Y > n_Z$

In this case, 3.56 gives $\mathcal{D} = n_X^2 \mathcal{A} - n_Y^2 n_Z^2$, since $\mathcal{D}$ is defined to be positive and $\mathcal{A}$ takes the values from n_Y^2 to n_Z^2 as m_Y goes from 0 to 1. Thus, the possible values of n from Eqs. 3.57 are

$$
n_s = n_X \qquad \text{and} \qquad n_f = \frac{n_Y\, n_Z}{(n_Y^2\, m_Y^2 + n_Z^2\, m_Z^2)^{1/2}}. \tag{3.116}
$$

As m_Y goes from 0 to 1, n_f decreases from n_Y to n_Z, and is always smaller than n_X. Since $m_X = 0$ and $n_f \neq n_X$, from the first lines of Eqns. 3.58 and 3.59 we obtain $e_{fX} = d_{fX} = 0$, i.e., the unit vectors $\hat{d}_f$ and $\hat{e}_f$ lie on the YZ plane, along with the propagation vector **k**. The unit vector $\hat{d}_s$, perpendicular to both **k** and to $\hat{d}_f$ must therefore lie along the X axis. Thus, $d_{sY} = d_{sZ} = 0$, which requires $\rho_s = 0$ (from Eqs. 3.59), which in turn requires $e_{sY} = e_{sZ} = 0$. Thus, for **k** along the YZ plane and $n_X > n_Y > n_Z$ the **D** and **E** vectors of the slow wave are along the X axis, and those of the fast wave lie on the YZ plane.

Since n_f goes from n_Z to n_Y, i.e., for this case n_f is less than n_Y and greater than n_Z, Eqns. 3.58 and 3.59 show that e_{fY} and d_{fY} are negative and e_{fZ} and d_{fZ} are positive, i.e., the vectors $\hat{d}_f$ and $\hat{e}_f$ lie in the second quadrant of the YZ plane as shown in Figure 3.31.

Since the angle between **k** and the Z axis is denoted by θ, then $\hat{d}_f$ makes an angle θ with the $-Y$ direction, so that $\tan\theta = |d_{fY}|/|d_{fZ}|$. If θ' is the angle between $\hat{e}_f$ and the $-Y$ direction, then $\tan\theta' = |\hat{e}_{fY}|/|\hat{e}_{fZ}|$.

Equation 3.112 shows that for n_Y bigger than n_Z, θ' is larger than θ, so that $\hat{e}_f$ points between the directions of $\hat{d}_f$ and Z, as shown in Figure 3.31.

Denoting by ρ_{f4} the angle between $\hat{e}_f$ and $\hat{d}_f$, we have $\theta' = \theta + \rho_{f4}$. Using Eqs. 3.112

$$
\tan(\rho_{f4} + \theta) = \left(\frac{n_Y}{n_Z}\right)^2 \tan\theta \tag{3.117}
$$

which can be solved for ρ_{f4}:

$$
\tan\rho_{f4} = \frac{(p_4 - 1)\tan\theta}{1 + p_4 \tan^2\theta} \tag{3.118}
$$

where $p_4 = (n_Y/n_Z)^2$ is greater than 1.

The unit vectors $\hat{d}_f$ and $\hat{e}_f$ are given (from Figure 3.31) by

$$\begin{aligned}
\hat{d}_f &= -\cos\theta \hat{Y} + \sin\theta \hat{Z} \\
\hat{e}_f &= -\cos(\theta + \rho_{f4})\hat{Y} + \sin(\theta + \rho_{f4})\hat{Z}.
\end{aligned}$$

$$(3.119)$$

Equations 3.114 and 3.118 show that the angles ρ_{s4} and ρ_{f4} can be denoted by the same symbol ρ_4, if ρ_4 is defined as

$$\rho_4 = \tan^{-1}\frac{|p_4 - 1|\tan\theta}{1 + p_4\tan^2\theta} = \tan^{-1}\frac{|n_Y^2 - n_Z^2|\sin\theta\cos\theta}{n_Y^2\sin^2\theta + n_Z^2\cos^2\theta} \qquad (3.120)$$

3.7.6 Propagation along the Principal Plane ZX

For propagation along the ZX plane an additional complication arises because of the choice of n_Y as having a value between n_X and n_Z. With $m_Y = 0$, Eqns. 3.55 and 3.56 give

$$\mathcal{A} = n_X^2 m_X^2 + n_Z^2 m_Z^2, \quad \mathcal{B} = n_X^2 n_Z^2 + n_Y^2 \mathcal{A} \quad \text{and} \quad \mathcal{C} = n_X^2 n_Y^2 n_Z^2 \quad (3.121)$$

and $\mathcal{D} = \pm(n_Y^2 \mathcal{A} - n_X^2 n_Z^2)$, where the sign chosen depends on the value of m_Z that makes $\mathcal{D}$ positive. For m_Z varying from 0 to 1, the two possible values of n are denoted by n_1 and n_2

$$n_1 = \frac{n_X\, n_Z}{(n_X^2\, m_X^2 + n_Z^2\, m_Z^2)^{1/2}} \qquad \text{and} \qquad n_2 = n_Y. \qquad (3.122)$$

To determine the assignment of n_1 and n_2 to n_s and n_f we consider the four combinations of cases $n_X < n_Y < n_Z$ and $n_X > n_Y > n_Z$ with the conditions $\theta < \Omega$ and $\theta > \Omega$.

3.7.7 k along ZX Plane, Case 1a: $n_X < n_Y < n_Z$, $\theta < \Omega$

As the angle θ goes from 0 to Ω, the value of n_1 increases from n_X to n_Y so n_2 is greater than n_1. Thus, for this range of θ, $n_s = n_2 = n_Y$ and $n_f = n_1 = \dfrac{n_X\, n_Z}{(n_X^2\, m_X^2 + n_Z^2\, m_Z^2)^{1/2}}$. At $\theta = \Omega$, n_1 and n_2 are equal to each other and light propagates as in an isotropic medium with no birefringence and with refractive index n_Y.

Since n_f goes from n_X to n_Y, i.e., for this case n_f is greater than n_X, Eqns. 3.58 and 3.59 show that e_{fX} and d_{fX} are positive. Since n_Z is larger than both n_X and n_Y, n_f is smaller than n_Z, so both e_{fZ} and d_{fZ} are negative. The vectors $\hat{d}_f$ and $\hat{e}_f$ therefore lie in the fourth quadrant of the XZ plane, as shown in Figure 3.32.

$$n_X < n_Y < n_Z$$
$$\theta < \Omega$$

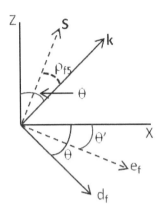

$$n_X < n_Y < n_Z$$
$$\theta > \Omega$$

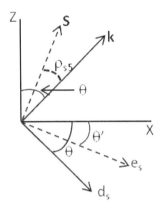

FIGURE 3.32

Directions of $\hat{d}$ and $\hat{e}$ for Case 1: $n_X < n_Y < n_Z$.

FIGURE 3.33

Directions of $\hat{d}$ and $\hat{e}$ for Case 2: $n_X > n_Y > n_Z$.

Since the angle between **k** and the Z axis is denoted by θ, $\hat{d}_f$ makes an angle θ with the X direction, so that $\tan \theta = |d_{fZ}|/|d_{fX}|$. If θ' is the angle between $\hat{e}_f$ and the X direction, then $\tan \theta' = |\hat{e}_{fZ}|/|\hat{e}_{fX}|$.

From Eqs. 3.32 we have

$$\frac{E_Z}{E_X} = \left(\frac{n_X}{n_Z}\right)^2 \frac{D_Z}{D_X} \tag{3.123}$$

so that

$$\tan \theta' = \left(\frac{n_X}{n_Z}\right)^2 \tan \theta. \tag{3.124}$$

Equation 3.124 shows that for n_X smaller than n_Z, θ' is smaller than θ, so that $\hat{e}_f$ points between the directions of $\hat{d}_f$ and X, as shown in Figure 3.32.

Denoting by ρ_{f5} the angle between $\hat{e}_f$ and $\hat{d}_f$, we have $\theta' = \theta - \rho_{f5}$. Using Eqs. 3.124

$$\tan(\theta - \rho_{f5}) = \left(\frac{n_X}{n_Z}\right)^2 \tan \theta \tag{3.125}$$

which can be solved for ρ_{f5}:

$$\tan \rho_{f5} = \frac{(1 - p_5) \tan \theta}{1 + p_5 \tan^2 \theta} \tag{3.126}$$

where $p_5 \equiv (n_X/n_Z)^2$ is less than 1.

The unit vectors $\hat{d}_f$ and $\hat{e}_f$ are given (from Figure 3.32) by

$$\hat{d}_f = \cos\theta\hat{X} - \sin\theta\hat{Z}$$
$$\hat{e}_f = \cos(\theta - \rho_{f5})\hat{X} - \sin(\theta - \rho_{f5})\hat{Z}.$$

$$(3.127)$$

3.7.8   k along ZX Plane, Case 1b: $n_X < n_Y < n_Z$, $\theta > \Omega$

For θ increasing from Ω to $90°$, n_1 increases from n_Y to n_Z, whereas $n_2 = n_Y$ so $n_1 > n_2$. Thus, for this range of θ, $n_s = n_1 = \dfrac{n_X\, n_Z}{(n_X^2\, m_X^2 + n_Z^2\, m_Z^2)^{1/2}}$ and $n_f = n_2 = n_Y$.

Since n_s goes from n_Y to n_Z, n_s is larger than n_X and smaller than n_Z. Eqns. 3.58 and 3.59 show that in this case e_{sX} and d_{sX} are positive and e_{sZ} and d_{sZ} are negative, i.e., the vectors $\hat{d}_s$ and $\hat{e}_s$ lie in the fourth quadrant of the XZ plane, as shown in Figure 3.33. The angle θ' between $\hat{e}_s$ and the X direction is equal to $\theta - \rho_{s5}$, where ρ_{s5} is given by Eqs. 3.130. The unit vectors $\hat{d}_s$ and $\hat{e}_s$ are then given (from Figure 3.33) by

$$\hat{d}_s = \cos\theta\hat{X} - \sin\theta\hat{Z}$$
$$\hat{e}_s = \cos(\theta - \rho_{s5})\hat{X} - \sin(\theta - \rho_{s5})\hat{Z}.$$

$$(3.128)$$

3.7.9   k along ZX Plane, Case 2a: $n_X > n_Y > n_Z$, $\theta < \Omega$

As the angle θ goes from 0 to Ω, the value of n_1 decreases from n_X to n_Y, whereas $n_2 = n_Y$, so n_1 is bigger than n_2. Thus, for this range of θ, $n_s = n_1 = \dfrac{n_X\, n_Z}{(n_X^2\, m_X^2 + n_Z^2\, m_Z^2)^{1/2}}$ and $n_f = n_2 = n_Y$. Again, at $\theta = \Omega$, n_1 and n_2 are equal to each other and light propagates as in an isotropic medium with no birefringence and with refractive index equal to n_Y.

Since n_s, which goes from n_X to n_Y, is smaller than n_X, we find from Eqns. 3.58 and 3.59 that e_{sX} and d_{sX} are negative. Since n_s is larger than n_Z, both e_{sZ} and d_{sZ} are positive, i.e., the vectors $\hat{d}_s$ and $\hat{e}_s$ lie in the second quadrant of the XZ plane, as shown in Figure 3.34.

Since the angle between $\mathbf{k}$ and the Z axis is denoted by θ, $\hat{d}_s$ makes an angle θ with the $-X$ direction, so that $\tan\theta = |d_{sZ}|/|d_{sX}|$. If θ' is the angle between $\hat{e}_s$ and the $-X$ direction, $\tan\theta' = |\hat{e}_{sZ}|/|\hat{e}_{sX}|$.

Equation 3.124 shows that for n_X bigger than n_Z, θ' is bigger than θ, so that $\hat{e}_s$ points between the directions of $\hat{d}_s$ and Z, as shown in Figure 3.34.

Denoting by ρ_{s5} the angle between $\hat{e}_s$ and $\hat{d}_s$, we have $\theta' = \theta + \rho_{s5}$. Using

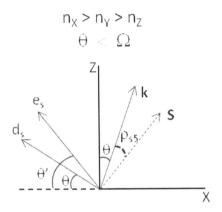

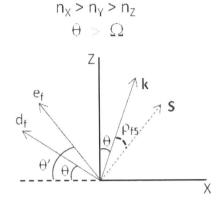

FIGURE 3.34
Directions of $\hat{d}$ and $\hat{e}$ for Case 1: $n_X < n_Y < n_Z$

FIGURE 3.35
Directions of $\hat{d}$ and $\hat{e}$ for Case 2: $n_X > n_Y > n_Z$

Eqs. 3.124

$$\tan(\rho_{s5} + \theta) = \left(\frac{n_X}{n_Z}\right)^2 \tan\theta \tag{3.129}$$

which can be solved for ρ_{s5}:

$$\tan\rho_{s5} = \frac{(p_5 - 1)\tan\theta}{1 + p_5 \tan^2\theta} \tag{3.130}$$

where $p_5 = (n_X/n_Z)^2$ is greater than 1.

The unit vectors $\hat{d}_s$ and $\hat{e}_s$ are given (from Figure 3.34) by

$$\hat{d}_s = -\cos\theta\hat{X} + \sin\theta\hat{Z}$$
$$\hat{e}_s = -\cos(\theta + \rho_{s5})\hat{X} + \sin(\theta + \rho_{s5})\hat{Z}$$
$$\tag{3.131}$$

and $\hat{d}_f = \hat{e}_f = \hat{Y}$.

3.7.10 k along ZX Plane, Case 2b: $n_X > n_Y > n_Z$, $\theta > \Omega$

For θ increasing from Ω to $90°$, n_1 decreases from n_Y to n_Z, whereas n_2 stays equal to n_Y, so that $n_2 > n_1$. Thus, for this range of θ, $n_s = n_2 = n_Y$ and $n_f = n_1 = \dfrac{n_X\, n_Z}{(n_X^2\, m_X^2 + n_Z^2\, m_Z^2)^{1/2}}$.

With n_f going from n_Y to n_Z, we have $n_f < n_X$, so that from Eqs. 3.58 and 3.59 we find that e_{fX} and d_{fX} are negative. Since n_f is bigger than n_Z, both e_{fZ} and d_{fZ} are positive, i.e., the vectors $\hat{d}_f$ and $\hat{e}_f$ lie in the second quadrant of the XZ plane, as shown in Figure 3.35.

The angle θ' between $\hat{e}_f$ and the X direction is equal to $\theta + \rho_{f5}$, where ρ_{f5} is given by Eqs. 3.126. The unit vectors $\hat{d}_f$ and $\hat{e}_f$ are then given (from Figure 3.35) by

$$
\begin{aligned}
\hat{d}_f &= -\cos\theta \hat{X} + \sin\theta \hat{Z} \\
\hat{e}_f &= -\cos(\theta + \rho_{f5})\hat{X} + \sin(\theta + \rho_{f5})\hat{Z}.
\end{aligned}
$$

(3.132)

Equations 3.126 and 3.130 show that the angles ρ_{f5} and ρ_{s5} can be denoted by the same symbol ρ_5, defined as

$$
\rho_5 = \tan^{-1}\frac{|p_5 - 1|\tan\theta}{1 + p_5 \tan^2\theta} = \tan^{-1}\frac{|n_Z^2 - n_X^2|\sin\theta\cos\theta}{n_X^2\sin^2\theta + n_Z^2\cos^2\theta}.
$$

(3.133)

3.7.11 Propagation along the Principal Plane XY

When the propagation vector $\mathbf{k}$ lies in the XY plane, $m_Z = 0$ and $m_X^2 + m_Y^2 = 1$, so that Eqns. 3.55 give

$$
\mathcal{A} = n_X^2 m_X^2 + n_Y^2 m_Y^2, \quad \mathcal{B} = n_X^2 n_Y^2 + n_Z^2 \mathcal{A} \quad \text{and} \quad \mathcal{C} = n_X^2 n_Y^2 n_Z^2. \tag{3.134}
$$

We consider the following two cases:

3.7.12 k along XY Plane, Case 1: $n_X < n_Y < n_Z$

In this case, from 3.56, we get $\mathcal{D} = n_Z^2 \mathcal{A} - n_X^2 n_Y^2$, since $\mathcal{D}$ is defined to be positive and $\mathcal{A}$ takes the values from n_X^2 to n_Y^2 as m_Y goes from 0 to 1. Thus, the possible values of n (from Eqs. 3.57) are

$$
n_s = n_Z \quad \text{and} \quad n_f = \frac{n_X \, n_Y}{(n_X^2 \, m_X^2 + n_Y^2 \, m_Y^2)^{1/2}}. \tag{3.135}
$$

As m_X goes from 0 to 1, n_f takes the values from n_X to n_Y, so that $n_f \geq n_X$. Since $m_Z = 0$ and $n_f \neq n_Z$, from the third lines of Eqns. 3.58 and 3.59 we obtain $e_{fZ} = d_{fZ} = 0$, i.e., the unit vectors $\hat{d}_f$ and $\hat{e}_f$ lie on the XY plane, along with the propagation vector $\mathbf{k}$. The unit vector $\hat{d}_s$, which is perpendicular to both $\mathbf{k}$ and to $\hat{d}_f$ must therefore lie along the Z axis. Thus $d_{sX} = d_{sY} = 0$, which requires $\rho_s = 0$ (from Eqs. 3.59), which in turn requires $e_{sX} = e_{sY} = 0$. So, for $\mathbf{k}$ along the XY plane and $n_X < n_Y < n_Z$ the $\mathbf{D}$ and $\mathbf{E}$ vectors of the slow wave are along the Z axis, and those of the fast wave lie on the XY plane.

Since n_f goes from n_X to n_Y, n_f is greater than n_X and smaller than n_Y, so Eqns. 3.58 and 3.59 show that e_{fX} and d_{fX} are positive and e_{fY} and d_{fY} are negative, i.e., the vectors $\mathbf{D}$ and $\mathbf{E}$ lie in the fourth quadrant of the XY plane as shown in Figure 3.36.

$$n_X < n_Y < n_Z$$

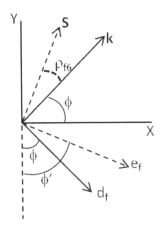

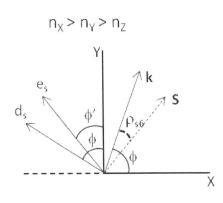

$$n_X > n_Y > n_Z$$

FIGURE 3.36
Directions of $\hat{d}_f$ and $\hat{e}_f$ for Case 1: $n_X < n_Y < n_Z$.

FIGURE 3.37
Directions of $\hat{d}_s$ and $\hat{e}_s$ for Case 2: $n_X > n_Y > n_Z$.

Since the angle between **k** and the X axis is denoted by ϕ, $\hat{d}_f$ makes an angle ϕ with the $-Y$ direction, so that $\tan\phi = |d_{fX}|/|d_{fY}|$. If ϕ' is the angle between $\hat{e}_f$ and the $-Y$ direction, then $\tan\phi' = |e_{fX}|/|e_{fY}|$.

From Eqs. 3.32 we have

$$\frac{E_X}{E_Y} = \left(\frac{n_Y}{n_X}\right)^2 \frac{D_X}{D_Y} \tag{3.136}$$

so that

$$\tan\phi' = \left(\frac{n_Y}{n_X}\right)^2 \tan\phi. \tag{3.137}$$

For n_X smaller than n_Y, ϕ' is larger than ϕ, so that $\hat{d}_f$ points between the directions of $\hat{e}_f$ and $-Y$, as shown in Figure 3.36.

Denoting the angle between $\hat{e}_f$ and $\hat{d}_f$ by ρ_{f6}, we have $\phi' = \phi + \rho_{f6}$. Using Eqs. 3.137

$$\tan(\phi + \rho_{f6}) = \left(\frac{n_Y}{n_X}\right)^2 \tan\phi \tag{3.138}$$

which can be solved for ρ_{f6}:

$$\tan\rho_{f6} = \frac{(p_6 - 1)\tan\phi}{1 + p_6 \tan^2\phi} \tag{3.139}$$

where $p_6 \equiv (n_Y/n_X)^2$ is greater than 1.

From Figure 3.36 the unit vectors $\hat{d}_f$ and $\hat{e}_f$ are given by

$$
\begin{aligned}
\hat{d}_f &= \sin\phi\hat{X} - \cos\phi\hat{Y} \\
\hat{e}_f &= \sin(\phi + \rho_{f6})\hat{X} - \cos(\phi + \rho_{f6})\hat{Y}.
\end{aligned}
$$

$$(3.140)$$

3.7.13 k along XY Plane, Case 2: $n_X > n_Y > n_Z$

In this case, $\mathcal{D} = n_X^2 n_Y^2 - n_Z^2 \mathcal{A}$ from 3.56, since $\mathcal{D}$ is defined to be positive and $\mathcal{A}$ takes the values from n_Y^2 to n_X^2 as m_X goes from 0 to 1. Thus, the possible values of n from Eqs. 3.57 are

$$
n_s = \frac{n_X \, n_Y}{(n_X^2 \, m_X^2 + n_Y^2 \, m_Y^2)^{1/2}} \qquad \text{and} \qquad n_f = n_Z.
$$

$$(3.141)$$

As m_X goes from 0 to 1, n_s takes the values from n_X to n_Y, and is always larger than n_Z. Since $m_Z = 0$ and $n_s \neq n_z$, from the third lines of Eqns. 3.58 and 3.59 we obtain $e_{sZ} = d_{sZ} = 0$, i.e., the unit vectors $\hat{d}_s$ and $\hat{e}_s$ lie on the XY plane, along with the propagation vector **k**. The unit vector $\hat{d}_f$, which is perpendicular to both **k** and to $\hat{d}_s$ must therefore lie along the Z axis. Thus $d_{fX} = d_{fY} = 0$, which requires $\rho_f = 0$ (from Eqs. 3.59), which in turn requires $e_{fX} = e_{fY} = 0$. So for **k** along the XY plane and $n_X > n_Y > n_Z$, the **D** and **E** vectors of the fast wave are along the Z axis, and those of the slow wave lie on the XY plane.

n_s goes from n_Y to n_X, which means $n_s \leq n_X$. Equations 3.58 and 3.59 then show that e_{sX} and d_{sX} are negative. Since $n_s \geq n_Y$, e_{sY} and d_{sY} are positive, i.e., the vectors $\hat{d}_s$ and $\hat{e}_s$ lie in the second quadrant of the XY plane as shown in Figure 3.37.

Since the angle between **k** and the X axis is denoted by ϕ, $\hat{d}_s$ makes an angle ϕ with the Y direction, so that $\tan\phi = |d_{sX}|/|d_{sY}|$. If ϕ' is the angle between $\hat{e}_s$ and the Y direction, then $\tan\phi' = |\hat{e}_{sX}|/|\hat{e}_{sY}|$.

Equation 3.137 shows that for n_Y smaller than n_X, ϕ' is smaller than ϕ, so that $\hat{e}_s$ points between the directions of $\hat{d}_s$ and Y, as shown in Figure 3.37.

Denoting by ρ_{s6} the angle between $\hat{e}_s$ and $\hat{d}_s$, we have $\phi' = \phi - \rho_{s6}$. Using Eqs. 3.137

$$
\tan(\phi - \rho_{s6}) = \left(\frac{n_Y}{n_X}\right)^2 \tan\phi
$$

$$(3.142)$$

which can be solved for ρ_{s6}:

$$
\tan\rho_{s6} = \frac{(1 - p_6)\tan\phi}{1 + p_6 \tan^2\phi}
$$

$$(3.143)$$

where $p_6 = (n_Y/n_X)^2$ is smaller than 1.

From Figure 3.37 the unit vectors $\hat{d}_s$ and $\hat{e}_s$ are given by

$$
\begin{aligned}
\hat{d}_s &= -\sin\phi\hat{X} + \cos\phi\hat{Y} \\
\hat{e}_s &= -\sin(\phi - \rho_{s6})\hat{X} + \cos(\phi - \rho_{s6})\hat{Y}.
\end{aligned}
$$

$$(3.144)$$

Equations 3.139 and 3.143 show that the angles ρ_{f6} and ρ_{s6} can be denoted by the same symbol ρ_6, if ρ_6 is defined as

$$
\rho_6 = \tan^{-1}\frac{|p_6 - 1|\tan\phi}{1 + p_5\tan^2\phi} = \tan^{-1}\frac{|n_X^2 - n_Y^2|\sin\phi\cos\phi}{n_Y^2\sin^2\phi + n_X^2\cos^2\phi}. \qquad (3.145)
$$

3.7.14 Summary of the Cases of Propagation along Principal Planes

Tables 3.3 and 3.4 summarize the results of this section. The values of the angle δ for the different cases are obtained from Eqs. 3.89. For each case, the n_s and n_f values are first calculated using Eqns. 3.57. For the θ, ϕ, and δ values for each case, Eqns. 3.91, 3.92, 3.93, and 3.94 are used to find the components $d_{sX}, d_{sY}, d_{sZ}, d_{fX}, d_{fY}$, and d_{fZ}. From these values of the components of the unit vectors $\hat{d}_s$ and $\hat{d}_f$, angles ρ_s and ρ_f are found using Eqns. 3.106. Using the values of ρ_s and ρ_f along with those of the components of the unit vectors $\hat{d}_s$ and $\hat{d}_f$ and the angles δ in Eqns. 3.96, 3.98, 3.100, and 3.102, components of the unit vectors $\hat{e}_s$ and $\hat{e}_f$ are obtained.

3.8 Uniaxial Crystals

When two of the principal refractive indices (by definition n_X and n_Y) are equal, Eqs. 3.73 and Figure 3.20 show that the two optic axes coalesce into one axis parallel to the Z direction. Such crystals with only *one* optic axis are of course called *uniaxial*. The two equal indices n_X and n_Y in uniaxial crystals are given the name n_o, so that

$$
D_X = n_o^2 E_X \qquad D_Y = n_o^2 E_Y \qquad D_Z = n_Z^2 E_Z. \qquad (3.146)
$$

A crystal with $n_Z > n_o$ is called *positive* uniaxial and if $n_Z < n_o$ it is called *negative* uniaxial. We saw in the last section that for a biaxial crystal, n_s and n_f take simple forms when the propagation vector $\hat{\mathbf{k}}$ is along special directions, such as along the principal axes or along the principal planes. For uniaxial crystals, n_s and n_f have such simple expressions for *arbitrary* directions of propagation, because of the additional symmetry condition. These expressions are similar to those for propagation along principal planes in biaxial crystals,

| | **k along YZ plane ($\phi = 90°$, $\delta = 0$)** | | **k along XY plane ($\theta = 90°$, $\delta = 0$)** | |
| | Case 1 | Case 2 | Case 1 | Case 2 |
	$n_X < n_Y < n_Z$	$n_X > n_Y > n_Z$	$n_X < n_Y < n_Z$	$n_X > n_Y > n_Z$
n_s	$(n_Y\,n_Z)/\sqrt{A_{YZ}}$	n_X	n_Z	$(n_X\,n_Y)/\sqrt{A_{XY}}$
n_f	n_X	$(n_Y\,n_Z)/\sqrt{A_{YZ}}$	$(n_X\,n_Y)/\sqrt{A_{XY}}$	n_Z
ρ_s	ρ_4	0	ρ_6	ρ_6
ρ_f	0	ρ_4	0	0
e_{sX}	0	-1	0	0
e_{sY}	$\cos(\theta - \rho_4)$	0	0	$-\sin(\phi - \rho_6)$
e_{sZ}	$-\sin(\theta - \rho_4)$	0	-1	$\cos(\phi - \rho_6)$
e_{fX}	1	0	$\sin(\phi + \rho_6)$	0
e_{fY}	0	$-\cos(\theta + \rho_4)$	$-\cos(\phi + \rho_6)$	0
e_{fZ}	0	$\sin(\theta + \rho_4)$	0	1

$$A_{YZ} = n_Y^2 \sin^2\theta + n_Z^2 \cos^2\theta; \quad A_{XY} = n_Y^2 \sin^2\phi + n_X^2 \cos^2\phi.$$

$$\rho_4 = \tan^{-1}\left(\frac{|n_Y^2 - n_Z^2|\sin\theta\cos\theta}{A_{YZ}}\right),$$

$$\rho_6 = \tan^{-1}\left(\frac{|n_X^2 - n_Y^2|\sin\phi\cos\phi}{A_{XY}}\right)$$

TABLE 3.3

n_s, n_f, ρ_s, ρ_f, $\hat{e}_s$, and $\hat{e}_f$ for **k** along the YZ and the XY planes. The components of the unit vector $\hat{d}$ are obtained from the corresponding components of the unit vector $\hat{e}$, with the walk-off angles ρ_4 or ρ_6 set equal to zero.

| | k along ZX plane ($\phi = 0$) | | | |
| | Case 1, $n_X < n_Y < n_Z$ | | Case 2, $n_X > n_Y > n_Z$ | |
	$\theta < \Omega, \delta = 90°$	$\theta > \Omega, \delta = 0°$	$\theta < \Omega, \delta = 90°$	$\theta > \Omega, \delta = 0°$
n_s	n_Y	n_Y	n_Y	n_Y
n_f	$(n_X\, n_Z)/\sqrt{A_{XZ}}$	$(n_X\, n_Z)/\sqrt{A_{XZ}}$	$(n_X\, n_Z)/\sqrt{A_{XZ}}$	$(n_X\, n_Z)/\sqrt{A_{XZ}}$
ρ_s	0	0	0	0
ρ_f	ρ_5	ρ_5	ρ_5	ρ_5
e_{sX}	0	0	0	0
e_{sY}	1	1	1	1
e_{sZ}	0	0	0	0
e_{fX}	$\cos(\theta - \rho_5)$	$\cos(\theta - \rho_5)$	$-\cos(\theta + \rho_5)$	$-\cos(\theta + \rho_5)$
e_{fY}	0	-1	-1	0
e_{fZ}	$-\sin(\theta - \rho_5)$	$-\sin(\theta - \rho_5)$	$\sin(\theta + \rho_5)$	$\sin(\theta + \rho_f)$

$$A_{XZ} = n_X^2 \sin^2\theta + n_Z^2 \cos^2\theta$$

$$\rho_5 = \tan^{-1}\left(\frac{|n_Z^2 - n_X^2|\,\sin\theta\cos\theta}{A_{XZ}}\right)$$

TABLE 3.4

n_s, n_f, ρ_s, ρ_f, $\hat{e}_s$, and $\hat{e}_f$ for **k** along the ZX plane. The components of the unit vector $\hat{d}$ are obtained from the corresponding components of the unit vector $\hat{e}$, with the walk-off angle ρ_5 set equal to zero.

	Positive uniaxial $n_Z > n_o$	Negative uniaxial $n_Z < n_o$
$\mathcal{D}$	$n_o^2(n_Z^2 - \mathcal{A})$	$n_o^2(\mathcal{A} - n_Z^2)$
n_s	$(n_o n_Z)/\sqrt{\mathcal{A}}$	n_o
n_f	n_o	$(n_o n_Z)/\sqrt{\mathcal{A}}$

$$\mathcal{A} = n_o^2 \sin^2\theta + n_Z^2 \cos^2\theta$$

TABLE 3.5
The values of $\mathcal{D}$, n_s, and n_f for positive and negative uniaxial crystals

but are sufficiently different in notation and usage, making it worthwhile to re-derive them in detail here.

With $n_X = n_Y = n_o$ inserted in Eqns. 3.55 we get

$$\mathcal{A} = n_o^2(m_X^2 + m_Y^2) + n_Z^2 m_Z^2, \quad \mathcal{B} = n_o^2(n_Z^2 + \mathcal{A}) \quad \text{and} \quad \mathcal{C} = n_o^4 n_Z^2 \quad (3.147)$$

from which we obtain

$$\mathcal{D}^2 = n_o^4(n_Z^2 - \mathcal{A})^2. \tag{3.148}$$

Since from Eqs. 3.147 $\mathcal{A} = (n_o^2 \sin^2\theta + n_Z^2 \cos^2\theta)$, $\mathcal{A}$ takes the values between n_Z^2 and n_o^2. With $\mathcal{D}$ defined to be positive, we have the following two cases shown in Table 3.5:

Defining

$$n_e(\theta) \equiv \frac{n_o n_Z}{\sqrt{\mathcal{A}}} = \frac{n_o n_Z}{\sqrt{n_o^2 \sin^2\theta + n_Z^2 \cos^2\theta}} \tag{3.149}$$

the two waves traveling with speeds c/n_o and $c/n_e(\theta)$ in the uniaxial medium are called the *ordinary wave* and the *extraordinary wave*, respectively. In a positive uniaxial crystal the extraordinary wave is the slow wave, and the ordinary wave is the fast wave. For negative uniaxial crystals the situation is reversed, with the extraordinary wave being the fast wave and the ordinary wave the slow wave. Field directions for these two waves are determined next.

The subscripts o and e are used on the field variables to designate the cases of the *ordinary* and the *extraordinary* waves, respectively. Thus $\hat{d}_o$, $\hat{e}_o$, $\hat{d}_e$, and $\hat{e}_e$ denote the unit vectors along the directions of the displacement and the electric field vectors, respectively. In a positive uniaxial crystal, $\hat{d}_s = \hat{d}_e$, $\hat{e}_s = \hat{e}_e$, $\hat{d}_f = \hat{d}_o$, $\hat{e}_f = \hat{e}_o$, and in negative uniaxial crystals, the situation is reversed, with $\hat{d}_s = \hat{d}_o$, $\hat{e}_s = \hat{e}_o$, $\hat{d}_f = \hat{d}_e$, $\hat{e}_f = \hat{e}_e$.

3.8.1   Field Directions of the D and E Vectors for Extraordinary and Ordinary Waves

To determine the values of the components of the unit vectors $\hat{d}_o$, $\hat{e}_o$, $\hat{d}_e$, and $\hat{e}_e$ along the principal dielectric axes, we start with the facts that **k** and $\hat{d}$ are

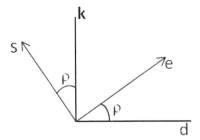

FIGURE 3.38
The unit vectors $\hat{e}$, $\hat{d}$, and $\hat{s}$ for the propagation vector $\mathbf{k}$. ρ is the walk-off angle.

perpendicular and that $\mathbf{k}$, $\hat{d}$, and $\hat{e}$ lie in one plane, as shown in Sects. 3.2.1 and 3.2.2. Denoting by ρ the angle between $\hat{d}$ and $\hat{e}$ we show below that two values of ρ are possible, one equal to 0 and the other non-zero. The non-zero value of ρ is found as a function of n_Z, n_o, and θ and it is shown that when $\rho \neq 0$, the refractive index n is equal to $n_e(\theta)$, i.e., for this case, $\hat{d} = \hat{d}_e$. The values of the components of d_e are found after that. Next, it is shown that when $\rho = 0$, n is equal to n_o, i.e., for this case, $\hat{d} = \hat{d}_o$. Lastly, the values of the components of d_o are found and the results are summarized in Table 3.6.

Redrawing Figures 3.26 and 3.27 as Figure 3.38 without the subscripts we rewrite Eqns. 3.95 as

$$\hat{e} = \hat{m}\sin\rho + \hat{d}\cos\rho \tag{3.150}$$

from which we obtain

$$e_X = m_X \, \sin\rho + d_X \, \cos\rho$$
$$e_Y = m_Y \, \sin\rho + d_Y \, \cos\rho$$
$$e_Z = m_Z \, \sin\rho + d_Z \, \cos\rho. \tag{3.151}$$

For a uniaxial crystal, we obtain from Eqs. 3.32

$$\frac{e_X}{e_Y} = \frac{d_X}{d_Y}$$
$$\frac{e_Z}{e_Y} = \frac{n_o^2 d_Z}{n_Z^2 d_Y} \tag{3.152}$$

and using Eqs. 3.151 we find

$$\sin\rho(m_X \, d_Y - m_Y \, d_X) = 0 \tag{3.153}$$

and

$$\tan\rho = \frac{(n_Z^2 - n_o^2) \, d_Z d_Y}{n_o^2 d_Z m_Y - n_Z^2 d_Y m_Z}. \tag{3.154}$$

There are two possible solutions of Eqs. 3.153: $\rho = 0$ and $\rho \neq 0$. We consider the case of nonzero ρ first.

3.8.2 $\rho \neq 0$ Case (Extraordinary Wave)

When $\rho \neq 0$, according to Eqs. 3.153,

$$m_X\, d_Y = m_Y\, d_X \tag{3.155}$$

which shows that the Z component of the cross product $\hat{m} \times \hat{d}$, given by $m_X\, d_Y - m_Y\, d_X$ is equal to 0. The vector $\hat{m} \times \hat{d}$ is therefore perpendicular to Z, as well as to $\hat{m}$ and $\hat{d}$. This means $\hat{m}$, $\hat{d}$, and Z are three vectors all perpendicular to the same vector—therefore $\hat{m}$, $\hat{d}$, and Z must be coplanar. By definition, vector u shown in Figure 3.3 also lies on the plane containing $\hat{m}$ and Z.

The plane containing the propagation vector $\mathbf{k}$ and the axis Z is defined as the *principal plane* in Ref. [2], although the planes XY, YZ, and ZX are also called the *principal planes* in the same book. To avoid possible confusion arising from the use of the same name, we will define the $\mathbf{k}$-Z plane here the $\mathbf{k}$-Z principal plane. The directions of the unit vectors d and e on the $\mathbf{k}$-Z principal plane are shown in Figures 3.39 and 3.40 for the cases of positive and negative axial crystals, respectively.

In Figures 3.39 and 3.40, the quadrants into which d and e point are determined by the following method. Since $\hat{m}$ and $\hat{d}$ are perpendicular to each other, we have

$$m_X\, d_X + m_Y\, d_Y + m_Z\, d_Z = 0. \tag{3.156}$$

Eliminating d_X in Eqns. 3.155 and 3.156, we obtain the relation between d_Y

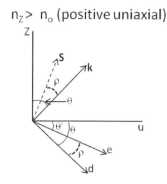

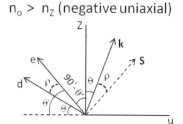

FIGURE 3.39

The orientation of the unit vectors $\hat{d}$ and $\hat{e}$ for the extraordinary wave traveling in a positive uniaxial crystal.

FIGURE 3.40

The orientation of the fields $\hat{d}$ and $\hat{e}$ for the extraordinary wave traveling in a negative uniaxial crystal.

and d_Z as

$$d_Y = -\frac{m_Y m_Z}{m_X^2 + m_Y^2} d_Z. \tag{3.157}$$

Following the same argument, we obtain the relation between d_X and d_Z as

$$d_X = -\frac{m_X m_Z}{m_X^2 + m_Y^2} d_Z. \tag{3.158}$$

For the **k** vector in the first octant of the XYZ coordinate axes, m_X, m_Y, and m_Z are all positive, so Eqs. 3.157 shows that d_Y and d_Z must have opposite signs. Equation 3.154 then shows that in a positive uniaxial crystal ($n_Z > n_o$), for $\tan \rho$ to be positive, d_Z must be negative and therefore d_Y must be positive. Similarly, in a negative uniaxial crystal ($n_Z < n_o$), for $\tan \rho$ to be positive, d_Z must be positive and therefore d_Y must be negative. These directions are indicated in Figures 3.39 and 3.40, showing that for the cases of positive and negative uniaxial crystals, $\hat{d}$ lies in the fourth and second quadrant of the $u - Z$ coordinate system, respectively.

The corresponding directions of the unit vector $\hat{e}$ are found as follows: From Eqs. 3.30 it can be shown (through a bit of algebra) that

$$\frac{e_Z}{e_u} = \frac{n_o^2}{n_Z^2} \frac{d_Z}{d_u} \tag{3.159}$$

where d_u and e_u are the components of $\hat{d}$ and $\hat{e}$ in the direction of the vector **u**. Since the angle between **k** and the Z axis is denoted by θ, $\hat{d}$ makes an angle θ with the **u** direction. If θ' is the angle between the vector $\hat{e}$ and the u direction, then Eqs. 3.159 shows that for a positive uniaxial crystal, i.e., with n_Z larger than n_o, θ is larger than θ', so that $\hat{e}$ points between the directions of $\hat{d}$ and $\hat{u}$, as shown in Figure 3.39.

The Poynting vector $\mathbf{S} = \mathbf{E} \times \mathbf{H}$ lies in the same plane as $\mathbf{E}$, $\mathbf{D}$, and **k**. The beam walk-off angle ρ between the vectors $\mathbf{S}$ and **k**, i.e., between the energy propagation direction and the direction of the propagation vector, is also the angle between $\hat{d}$ and $\hat{e}$, and we have $\rho = \theta - \theta'$. Using Eqs. 3.159

$$\tan \theta' = \tan(\theta - \rho) = \left(\frac{n_o}{n_Z}\right)^2 \tan \theta \tag{3.160}$$

which can be solved for ρ:

$$\tan \rho = \frac{(1 - p_1) \tan \theta}{1 + p_1 \tan^2 \theta} \tag{3.161}$$

where $p_1 \equiv (n_o/n_Z)^2$.

For the negative uniaxial crystal case, Eqs. 3.159 shows that for n_o larger than n_Z, θ is smaller than θ', so that $\hat{e}$ points between the directions of $\hat{d}$ and Z, as shown in Figure 3.40. In this case the walk-off angle $\rho = \theta' - \theta$, and using Eqs. 3.159

$$\tan \theta' = \tan(\theta + \rho) = p_1 \tan \theta \tag{3.162}$$

which can be solved for ρ:

$$\tan \rho = \frac{(p_1 - 1)\tan \theta}{1 + p_1 \tan^2 \theta} \tag{3.163}$$

3.8.3 Another Expression Relating ρ and θ

With the angles θ and θ' defined as above, the walk-off angle ρ is equal to $\pm(\theta - \theta')$, with the upper and lower signs applicable for positive and negative uniaxial crystals, respectively. Defining a sign parameter s, such that $s = +1$ for a positive and $s = -1$ for a negative uniaxial crystal, we have

$$\rho s = \theta - \theta' \quad \text{i.e.,} \quad \theta' = \theta - \rho s \tag{3.164}$$

so that

$$\tan \theta' = \tan(\theta - \rho s) = p_1 \tan \theta \tag{3.165}$$

which can be solved for ρ:

$$\tan \rho = s\frac{(1 - p_1)\tan \theta}{1 + p_1 \tan^2 \theta}. \tag{3.166}$$

ρ can also be expressed as

$$\rho = \tan^{-1}\frac{|n_Z^2 - n_o^2|\tan \theta}{1 + p_1 \tan^2 \theta} = \tan^{-1}\frac{|n_Z^2 - n_o^2|\sin \theta \cos \theta}{n_o^2 \sin^2 \theta + n_Z^2 \cos^2 \theta} \tag{3.167}$$

Using the relation $\cos^2 \rho = 1/(1 + \tan^2 \rho)$ we obtain from Eqs. 3.166

$$\cos \rho = \frac{n_Z^2 \cos^2 \theta + n_o^2 \sin^2 \theta}{(n_Z^4 \cos^2 \theta + n_o^4 \sin^2 \theta)^{1/2}}. \tag{3.168}$$

The refractive index n for the case of $\rho \neq 0$

With D and E denoting the magnitudes of the **D** and **E** vectors, Eqs. 3.32 can be re-written for uniaxial crystals as

$$\begin{aligned}
D\, d_X &= \varepsilon_0 n_o^2 E e_X \\
D\, d_Y &= \varepsilon_0 n_o^2 E e_Y \\
D\, d_Z &= \varepsilon_0 n_Z^2 E e_Z
\end{aligned} \tag{3.169}$$

which can be rewritten as

$$\begin{aligned}
e_X &= \frac{D}{\varepsilon_0 E}\frac{d_X}{n_o^2} \\
e_Y &= \frac{D}{\varepsilon_0 E}\frac{d_Y}{n_o^2} \\
e_Z &= \frac{D}{\varepsilon_0 E}\frac{d_Z}{n_Z^2}.
\end{aligned} \tag{3.170}$$

Squaring and summing the components of the unit vector $\hat{e}$ in Eqs. 3.170 we obtain

$$
\begin{aligned}
1 &= \frac{D^2}{\varepsilon_0^2 E^2} \left\{ \frac{d_X^2 + d_Y^2}{n_o^4} + \frac{d_Z^2}{n_Z^4} \right\} \\
&= \frac{D^2}{\varepsilon_0^2 E^2} \left\{ \frac{\cos^2 \theta}{n_o^4} + \frac{\sin^2 \theta}{n_Z^4} \right\}
\end{aligned}
\tag{3.171}
$$

where we have used Eqns. 3.157 and 3.158, with $d_Z = \pm \sin \theta$ as shown above to obtain

$$
d_X^2 + d_Y^2 = \cos^2 \theta.
\tag{3.172}
$$

From Eqs. 3.171 we obtain

$$
\frac{D}{\varepsilon_0 E} = \frac{n_o^2 n_Z^2}{(n_Z^4 \cos^2 \theta + n_o^4 \sin^2 \theta)^{1/2}}.
\tag{3.173}
$$

Using Eqns. 3.50, 3.168, and 3.173 we obtain the refractive index n

$$
\begin{aligned}
n^2 &= \frac{D}{\varepsilon_0 E} \frac{1}{\cos \rho} \\
&= \frac{n_Z^2 n_o^2}{n_Z^2 \cos^2 \theta + n_o^2 \sin^2 \theta}
\end{aligned}
\tag{3.174}
$$

so that from the definition of $n_e(\theta)$ in Eqs. 3.149 we get $n = n_e(\theta)$ when the walk-off angle ρ is not equal to zero.

The **D** *and* **E** *components of the extraordinary wave*

Figure 3.39 shows that for positive uniaxial crystals,

$$
\begin{aligned}
\hat{e} &= \hat{u} \cos \theta' - \hat{Z} \sin \theta' \\
\hat{d} &= \hat{u} \cos \theta - \hat{Z} \sin \theta
\end{aligned}
\tag{3.175}
$$

and using Eqs. 3.74 we obtain for positive uniaxial crystals

$$
\begin{aligned}
\hat{e} &= \hat{X} \cos \theta' \cos \phi + \hat{Y} \cos \theta' \sin \phi - \hat{Z} \sin \theta' \\
\hat{d} &= \hat{X} \cos \theta \cos \phi + \hat{Y} \cos \theta \sin \phi - \hat{Z} \sin \theta.
\end{aligned}
\tag{3.176}
$$

Similarly, Figure 3.40 shows that for negative uniaxial crystals,

$$
\begin{aligned}
\hat{e} &= -\hat{u} \cos \theta' + \hat{Z} \sin \theta' \\
\hat{d} &= -\hat{u} \cos \theta + \hat{Z} \sin \theta
\end{aligned}
\tag{3.177}
$$

and using Eqs. 3.74 again, we obtain for negative uniaxial crystals

$$
\begin{aligned}
\hat{e} &= -\hat{X} \cos \theta' \cos \phi - \hat{Y} \cos \theta' \sin \phi + \hat{Z} \sin \theta' \\
\hat{d} &= -\hat{X} \cos \theta \cos \phi - \hat{Y} \cos \theta \sin \phi + \hat{Z} \sin \theta.
\end{aligned}
\tag{3.178}
$$

Thus the unit vectors in the directions of the fields $\mathbf{D}_e$ and $\mathbf{E}_e$ can be written in a compact form as

$$\hat{e}_e = (s\cos\theta'\cos\phi, s\cos\theta'\sin\phi, -s\sin\theta')$$
$$\hat{d}_e = (s\cos\theta\cos\phi, s\cos\theta\sin\phi, -s\sin\theta) \qquad (3.179)$$

where s is equal to $+1$ or -1 for the positive and negative uniaxial crystal cases, respectively, and $\theta' = \theta - s\rho$.

3.8.4 $\rho = 0$ Case (Ordinary Wave)

Equation 3.154 shows that d_Z is equal to 0 when $\rho = 0$. From Eqs. 3.30, e_Z must then also be equal to 0, and the $\hat{d}$ and $\hat{e}$ unit vectors lie on the XY plane. Since from Eqs. 3.30

$$\hat{e} = e_X\hat{X} + e_Y\hat{Y},$$
$$\hat{d} = \varepsilon_0 n_o^2(e_X\hat{X} + e_Y\hat{Y}) \qquad (3.180)$$

$\hat{d}$ and $\hat{e}$ are parallel.

From Eqs. 3.170 with $e_Z = 0$, we obtain $D = \varepsilon_0 n_o^2 E$, whereas from Eqs. 3.50 with $\rho = 0$, $D = \varepsilon_0 n^2 E$, thus for this case $n = n_o$ and light propagates as the *ordinary wave*. To summarize, for the ordinary wave, the unit vectors $\hat{d}$ and $\hat{e}$, designated $\hat{d}_o$ and $\hat{e}_o$, are parallel (walk-off angle is zero) and they lie on the XY plane as shown in Figure 3.41.

Since $\hat{d}_o$ is perpendicular to $\mathbf{k}$, when $\hat{k}_u$ lies in the first quadrant of the XY coordinate plane, $\hat{d}_o$ can be in the second or the fourth quadrant, as shown in Figure 3.42 by the dashed and solid arrows, respectively. Equations 3.58 do not provide any guidance for the selection of the quadrant here, as they did in the case of propagation along the principal planes of the biaxial crystal. Following the choice established in earlier work (Sec. 3.5 of Ref. [9]), the direction of $\hat{d}$ is chosen here to be in the fourth quadrant so that for the wave with propagation vector $\mathbf{k}$ traveling with speed c/n_o, the unit vectors along $\hat{e}$ and $\hat{d}$ are given by

$$\hat{e}_o = (\sin\phi, -\cos\phi, 0)$$
$$\hat{d}_o = (\sin\phi, -\cos\phi, 0). \qquad (3.181)$$

3.8.5 Two Special Cases: $\theta = 0$ and $\theta = 90°$

From Eqns. 3.161 and 3.163, when $\mathbf{k}$ is along the Z axis, i.e., for $\theta = 0$, the walk-off angle ρ is also equal to 0, and the $\hat{d}$ and $\hat{e}$ vectors are parallel. With $\theta = 0$, $n_e(\theta) = n_o$ and the angle ϕ becomes indeterminate. The unit vector $\hat{d}$ can be along any direction on the XY plane and the crystal behaves as an isotropic material with refractive index n_o. This is in contrast with the case of a biaxial crystal, in which $\hat{d}$ is along the X or Y axes when $\mathbf{k}$ is along Z.

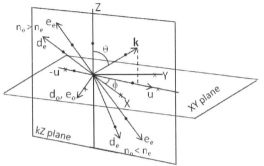

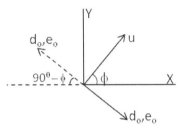

FIGURE 3.41

Polarization directions of unit vectors $\hat{d}$ and $\hat{e}$ of the ordinary and extraordinary waves. The *ordinary* fields lie on the XY plane and are shown by the cross-marks on the lines. The *extraordinary* fields lie on the $\mathbf{k} - Z$ plane and are shown by dots on the lines.

FIGURE 3.42

Possible directions of unit vectors $\hat{d}$ and $\hat{e}$ of the ordinary wave on the XY plane.

When $\mathbf{k}$ lies on the XY plane, i.e., $\theta = 90°$, the walk-off angle ρ for the extraordinary wave is equal to 0, from Eqs. 3.168. From Eqns. 3.179, the unit vector $\hat{d}_e$ for the extraordinary wave is along the positive or negative Z axis, depending on whether it is a negative or positive uniaxial crystal, respectively. The value of $n_e(\theta)$ is n_Z, irrespective of the angle ϕ. In a biaxial crystal, for $\mathbf{k}$ lying on the XY plane, the walk-off angle is not zero except for $\phi = 0$ or $90°$.

Results for the slow and fast components of the electric fields for a uni-axial medium with propagation vector $\mathbf{k}$ at an angle θ with the Z axis are summarized in Table 3.6.

3.9 Propagation Equation in Presence of Walk-Off

The propagation equation for the electric field component of light is expressed conveniently in a laboratory coordinate system (xyz) with the z direction along the propagation vector $\mathbf{k}$ and the transverse variation of the field being on the xy plane. Polarization components of the electric fields have thus far been determined with respect to the principal dielectric axes (XYZ) of the crystal. To derive the equation describing light propagation in the presence of beam walk-off, it is necessary to perform a transformation between the xyz and XYZ coordinate systems.

	Positive uniaxial $n_Z > n_o$	Negative uniaxial $n_Z < n_o$
n_s	$(n_Z\, n_o)/\sqrt{\mathcal{A}}$	n_o
n_f	n_o	$(n_Z\, n_o)/\sqrt{\mathcal{A}}$
ρ	$\tan^{-1}\dfrac{\lvert n_Z^2 - n_o^2\rvert \sin\theta\cos\theta}{\mathcal{A}}$	
e_{sX}	$\cos(\theta-\rho)\cos\phi$	$\sin\phi$
e_{sY}	$\cos(\theta-\rho)\sin\phi$	$-\cos\phi$
e_{sZ}	$-\sin(\theta-\rho)$	0
e_{fX}	$\sin\phi$	$-\cos(\theta+\rho)\cos\phi$
e_{fY}	$-\cos\phi$	$-\cos(\theta+\rho)\sin\phi$
e_{fZ}	0	$\sin(\theta+\rho)$

$$\mathcal{A} - n_o^2\sin^2\theta + n_Z^2\cos^2\theta$$

TABLE 3.6

n_s, n_f, ρ, $\hat{e}_s$, and $\hat{e}_f$ for **k** an uniaxial crystal. The components of the unit vector $\hat{d}$ are obtained from the corresponding components of the unit vector $\hat{e}$, with the walk-off angle ρ set equal to zero.

3.9.1 Transformation between Laboratory and Crystal Coordinate Systems

For light propagating with propagation vector **k** having polar coordinates (θ, ϕ) with respect to the principal dielectric axes in an anisotropic crystal with known values of n_X, n_Y, and n_Z, components of the vectors $\hat{m}$, $\hat{d}$, and $\hat{e}$ along the X, Y, and Z directions for the two allowed polarization directions (denoted by indices s and f) are known from Eqns. 3.64, 3.91, 3.92, 3.93, 3.94, and 3.95. For a given beam of light, the laboratory coordinate system (xyz) can be defined with $\hat{z}$ equal to $\hat{m}$, and x equal to the unit vector $\hat{d}$ for one of the polarizations. The unit vector $\hat{y}$ can then be determined from the cross product $\hat{z} \times \hat{x}$, i.e., we have

$$\begin{aligned}
\hat{x} &= \hat{d} = d_X\hat{X} + d_Y\hat{Y} + d_Z\hat{Z} \\
\hat{z} &= \hat{m} = m_X\hat{X} + m_Y\hat{Y} + m_Z\hat{Z} \\
\hat{y} &= (m_Y\, d_z - m_Z\, d_Y)\,\hat{X} + (m_Z\, d_X - m_X\, d_Z)\,\hat{Y} \\
&\quad + (m_X\, d_Y - m_Y\, d_X)\,\hat{Z}
\end{aligned} \tag{3.182}$$

and from Eqs. 3.95

$$\hat{e} = \cos\rho\,\hat{d} + \sin\rho\,\hat{m} = \cos\rho\,\hat{x} + \sin\rho\,\hat{z} \tag{3.183}$$

with the subscripts s or f added to the unit vectors $\hat{d}$ or $\hat{e}$ as appropriate. For example, if $n_X < n_Y < n_Z$ and the slow component is being considered, the values of d_X, d_Y, and d_Z will be given by d_{sX}, d_{sY}, and d_{sZ} in Eqs. 3.91.

3.9.2 The Propagation Equation in Presence of Walk-Off

Since $\mathbf{k}$ has been chosen to be in the z direction, $\mathbf{k} \cdot \mathbf{r} = kz$, so Eqns. 3.12 and 3.13 can be rewritten as

$$
\begin{aligned}
\widetilde{\mathbf{E}} &= \mathbf{E}e^{-i\omega t} = \mathbf{A}\psi e^{-i\omega t} \\
\widetilde{\mathbf{D}} &= \mathbf{D}e^{-i\omega t} = \mathfrak{D}\psi e^{-i\omega t}
\end{aligned}
\tag{3.184}
$$

where

$$
\psi = e^{ikz}, \quad \text{i.e.,} \quad \frac{\partial \psi}{\partial z} = ik\psi \quad \text{and} \quad \nabla\psi = ik\psi\hat{z}.
\tag{3.185}
$$

Inserting the identity $\nabla \times (\nabla \times \mathbf{E}) = \nabla(\nabla \cdot \mathbf{E}) - \nabla^2\mathbf{E}$ in Eqs. 3.39 we obtain

$$
\nabla(\nabla \cdot \mathbf{E}) - \nabla^2\mathbf{E} = \mu_0\omega^2\mathbf{D}.
\tag{3.186}
$$

Using $\mathbf{E} = \psi\mathbf{A}$ from Eqs. 3.184,

$$
\begin{aligned}
\nabla \cdot \mathbf{E} = \nabla \cdot (\psi\mathbf{A}) &= \psi(\nabla \cdot \mathbf{A} + ikA_z) \\
&= \psi(f + ikA_z)
\end{aligned}
\tag{3.187}
$$

where we have defined

$$
\begin{aligned}
f &\equiv \nabla \cdot \mathbf{A} \\
&= \frac{\partial A_x}{\partial x} + \frac{\partial A_z}{\partial z}
\end{aligned}
\tag{3.188}
$$

since $\mathbf{A}$ lies on the xz plane (Eqs. 3.183). Since $\mathbf{D}$ is along the x direction, and from Eqs. 3.184, $\mathbf{D}$ is parallel to $\mathfrak{D}$, the z component of $\mathfrak{D}$, i.e., $\mathfrak{D}_z$ is equal to 0.

Thus

$$
\begin{aligned}
\nabla \cdot \mathbf{D} &= \nabla \cdot (\psi\mathfrak{D}) \\
&= \psi(\nabla \cdot \mathfrak{D}) + \mathfrak{D} \cdot \nabla\psi \\
&= \psi(\nabla \cdot \mathfrak{D}) + ik\mathfrak{D}_z \\
&= \psi(\nabla \cdot \mathfrak{D}).
\end{aligned}
\tag{3.189}
$$

Thus the Maxwell's equation $\nabla \cdot \mathbf{D} = 0$ requires $\nabla \cdot \mathfrak{D}$ to be zero as well.

In an isotropic medium, $\mathbf{A}$ is parallel to $\mathfrak{D}$, thus $\nabla \cdot \mathfrak{D} = 0$ requires $\nabla \cdot \mathbf{A}$ to be equal to zero, so that $f = \nabla \cdot \mathbf{A} = 0$. In anisotropic media, in which $\mathfrak{D}$ and $\mathbf{A}$ are in general not parallel to each other, f is non-zero. However, for small anisotropy, the value of f is small, and the gradient of f can usually be ignored, i.e., $\nabla f \approx 0$. This is the key assumption in the derivation of the propagation equation below.

Using $\nabla f \approx 0$ and Eqns. 3.187 and 3.185

$$
\begin{aligned}
\nabla(\nabla \cdot \mathbf{E}) = \nabla(\nabla \cdot (\psi\mathbf{A})) &= \nabla(\psi f) + ik\nabla(\psi A_z) \\
&\approx (f + ikA_z)\nabla\psi + ik\psi\nabla A_z \\
&= ik\psi\{(f + ikA_z)\hat{z} + \nabla A_z\}.
\end{aligned}
\tag{3.190}
$$

Under the paraxial approximation, i.e., ignoring the $\partial^2/\partial z^2$ term, we get

$$\nabla^2 \mathbf{E} = \nabla^2(\psi \mathbf{A}) = \psi \left(\nabla_T^2 \mathbf{A} + 2ik\frac{\partial \mathbf{A}}{\partial z} - k^2 \mathbf{A} \right). \qquad (3.191)$$

If A and $\mathfrak{D}$ denote the magnitudes of the vectors $\mathbf{A}$ and $\mathfrak{D}$ respectively, we have (using Eqs. 3.183)

$$
\begin{aligned}
\mathbf{A} &= \hat{x}A_x + \hat{z}A_z \\
&= (\hat{x}\cos\rho + \hat{z}\sin\rho)A, \qquad (3.192)
\end{aligned}
$$

and

$$f = \cos\rho\frac{\partial A}{\partial x} + \sin\rho\frac{\partial A}{\partial z}. \qquad (3.193)$$

Inserting Eqns. 3.190 and 3.191 in Eqs. 3.186 (and canceling the common ψ) we get

$$
\begin{aligned}
ik\{(f + ikA_z)\hat{z} \quad &+ \quad \nabla A_z\} \\
&- \quad (\nabla_T^2 \mathbf{A} + 2ik\frac{\partial \mathbf{A}}{\partial z} - k^2 \mathbf{A}) = \mu_0\omega^2\mathfrak{D}. \qquad (3.194)
\end{aligned}
$$

Taking the dot products of the vectors $\mathbf{A}$, $\mathfrak{D}$, and ∇A_z with unit vector $\hat{e}$ given in Eqs. 3.183 we get

$$\hat{e} \cdot \mathbf{A} = A, \quad \hat{e} \cdot \mathfrak{D} = \mathfrak{D}\cos\rho \qquad (3.195)$$

and

$$
\begin{aligned}
\hat{e} \cdot \nabla A_z &= \cos\rho\frac{\partial A_z}{\partial x} + \sin\rho\frac{\partial A_z}{\partial z} \\
&= \cos\rho\sin\rho\frac{\partial A}{\partial x} + \sin^2\rho\frac{\partial A}{\partial z}. \qquad (3.196)
\end{aligned}
$$

Using Eqns. 3.196 and 3.195 in Eqs. 3.194 we get

$$
\begin{aligned}
ik\sin\rho(f + ikA_z) \quad &+ \quad ik\left(\cos\rho\sin\rho\frac{\partial A}{\partial x} + \sin^2\rho\frac{\partial A}{\partial z} \right) \\
&- \left(\nabla_T^2 A + 2ik\frac{\partial A}{\partial z} - k^2 A \right) = \mu_0\omega^2 \hat{e} \cdot \mathfrak{D} \quad (3.197)
\end{aligned}
$$

where the right-hand side of Eqs. 3.197 is intentionally left in this form.

With $A_x = A\cos\rho$ and $A_z = A\sin\rho$ (from Eqs. 3.192), and inserting Eqs. 3.193 in Eqs. 3.197 we obtain

$$
\begin{aligned}
ik\sin\rho(\cos\rho\frac{\partial A}{\partial x} \quad &+ \quad \sin\rho\frac{\partial A}{\partial z} + ik\sin\rho A) \\
&+ \quad ik\sin\rho(\cos\rho\frac{\partial A}{\partial x} + \sin\rho\frac{\partial A}{\partial z}) \\
&- (\nabla_T^2 A + 2ik\frac{\partial A}{\partial z} - k^2 A) = \mu_0\omega^2\hat{e}\cdot\mathfrak{D}. \quad (3.198)
\end{aligned}
$$

Rearranging and collecting the terms in Eqs. 3.197

$$
\begin{aligned}
2ik\sin\rho\cos\rho\frac{\partial A}{\partial x} \quad &- \quad 2ik\cos^2\rho\frac{\partial A}{\partial z} - \nabla_T^2 A \\
&= \quad \mu_0\omega^2\hat{e}\cdot\mathfrak{D} - k^2 A\cos^2\rho \\
&= \quad 0 \qquad\qquad\qquad\qquad (3.199)
\end{aligned}
$$

where the last equality follows from Eqs. 3.50 with $k = n\omega/c$ and the second relation in Eqs. 3.195. Rewriting Eqs. 3.199 with the terms rearranged, we get

$$
\frac{\partial A}{\partial z} = \frac{i}{2k\cos^2\rho}\nabla_T^2 A + \tan\rho\frac{\partial A}{\partial x}, \qquad\qquad (3.200)
$$

which is the linear propagation equation in presence of walk-off.

Bibliography

[1] M. Born and E. Wolf, *Principles of optics, 7th Edition*, Cambridge University Press, Cambridge, 1999.

[2] V. G. Dmitriev, G. G. Gurzadyan, and D. N. Nikogosyan, *Handbook of Nonlinear Optical Crystals*, Springer, Berlin, 1999.

[3] J. D. Jackson, *Classical Electrodynamics*, John Wiley and Sons, Inc., New York, 1975.

[4] S. A. Akhmanov and S. Yu. Nikitin, *Physical Optics*, Clarendon Press, Oxford, 1997.

[5] A. Yariv, *Quantum Electronics*, John Wiley and Sons, Inc., 1978.

[6] J. F. Nye, *Physical properties of crystals: their representation by tensors and matrices*, Imprint Oxford [Oxfordshire] Clarendon Press, New York, Oxford University Press, 1985.

[7] H. Ito and H. Inaba, Optical Properties and UV N_2 Laser-Pumped Parametric Fluorescence in $LiCOOH\cdot H_2O$, *IEEE J. Quantum Electron.* **8**, 612, 1972.

[8] O. I. Lavrovskaya, N. I. Pavlova, and A. V. Tarasov, Second harmonic generation of light from an YAG:Nd^{3+} laser in an optically biaxial crystal $KTiOPO_4$, *Sov. Phys. Crystallogr.* **31**, 678, 1986.

[9] F. Zernike and J. E. Midwinter, *Applied Nonlinear Optics*, John Wiley and Sons, New York, 1973.

4

Wave Propagation across the Interface of Two Homogeneous Media

In this chapter we continue our mathematical treatment of the propagation of electromagnetic waves through linear optical media begun in Chapter 2, where there are no boundary conditions for the volume, nor diffractive or spatially limiting elements within the volume.

The major difference between the mathematical models presented in this chapter and those of Chapter 2 is that we now allow the volume through which the light is propagating to be composed of more than one linear optical medium. However, we do have the following restrictions on the interface between the two different optical media:

- the shape of the interface between the media is that of a plane, and

- there are no other restrictive or diffractive elements within the plane between the media.

Treating the light as a vector electromagnetic field, we derive the Fresnel reflection and transmission coefficients to determine the reflected and transmitted light fields for the s- and p-polarization components of the incident light. We present the case in which the interface is along the x-y plane of a Cartesian coordinate system, and we show how to find the coordinate transformation for treating the case in which the interface is along any other plane.

4.1 Reflection and Refraction at a Planar Interface

The phenomena of reflection and refraction take place when light traveling in one medium falls on a surface separating that medium from another one with different optical properties. Suppose a beam of light traveling as a plane wave with wave vector $\mathbf{k}_i$ in one homogeneous and isotropic medium is incident upon an interface with another homogeneous and isotropic medium. The electric and magnetic fields of the incident light beam can be expressed as

$$\widetilde{\mathbf{E}}_i\left(\mathbf{r}, t\right) = E_o e^{i(\mathbf{k}_i \cdot \mathbf{r} - \omega t)} \hat{e}_i, \qquad (4.1)$$

and

$$\widetilde{\mathbf{H}}_i\left(\mathbf{r}, t\right) = H_o e^{i(\mathbf{k}_i \cdot \mathbf{r} - \omega t)} \hat{h}_i, \tag{4.2}$$

where $\hat{e}_i$ and $\hat{h}_i$ represent the unit vectors in the directions of the polarization of the electric and magnetic fields, respectively. Upon interacting with the boundary between media of two different refractive indices the fields are split into a reflected wave and a transmitted (refracted) wave. The reflected and transmitted electric fields can be represented as

$$\widetilde{\mathbf{E}}_r\left(\mathbf{r}, t\right) = R E_o e^{i(\mathbf{k}_r \cdot \mathbf{r} - \omega t)} \hat{e}_r, \tag{4.3}$$

and

$$\widetilde{\mathbf{E}}_t\left(\mathbf{r}, t\right) = T E_o e^{i(\mathbf{k}_t \cdot \mathbf{r} - \omega t)} \hat{e}_t, \tag{4.4}$$

where R and T are the coefficients of reflection and transmission, $\mathbf{k}_r$ and $\mathbf{k}_t$ are the wave vectors, and $\hat{e}_r$ and $\hat{e}_t$ are the unit vectors of the polarization direction of the reflected and transmitted waves, respectively.

To find the relationships of the directions of the reflected and transmitted waves to that of the incident wave, we compare their phases. Without loss of generality, the interface plane is assumed to be the x-y plane $(x, y, 0)$, with the incident, reflected and transmitted wave vectors $\mathbf{k}_i$, $\mathbf{k}_r$, and $\mathbf{k}_t$ as illustrated in Figure 4.1. Because the boundary is an infinite plane, the relationship of the reflected and transmitted amplitudes to the incident amplitude cannot depend on the position $\mathbf{r}$ on the boundary, and it also cannot depend on the time. This leads to the condition that the phases of the incident, reflected, and transmitted fields must be equal, so

$$\mathbf{k}_i \cdot \mathbf{r} - \omega t = \mathbf{k}_r \cdot \mathbf{r} - \omega t = \mathbf{k}_t \cdot \mathbf{r} - \omega t. \tag{4.5}$$

Since at the boundary $z = 0$, this can be simplified to

$$k_{ix} x + k_{iy} y = k_{rx} x + k_{ry} y = k_{tx} x + k_{ty} y. \tag{4.6}$$

Since Eq. 4.6 holds for all values of x and y, we have (setting $x = 0$ and $y = 0$, respectively)

$$k_{ix} = k_{rx} = k_{tx} \tag{4.7}$$

and

$$k_{iy} = k_{ry} = k_{ty}. \tag{4.8}$$

The plane containing the wave vector of the incident beam and the normal to the surface (i.e., $\mathbf{k}_i$ and z in this case) is called the *plane of incidence*. Using Eqs. 4.7 and 4.8 it can be shown that the wave vectors of the reflected and the transmitted beams, i.e., $\mathbf{k}_r$ and $\mathbf{k}_t$, also lie in this plane. One way to show this is to find a vector $\mathbf{N}_i$ that is normal to the plane of incidence, such as

$$\begin{aligned} \mathbf{N}_i &\equiv \mathbf{k}_i \times \hat{z} \\ &= \hat{x} k_{iy} - \hat{y} k_{ix}. \end{aligned} \tag{4.9}$$

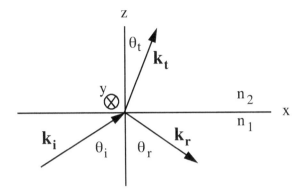

FIGURE 4.1
Illustration of the incident, reflected, and transmitted wave vectors of a light
wave incident upon an interface between two different media along the x-y
plane and their corresponding angles with respect to the normal direction,
the z-direction.

A wave vector then lies in the plane of incidence if its scalar product with $\mathbf{N}_i$
vanishes. Using Eqs. 4.7 and 4.8, we find that $\mathbf{k}_r \cdot \mathbf{N}_i = \mathbf{k}_t \cdot \mathbf{N}_i = 0$, so $\mathbf{k}_r$ and
$\mathbf{k}_t$ both lie in the plane of incidence.

From Eq. 4.7 we obtain

$$k_i \sin\theta_i = k_r \sin\theta_r = k_t \sin\theta_t \tag{4.10}$$

where θ_i, θ_r, and θ_t are the angles of incidence, reflection, and transmission,
respectively. As illustrated in Figure 4.1, θ_i and θ_t are measured from the positive z-axis, and θ_r is measured from the negative z-axis. Using the definition
of the wave number, $k \equiv 2\pi n/\lambda$, where λ is the wavelength of light in vacuum,
and designating the refractive index, n, of the first medium as n_1 and that of
the second medium as n_2, Eq. 4.10 simplifies to

$$n_1 \sin\theta_i = n_1 \sin\theta_r = n_2 \sin\theta_t. \tag{4.11}$$

From this we find that

$$n_1 \sin\theta_i = n_2 \sin\theta_t \tag{4.12}$$

and

$$\sin\theta_i = \sin\theta_r. \tag{4.13}$$

Because θ_i and θ_r have to be between 0 and $\pi/2$, solutions for this equation
are single valued and

$$\theta_i = \theta_r. \tag{4.14}$$

Equations 4.12 and 4.14 are known as Snell's laws of refraction and reflection,
respectively.

4.2   Fresnel Reflection and Transmission Coefficients

To find the reflection and transmission coefficients R and T, we must distinguish the electric and magnetic field components that are parallel and perpendicular to the plane of incidence. The orientation of the Cartesian coordinate system is chosen such that the z-axis is normal to the planar interface between the media, and the x-axis is chosen to lie in the plane of incidence. Therefore, the y-axis is parallel to the interface and perpendicular to the plane of incidence. Electric field components that are parallel to the plane of incidence (the x-z plane) are labeled "p" (for parallel), and the components that are perpendicular to the plane of incidence (the y-direction) are labeled "s" (from the German "senkrecht" for perpendicular).

The amplitude and polarization directions of $\widetilde{\mathbf{E}}_i$ and $\widetilde{\mathbf{H}}_i$ from Eqs. 4.1 and 4.2 can be written in component form in terms of s-polarization and p-polarization components as

$$E_o \hat{e}_i = E_p \hat{p}_E + E_s \hat{s}_E \tag{4.15}$$

and

$$H_o \hat{h}_i = H_p \hat{p}_H + H_s \hat{s}_H. \tag{4.16}$$

Now the incident electric and magnetic fields can be expressed as

$$\widetilde{\mathbf{E}}_i \left(\mathbf{r}, t \right) = e^{i(\mathbf{k}_i \cdot \mathbf{r} - \omega t)} \left(E_p \hat{p}_E + E_s \hat{s}_E \right) \tag{4.17}$$

and

$$\widetilde{\mathbf{H}}_i \left(\mathbf{r}, t \right) = e^{i(\mathbf{k}_i \cdot \mathbf{r} - \omega t)} \left(H_p \hat{p}_H + H_s \hat{s}_H \right), \tag{4.18}$$

where $\hat{p}_E$ and $\hat{p}_H$ are the unit vectors for the electric and magnetic fields that comprise a p-polarized light wave. Similarly, $\hat{s}_E$ and $\hat{s}_H$ are the unit vectors for the electric and magnetic fields that would comprise an s-polarized light wave. Thus, p-polarized light is composed of only the first term of Eqs. 4.17 and 4.18, and s-polarized light is composed of only the second term of Eqs. 4.17 and 4.18. Any linear light polarization can be decomposed in this way.

For what follows it is convenient to define the fractions

$$e_p \equiv \frac{E_p}{E_o} \tag{4.19}$$

$$h_p \equiv \frac{H_p}{H_o} \tag{4.20}$$

$$e_s \equiv \frac{E_s}{E_o} \tag{4.21}$$

$$h_s \equiv \frac{H_s}{H_o}. \tag{4.22}$$

From Eqs. 4.15 and 4.16, it can be shown that

$$e_p^2 + e_s^2 = 1 \tag{4.23}$$

and

$$h_p^2 + h_s^2 = 1. \tag{4.24}$$

With these, we find

$$\widetilde{\mathbf{E}}_i\left(\mathbf{r}, t\right) = E_o e^{i(\mathbf{k}_i \cdot \mathbf{r} - \omega t)}\left(e_p \hat{p}_E + e_s \hat{s}_E\right) \tag{4.25}$$

and

$$\widetilde{\mathbf{H}}_i\left(\mathbf{r}, t\right) = H_o e^{i(\mathbf{k}_i \cdot \mathbf{r} - \omega t)}\left(h_p \hat{p}_H + h_s \hat{s}_H\right). \tag{4.26}$$

The reflected fields can be written as

$$\widetilde{\mathbf{E}}_r\left(\mathbf{r}, t\right) = E_o e^{i(\mathbf{k}_r \cdot \mathbf{r} - \omega t)}\left(R_p e_p \hat{p}_{Er} + R_s e_s \hat{s}_{Er}\right) \tag{4.27}$$

and

$$\widetilde{\mathbf{H}}_r\left(\mathbf{r}, t\right) = H_o e^{i(\mathbf{k}_r \cdot \mathbf{r} - \omega t)}\left(R_p h_p \hat{p}_{Hr} + R_s h_s \hat{s}_{Hr}\right), \tag{4.28}$$

where R_p and R_s are the reflection coefficients representing the fraction of the incident amplitude of a given polarization direction reflected back into the first medium. Note that the amount of light that is reflected depends upon whether the light is s- or p-polarized, so two different coefficients are needed. The vectors $\hat{p}_{Er}$, $\hat{s}_{Er}$, $\hat{p}_{Hr}$, and $\hat{s}_{Hr}$ are the unit vectors of the polarization directions of the reflected fields.

Likewise, the transmitted fields can be written as

$$\widetilde{\mathbf{E}}_t\left(\mathbf{r}, t\right) = E_o e^{i(\mathbf{k}_t \cdot \mathbf{r} - \omega t)}\left(T_p e_p \hat{p}_{Et} + T_s e_s \hat{s}_{Et}\right) \tag{4.29}$$

and

$$\widetilde{\mathbf{H}}_t\left(\mathbf{r}, t\right) = H_o e^{i(\mathbf{k}_t \cdot \mathbf{r} - \omega t)}\left(T_p h_p \hat{p}_{Ht} + T_s h_s \hat{s}_{Ht}\right), \tag{4.30}$$

where T_p and T_s are the transmission coefficients representing the fraction of the incident amplitude transmitted into the second medium for the p- and s-polarized components, respectively.

All of the light propagation directions, unit vectors, and angles are illustrated in Figure 4.2, where the angles of incidence, reflection, and transmission are θ_i, θ_r, and θ_t, respectively. Figure 4.2 also uses the same Cartesian coordinate system as the previous section where the planar interface between the two media is the x-y plane. For a given direction of $\mathbf{k}$ and $\widetilde{\mathbf{E}}$, the direction of the corresponding $\mathbf{H}$ is obtained from the relation

$$\widetilde{\mathbf{H}} = \frac{1}{\omega \mu_o} \mathbf{k} \times \widetilde{\mathbf{E}}. \tag{4.31}$$

Reverting from s- and p-directions to Cartesian coordinates is necessary here in order to apply the boundary conditions for electromagnetic fields encountering a boundary between two different media. The components of each field that reside in the plane of incidence (p-polarization for the electric field and s-polarization for the magnetic field) must be further split into components that are parallel and perpendicular to the surface. Using Figure 4.2 it

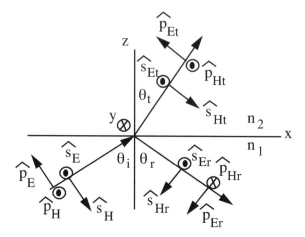

FIGURE 4.2

Illustration of the direction of the components of the incident electric and magnetic fields expressed in the p-polarization, $\hat{p}_E$ and $\hat{p}_H$, and s-polarization, $\hat{s}_E$ and $\hat{s}_H$, directions for an incident field with the plane of incidence in the x-z plane.

can be deduced that the unit vectors of the incident electric and magnetic fields are

$$
\begin{aligned}
\hat{p}_E &= -\cos\theta_i\hat{x} + \sin\theta_i\hat{z} \\
\hat{p}_H &= -\hat{y} \\
\hat{s}_E &= -\hat{y} \\
\hat{s}_H &= \cos\theta_i\hat{x} - \sin\theta_i\hat{z}.
\end{aligned}
\tag{4.32}
$$

The reflected unit vectors are also deduced to be

$$
\begin{aligned}
\hat{p}_{Er} &= -\cos\theta_r\hat{x} - \sin\theta_r\hat{z} \\
\hat{p}_{Hr} &= \hat{y} \\
\hat{s}_{Er} &= -\hat{y} \\
\hat{s}_{Hr} &= -\cos\theta_r\hat{x} - \sin\theta_r\hat{z}.
\end{aligned}
\tag{4.33}
$$

The transmitted fields maintain the same sign as the incident fields for all unit vector components, and are now functions of the angle of transmission, or

$$
\begin{aligned}
\hat{p}_{Et} &= -\cos\theta_t\hat{x} + \sin\theta_t\hat{z} \\
\hat{p}_{Ht} &= -\hat{y} \\
\hat{s}_{Et} &= -\hat{y} \\
\hat{s}_{Ht} &= \cos\theta_t\hat{x} - \sin\theta_t\hat{z}.
\end{aligned}
\tag{4.34}
$$

Using the relationship between the amplitudes of $\mathbf{E}$ and $\mathbf{H}$ in a non-magnetic dielectric medium,

$$H_o = \frac{n}{\mu_o c} E_o, \tag{4.35}$$

and plugging in the incident field unit vectors, the incident electric and magnetic fields can be written in the Cartesian coordinate system as

$$\begin{aligned}
\widetilde{\mathbf{E}}_i\left(\mathbf{r},t\right) = \quad & E_o e^{i(\mathbf{k}_i\cdot\mathbf{r}-\omega t)} \\
& \times \quad \left(-e_p \cos\theta_i \hat{x} - e_s \hat{y} + e_p \sin\theta_i \hat{z}\right)
\end{aligned} \tag{4.36}$$

and

$$\begin{aligned}
\widetilde{\mathbf{H}}_i\left(\mathbf{r},t\right) = \quad & \frac{n_1}{\mu_o c} E_o e^{i(\mathbf{k}_i\cdot\mathbf{r}-\omega t)} \\
& \times \quad \left(h_s \cos\theta_i \hat{x} - h_p \hat{y} - h_s \sin\theta_i \hat{z}\right),
\end{aligned} \tag{4.37}$$

where n_1 is the refractive index of the medium containing the incident and reflected light fields, and c is the speed of light in vacuum.

Using the unit vectors for the polarization of the reflected fields we can write the reflected light fields as

$$\begin{aligned}
\widetilde{\mathbf{E}}_r\left(\mathbf{r},t\right) = \quad & -E_o e^{i(\mathbf{k}_r\cdot\mathbf{r}-\omega t)} \\
& \times \quad \left(R_p e_p \cos\theta_r \hat{x} + R_s e_s \hat{y} + R_p e_p \sin\theta_r \hat{z}\right)
\end{aligned} \tag{4.38}$$

and

$$\begin{aligned}
\widetilde{\mathbf{H}}_r\left(\mathbf{r},t\right) = \quad & \frac{n_1}{\mu_o c} E_o e^{i(\mathbf{k}_r\cdot\mathbf{r}-\omega t)} \\
& \times \quad \left(-R_s h_s \cos\theta_r \hat{x} + R_p h_p \hat{y} - R_s h_s \sin\theta_r \hat{z}\right).
\end{aligned} \tag{4.39}$$

Applying the unit vectors for the transmitted fields, the transmitted $\mathbf{E}$ and $\mathbf{H}$ become

$$\begin{aligned}
\widetilde{\mathbf{E}}_t\left(\mathbf{r},t\right) = \quad & E_o e^{i(\mathbf{k}_t\cdot\mathbf{r}-\omega t)} \\
& \times \quad \left(-T_p e_p \cos\theta_t \hat{x} - T_s e_s \hat{y} + T_p e_p \sin\theta_t \hat{z}\right)
\end{aligned} \tag{4.40}$$

and

$$\begin{aligned}
\widetilde{\mathbf{H}}_t\left(\mathbf{r},t\right) = \quad & \frac{n_2}{\mu_o c} E_o e^{i(\mathbf{k}_t\cdot\mathbf{r}-\omega t)} \\
& \times \quad \left(T_s h_s \cos\theta_t \hat{x} - T_p h_p \hat{y} - T_s h_s \sin\theta_t \hat{z}\right),
\end{aligned} \tag{4.41}$$

where n_2 is the refractive index of the second medium.

The electromagnetic boundary conditions require that the tangential components of the fields be continuous across the boundary; i.e., the sum of the fields in the region $z < 0$ (incident and reflected) must be equal to the field in the region $z > 0$ (transmitted) at the boundary, or

$$\begin{aligned}
E_{ix} + E_{rx} &= E_{tx} \\
E_{iy} + E_{ry} &= E_{ty} \\
H_{ix} + H_{rx} &= H_{tx} \\
H_{iy} + H_{ry} &= H_{ty}
\end{aligned} \tag{4.42}$$

for all points along the boundary $(z = 0)$.

Recalling Eq. 4.5, substituting the x- and y-components of Eqs. 4.36–4.41 into Eq. 4.42, and using the fact that $\theta_i = \theta_r$ it can be shown that

$$\begin{aligned}
(1 + R_p) \cos \theta_i &= T_p \cos \theta_t \\
1 + R_s &= T_s \\
(1 - R_s) n_1 \cos \theta_i &= T_s n_2 \cos \theta_t \\
(1 - R_p) n_1 &= T_p n_2.
\end{aligned} \tag{4.43}$$

Solving Eq. 4.43 for the various reflection and transmission coefficients results in

$$\begin{aligned}
R_p &= \frac{-n_2 \cos \theta_i + n_1 \cos \theta_t}{n_2 \cos \theta_i + n_1 \cos \theta_t} \\
R_s &= \frac{n_1 \cos \theta_i - n_2 \cos \theta_t}{n_1 \cos \theta_i + n_2 \cos \theta_t}
\end{aligned} \tag{4.44}$$

and

$$\begin{aligned}
T_p &= \frac{2 n_1 \cos \theta_i}{n_2 \cos \theta_i + n_1 \cos \theta_t} \\
T_s &= \frac{2 n_1 \cos \theta_i}{n_1 \cos \theta_i + n_2 \cos \theta_t}.
\end{aligned} \tag{4.45}$$

Equations 4.44 and 4.45 are known as the Fresnel formulae, Fresnel equations, or Fresnel relations, the reflection ratios (R_s, R_p) and transmission ratios (T_s, T_p) of the s- and p-polarized waves are called the Fresnel coefficients, and the phenomenon of reflection of light from a specular interface between two media is called Fresnel reflection.[1]

---

[1] Note that the sign of R_p in Eq. 4.44 is the opposite of that found in Born and Wolf, and Jackson (Refs. [1] and [3] of Ch. 3) because the direction of $\hat{p}_{Er}$ chosen here is the opposite of that chosen in those books.

4.3 Reflection and Refraction at an Interface Not Normal to a Cartesian Axis

In the last section the Fresnel equations were derived assuming that the interface was oriented along the Cartesian x-y plane, i.e., with the normal to the interface along the z-direction. For situations in which the Cartesian coordinate directions are already fixed and the normal to an interface is not along the z-direction, a coordinate transformation must be done to determine the appropriate Fresnel coefficients. Here the general procedure to be followed for such a coordinate transformation is outlined.

Suppose a plane wave of light with wave vector $\mathbf{k}_i$ in a direction with unit vector $\hat{m}$ is incident on a plane surface with the surface normal along a unit vector $\hat{n}$. The plane of incidence is then the plane containing the vectors $\hat{m}$ and $\hat{n}$. We write the vectors $\hat{m}$, $\hat{n}$ and the incident electric field $\mathbf{E}_i$ in the Cartesian coordinate system xyz as

$$\begin{aligned} \hat{m} &= m_x\hat{x} + m_y\hat{y} + m_z\hat{z} \\ \hat{n} &= n_x\hat{x} + n_y\hat{y} + n_z\hat{z} \\ \mathbf{E}_i &= E_o(e_x\hat{x} + e_y\hat{y} + e_z\hat{z}). \end{aligned} \tag{4.46}$$

The values of m_x, m_y, m_z, n_x, n_y, n_z, and e_x, e_y, e_z are all assumed to be known.

A new Cartesian coordinate sytem (x_1, y_1, z_1) is now constructed such that

$$\begin{aligned} \hat{z}_1 &= \hat{n} \\ \hat{y}_1 &= \hat{n} \times \hat{m} \\ \hat{x}_1 &= \hat{y}_1 \times \hat{z}_1 = (\hat{n} \times \hat{m}) \times \hat{n}. \end{aligned} \tag{4.47}$$

It can be shown that $\hat{x}_1$ lies on the plane containing the unit vectors $\hat{m}$ and $\hat{n}$, i.e., the plane of incidence. Thus in the new coordinate system, the z_1 axis is aligned with the surface normal $\hat{n}$, the y_1 axis is perpendicular to the plane of incidence, and the x_1 axis lies in the plane of incidence. The $(x_1 y_1 z_1)$ coordinate system then plays the role of the (xyz) coordinates of the last sections. The transformation equations between the two coordinate systems are

$$\begin{aligned} \hat{z}_1 &= n_x\hat{x} + n_y\hat{y} + n_z\hat{z} \\ \hat{y}_1 &= (n_y m_z - m_y n_z)\hat{x} + (m_x n_z - n_x m_z)\hat{y} + (n_x m_y - m_x n_y)\hat{z} \\ \hat{x}_1 &= \{(m_x n_z - n_x m_z)n_z - (n_x m_y - m_x n_y)n_y\}\hat{x} \\ &\quad + \{(n_x m_y - m_x n_y)n_x - (n_y m_z - m_y n_z)n_z\}\hat{y} \\ &\quad + \{(n_y m_z - m_y n_z)n_y - (m_x n_z - n_x m_z)n_x\}\hat{z}, \end{aligned}$$

$$\tag{4.48}$$

from which the inverse transformation equations relating the (x,y,z) coordinates to the (x_1,y_1,z_1) coordinates can be obtained.

The angle of incidence θ_i is the angle between the normal to the surface, $\hat{n}$, and the unit vector along the propagation direction, $\hat{m}$, or

$$\theta_i = \arccos(\hat{n} \cdot \hat{m}). \tag{4.49}$$

The angle θ_r is obtained from the law of reflection

$$\theta_r = \theta_i \tag{4.50}$$

where θ_r is the angle between the propagation vector $\mathbf{k}_r$ of the reflected beam and the *negative* z_1 direction, as shown in Figure 4.1. Using Snell's law, the angle of refraction θ_t between the propagation vector $\mathbf{k}_t$ of the transmitted beam with respect to $\hat{z}_1$ is

$$\theta_t = \arcsin\left(\frac{n_1}{n_2}\sin\theta_i\right). \tag{4.51}$$

Using these values of θ_i, θ_r, and θ_t, the Fresnel coefficients R_s, T_s, R_p, and T_p for the s and p polarizations can be obtained from Eqs. 4.44 or 4.45. The reflected and transmitted electric field amplitudes can then be written in terms of the Cartesian components along the $(x_1y_1z_1)$ axes as

$$\mathbf{E}_r = E_o\left(-R_p\sqrt{e_x^2 + e_z^2}\cos\theta_r\;\hat{x}_1 + R_s e_y\;\hat{y}_1 - R_p\sqrt{e_x^2 + e_z^2}\sin\theta_r\;\hat{z}_1\right) \tag{4.52}$$

and

$$\mathbf{E}_t = E_o\left(-T_p\sqrt{e_x^2 + e_z^2}\cos\theta_t\;\hat{x}_1 + T_s e_y\;\hat{y}_1 + T_p\sqrt{e_x^2 + e_z^2}\sin\theta_t\;\hat{z}_1\right). \tag{4.53}$$

From the transformation relations between $(x_1y_1z_1)$ and (xyz) coordinates (obtained from the inverse solutions of the Eqs. 4.48) the values of the electric field components of the reflected and transmitted beams in the original (xyz) coordinate system can then be obtained.

5

Light Propagation in a Dielectric Waveguide

In this chapter we present mathematical models for light propagation within a two-dimensional slab waveguide. First, the conditions necessary for a light wave to propagate within the waveguide are explained. Then we derive the equations for the electric and magnetic field amplitudes of transverse electric and transverse magnetic modes and discuss the properties of the evanescent waves in the media just outside of the waveguide.

5.1 Conditions for Guided Waves

In Chapter 4 we explored the situation where the light propagating within one medium encountered a single planar interface to a different medium. Here we will extend our investigation to the situation where three different media are present. It is assumed that the two boundaries between the three media are parallel planes separated by some given distance. This specific arrangement is known as a "slab waveguide" (sometimes also referred to as a planar or two-dimensional waveguide).

Figure 5.1 shows the cross section of such a waveguide. Here, the coordinate system has been changed from Chapter 4 such that the z-direction is now the direction of propagation of the guided wave, and y is normal to the waveguide surfaces. We will derive the guided wave conditions and field amplitudes for slab waveguides here. For ridge (one-dimensional) waveguides, which are finite in size along the x-direction, no analytic solution for the guided wave fields exists, so they must be calculated from numerical methods. For alternative derivations and additional details see [1, 2].

First, total internal reflection must occur at both interfaces. For this, the refractive index n_2 of the second material must exceed the refractive indices, n_1 and n_3, of the other two materials, or

$$n_1 < n_2 > n_3. \tag{5.1}$$

In addition, the incident angle, θ_t , of the wave transmitted through the first

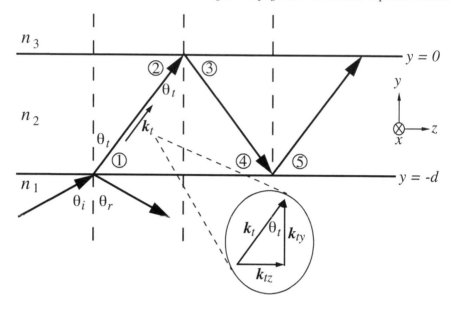

FIGURE 5.1

A y-z cross section of a slab dielectric waveguide oriented in the x-z-plane. An incident plane wave is partially transmitted at the n_1-n_2 boundary, and then undergoes total internal reflection at both boundaries to form a guided wave along z if the guided wave conditions are met.

interface must exceed the critical angle of both interfaces, resulting in

$$\arcsin\left(\frac{n_1}{n_2}\right) < \theta_t > \arcsin\left(\frac{n_3}{n_2}\right). \tag{5.2}$$

Under these conditions, a portion of the wave will be transmitted into the waveguide and will keep getting reflected completely and travel along the waveguide. However, these conditions alone are not sufficient to create a sustained wave along the waveguide. For the wave to travel unattenuatedly along the waveguide, consecutive reflections of the wave have to constructively interfere. This will be the case only for discrete incident angles, for which the phase difference due to the transverse propagation of the wave acquired in one complete bounce is an integer multiple of 2π. The traveling waves created by light incident at these angles are called the modes of the waveguide. Because the phase shift upon reflection at an interface differs for s-polarization and p-polarization, each of these polarizations has a different set of modes, with different incident angles. The modes for s-polarization are referred to as transverse electric ("TE") modes, because there is only one electric field component, which is pointed along the transverse direction, the x-direction in Figure 5.1. Similarly, p-polarization is referred to as transverse magnetic

("TM") mode, because there is only one component of magnetic field, and it is along the transverse direction.

To find the expression for the total phase acquired due to the transverse propagation, we follow a wave front through one complete bounce as shown in Figure 5.1. The transverse component of the wave vector is

$$k_{t_y} = k_t \cos \theta_t = \frac{2\pi n_2}{\lambda} \cos \theta_t, \tag{5.3}$$

where λ is the wavelength of the light in vacuum. From point 1 to point 2 in Figure 5.1, the transverse phase acquired by the wave front is

$$\phi_{12} = k_{t_y} y = \frac{2\pi n_2}{\lambda} \cos \theta_t d, \tag{5.4}$$

where d is the thickness of the intermediate layer of the waveguide. Going from point 2 (before reflection) to point 3 (after reflection), we need to distinguish between s- and p-polarization. The derivation for the phase difference upon reflection from a dielectric interface for different angles and polarizations can be found in Section 23-3 of [3]. For s-polarization (TE modes) the phase difference upon reflection at an interface to a lower index material ("internal reflection") for angles exceeding the critical angle is

$$\phi_{23_{TE}} = -2 \arctan \frac{\sqrt{\sin^2 \theta_t - \left(\frac{n_3}{n_2}\right)^2}}{\cos \theta_t}. \tag{5.5}$$

For p-polarization (TM modes) the phase difference upon internal reflection for angles exceeding the critical angle is

$$\phi_{23_{TM}} = -2 \arctan \frac{\sqrt{\sin^2 \theta_t - \left(\frac{n_3}{n_2}\right)^2}}{\left(\frac{n_3}{n_2}\right)^2 \cos \theta_t} + \pi. \tag{5.6}$$

From point 3 to point 4, both the wave vector component and the displacement along y are negative, resulting in a phase difference of

$$\phi_{34} = \frac{2\pi n_2}{\lambda} \cos \theta_t d, \tag{5.7}$$

the same as ϕ_{12}.

Finally, the phase differences due to the reflection at the first interface, going from point 4 (before reflection) to point 5 (after reflection), are

$$\phi_{45_{TE}} = -2 \arctan \frac{\sqrt{\sin^2 \theta_t - \left(\frac{n_1}{n_2}\right)^2}}{\cos \theta_t} \tag{5.8}$$

and

$$\phi_{45_{TM}} = -2 \arctan \frac{\sqrt{\sin^2 \theta_t - \left(\frac{n_1}{n_2}\right)^2}}{\left(\frac{n_1}{n_2}\right)^2 \cos \theta_t} + \pi. \tag{5.9}$$

We add all of these phases to obtain the total transverse phase difference and set it equal to an integer multiple of 2π, so

$$\phi = \phi_{12} + \phi_{23} + \phi_{34} + \phi_{45} = m2\pi, \tag{5.10}$$

where m is a positive integer, called the mode number. Plugging in the above expressions and simplifying, the condition for TE modes then becomes

$$m\pi = \frac{2\pi n_2}{\lambda} \cos \theta_t d$$

$$- \arctan \frac{\sqrt{\sin^2 \theta_t - \left(\frac{n_3}{n_2}\right)^2}}{\cos \theta_t} - \arctan \frac{\sqrt{\sin^2 \theta_t - \left(\frac{n_1}{n_2}\right)^2}}{\cos \theta_t}. \tag{5.11}$$

Similarly, we get the condition for TM modes to be

$$(m-1)\pi = \frac{2\pi n_2}{\lambda} \cos \theta_t d$$

$$- \arctan \frac{\sqrt{\sin^2 \theta_t - \left(\frac{n_3}{n_2}\right)^2}}{\left(\frac{n_3}{n_2}\right)^2 \cos \theta_t} - \arctan \frac{\sqrt{\sin^2 \theta_t - \left(\frac{n_1}{n_2}\right)^2}}{\left(\frac{n_1}{n_2}\right)^2 \cos \theta_t}. \tag{5.12}$$

Thus, only incident angles leading to transmitted angles that match one of these conditions lead to a guided wave. The discrete transmitted angles θ_t for each mode can be obtained from these equations through numerical methods. To distinguish the valid solutions of θ_t for different integers m, we will use the notation θ_m for the transmitted angle of guided mode m in what follows.

With increasing mode number m, the angles θ_m get smaller and smaller, so eventually θ_m will drop below the critical angle at one of the interfaces. Thus, for each combination of refractive indices and waveguide thickness, the number of guided modes is limited.

Overall, for the angles θ_m, a wave traveling along the z-direction of the waveguide with an effective propagation constant k_m results. Generally, the z-component of the wave vector in layer 2 is

$$k_{tz} = k_t \sin \theta_t. \tag{5.13}$$

For mode m, this is

$$k_m = \frac{2\pi n_2}{\lambda} \sin \theta_m. \tag{5.14}$$

Another way to express this is by defining the effective refractive index of the mode,

$$n_m \equiv n_2 \sin \theta_m, \tag{5.15}$$

so that

$$k_m = \frac{2\pi n_m}{\lambda}. \tag{5.16}$$

With this, the cutoff condition for guided modes can be written as

$$n_1 < n_m > n_3, \tag{5.17}$$

and we see that

$$n_1, n_3 < n_m < n_2. \tag{5.18}$$

Because a significant amount of the incident light is reflected at the first interface, it is usually more efficient or in some instances necessary to use other methods of coupling light into the waveguide, such as grating coupling, prism coupling, or end coupling.

5.2 Field Amplitudes for Guided Waves

The electric and magnetic fields of a waveguide mode can be found by solving the wave equation for dielectric media for the boundary conditions of electric and magnetic fields at an interface between two dielectric media.[1] Because guided waves are the sum of all the reflections of the transmitted wave, they must obey the wave equations, Eqs. 2.31 and 2.32. However, some properties of single wave solutions to the wave equations do not hold here. In particular, the wave sum can and does have electric and magnetic field components along the effective direction of propagation, so E_z and H_z can be non-zero.

In general, the expression for the electric field of an electromagnetic wave traveling along the z-direction is

$$\widetilde{\mathbf{E}}(\mathbf{r}, t) = \mathbf{E_o}(x, y) e^{i(kz - \omega t + \phi)}, \tag{5.19}$$

where $\mathbf{E_o}$ is the electric field amplitude vector, and ϕ is the initial phase. As shown above, guided wave modes travel along z with an effective propagation constant of k_m. For a wave traveling along z in a slab waveguide with infinite extent along x, the electric field cannot depend on x, simplifying the electric field to

$$\widetilde{\mathbf{E}}(\mathbf{r}, t) = \mathbf{E_o}(y) e^{i(k_m z - \omega t + \phi)}. \tag{5.20}$$

The corresponding expression for the magnetic field is

$$\widetilde{\mathbf{H}}(\mathbf{r}, t) = \mathbf{H_o}(y) e^{i(k_m z - \omega t + \phi)}. \tag{5.21}$$

---

[1]For a detailed discussion of the boundary conditions see Section 7.3.6 of [4].

Because all components of all fields satisfy wave equations of the same form, we will explicitly show the common steps of the derivation of the solution for the field amplitudes for the x-component of the electric field, E_x, only, knowing that E_y, E_z, H_x, H_y, and H_z have the same functional form.

The x-component of the electric field is

$$E_x(\mathbf{r}, t) = E_{ox}(y) e^{i(k_m z - \omega t + \phi)}, \tag{5.22}$$

where E_{ox} is the x-component of the electric field amplitude $\mathbf{E_o}$. Recalling the wave equation in a dielectric medium, Eq. 2.31, the relationship between electric field and electric displacement, Eq. 2.22, and plugging in this Ansatz yields

$$\nabla^2 E_{ox}(y) e^{i(k_m z - \omega t + \phi)} - \mu\epsilon \frac{\partial^2}{\partial t^2} E_{ox}(y) e^{i(k_m z - \omega t + \phi)} = 0. \tag{5.23}$$

Taking the derivatives and simplifying results in

$$\frac{d^2}{dy^2} E_{ox}(y) - k_m{}^2 E_{ox}(y) + \mu\epsilon\omega^2 E_{ox}(y) = 0, \tag{5.24}$$

where the partial derivative has been replaced because E_{ox} only depends on y.

Using the relationship $k = \omega\sqrt{\epsilon\mu}$, Eq. 5.24 simplifies to

$$\frac{d^2}{dy^2} E_{ox}(y) - \left(k_m{}^2 - k^2\right) E_{ox}(y) = 0, \tag{5.25}$$

where k is the propagation constant of the corresponding layer of dielectric material.

Defining

$$\delta_m \equiv \sqrt{k_m{}^2 - k^2} \tag{5.26}$$

the above equation can be written as

$$\frac{d^2}{dy^2} E_{ox}(y) - \delta_m{}^2 E_{ox}(y) = 0. \tag{5.27}$$

This is an ordinary second order differential equation with solutions consisting of exponential functions or sine and cosine functions, depending on whether δ_m is real or imaginary. In what follows, subscripts 1, 2, and 3 refer to the waveguide layers with n_1, n_2, and n_3, respectively.

In region 3 ($0 \leq y$), $n_3 < n_m$, so $k_m{}^2 - k_3{}^2 > 0$, and thus δ_{3m}, the value of δ_m in region 3, is real. The general solution for this case is

$$E_{ox3}(y) = E'_{x3} e^{\delta_{3m} y} + E''_{x3} e^{-\delta_{3m} y}. \tag{5.28}$$

E'_{x3} and E''_{x3} are coefficients that are determined by the power of the incident wave and the boundary conditions, as are all the prime and double-prime coefficients in what follows.

The same is true for region 1 ($y < -d$), with $n_1 < n_m$, so $k_m^2 - k_1^2 > 0$, and δ_{1m}, the value of δ_m in region 1, being real, making the general solution

$$E_{ox1}(y) = E_{x1}' e^{\delta_{1m}(y+d)} + E_{x1}'' e^{-\delta_{1m}(y+d)}. \tag{5.29}$$

Here, we chose to write a portion of each coefficient as $e^{\delta_{1m}d}$ and $e^{-\delta_{1m}d}$, respectively, for convenience when processing the boundary conditions at $y = -d$.

In region 2, $n_2 > n_m$, so $k_m^2 - k_2^2 < 0$ and thus δ_{2m} is imaginary. Note that k_2, the propagation constant in material 2, is the same as k_t from the previous section. Defining the parameter κ_{2m} with

$$\delta_{2m} \equiv i\kappa_{2m}, \tag{5.30}$$

and

$$\kappa_{2m} = \sqrt{k_2^2 - k_m^2}, \tag{5.31}$$

the differential equation becomes

$$\frac{d^2}{dy^2} E_{ox2}(y) + \kappa_{2m}^2 E_{ox2}(y) = 0. \tag{5.32}$$

The general solution for this case is

$$E_{ox2} = E_{x2}' \sin \kappa_{2m} y + E_{x2}'' \cos \kappa_{2m} y. \tag{5.33}$$

The solutions for all components of all fields ($E_{x,y,z}$, $H_{x,y,z}$) have this form. Because the energy carried by these waves cannot be infinite, all components of all fields also meet the same boundary conditions for $y = \infty$ and $y = -\infty$. The boundary conditions are

$$E_{ox1}(-\infty) = E_{ox3}(\infty) = 0. \tag{5.34}$$

For this to be the case,

$$E_{x1}'' = E_{x3}' = 0, \tag{5.35}$$

leaving

$$E_{ox} = \begin{cases} E_{x1}' e^{\delta_{1m}(y+d)} & \text{for } y < -d \\ E_{x2}' \sin \kappa_{2m} y + E_{x2}'' \cos \kappa_{2m} y & \text{for } -d \leq y \leq 0 \\ E_{x3}'' e^{-\delta_{3m} y} & \text{for } 0 < y \end{cases}. \tag{5.36}$$

Thus far, the equations were the same for all components of all fields. However, the boundary conditions at the interfaces differ for electric and magnetic field components parallel to the interface, and those perpendicular to the interface. Thus, we need to now distinguish TE modes, which have only an x-component of electric field, and y- and z-components of magnetic field, and TM modes, which have only an x-component of magnetic field, and y- and z-components of electric field.

5.2.1 TE Modes

The x-component of the electric field is parallel to the interfaces and thus must be continuous across the interfaces, so the boundary conditions at the two interfaces, $y = -d$ and $y = 0$, are

$$E_{ox1}(-d) = E_{ox2}(-d) \tag{5.37}$$

and

$$E_{ox2}(0) = E_{ox3}(0). \tag{5.38}$$

From these, we find

$$E'_{x1} = -E'_{x2} \sin \kappa_{2m} d + E''_{x2} \cos \kappa_{2m} d \tag{5.39}$$

and

$$E''_{x2} = E''_{x3}. \tag{5.40}$$

Thus,

$$E_{ox} = \begin{cases} (-E'_{x2} \sin \kappa_{2m} d + E''_{x2} \cos \kappa_{2m} d)\, e^{\delta_{1m}(y+d)} & \text{for } y < -d \\ E'_{x2} \sin \kappa_{2m} y + E''_{x2} \cos \kappa_{2m} y & \text{for } -d \leq y \leq 0 \\ E''_{x2} e^{-\delta_{3m} y} & \text{for } 0 < y \end{cases}. \tag{5.41}$$

Of the two parameters, E'_{x2} and E''_{x2}, one will be determined by the overall power of the incident wave. The remaining constraint that determines the relationship of the two can be found from the solution for H_{oz}.

Since the z-component of the magnetic field amplitude $\mathbf{H_o}$ is also parallel to the interfaces and therefore must be continuous across the boundaries, it has the same form as E_{ox}, so

$$H_{oz} = \begin{cases} (-H'_{z2} \sin \kappa_{2m} d + H''_{z2} \cos \kappa_{2m} d)\, e^{\delta_{1m}(y+d)} & \text{for } y < -d \\ H'_{z2} \sin \kappa_{2m} y + H''_{z2} \cos \kappa_{2m} y & \text{for } -d \leq y \leq 0 \\ H''_{z2} e^{-\delta_{3m} y} & \text{for } 0 < y \end{cases}. \tag{5.42}$$

The coefficients in this equation are all related to the coefficients of the x-component of the electric field by one of Maxwell's equations, Eq. 2.19. Specifically, the z-component of this equation is

$$\frac{\partial E_y}{\partial x} - \frac{\partial E_x}{\partial y} = -\frac{\partial B_z}{\partial t} \tag{5.43}$$

or

$$\frac{\partial E_y}{\partial x} - \frac{\partial E_x}{\partial y} = -\mu \frac{\partial H_z}{\partial t}. \tag{5.44}$$

Since the slab waveguide is infinite along the x-direction, the electric field cannot depend on x and the first term is zero. Carrying out the derivative on the wave expression for H_z corresponding to the expression for E_x shown in Eq. 5.22 results in

$$-\frac{\partial E_x}{\partial y} = i\mu\omega H_z. \tag{5.45}$$

For region 3, this means

$$\delta_{3m} E''_{x2} = i\mu_3\omega H''_{z2} \tag{5.46}$$

or

$$H''_{z2} = -i\frac{\delta_{3m}}{\mu_3\omega} E''_{x2}. \tag{5.47}$$

In region 2, we find

$$-\kappa_{2m}(E'_{x2}\cos\kappa_{2m}y - E''_{x2}\sin\kappa_{2m}y)$$
$$= i\mu_2\omega\,(H'_{z2}\sin\kappa_{2m}y + H''_{z2}\cos\kappa_{2m}y). \tag{5.48}$$

Since sine and cosine are linearly independent and this equation must hold for all values of y between $-d$ and 0, the coefficients of the two sine terms must be equal, as are the coefficients of the cosine terms, giving the relationships

$$\kappa_{2m} E''_{x2} = i\mu_2\omega H'_{z2} \tag{5.49}$$

or

$$H'_{z2} = -i\frac{\kappa_{2m}}{\mu_2\omega} E''_{x2} \tag{5.50}$$

and

$$-\kappa_{2m} E'_{x2} = i\mu_2\omega H''_{z2} \tag{5.51}$$

or

$$H''_{z2} = i\frac{\kappa_{2m}}{\mu_2\omega} E'_{x2}. \tag{5.52}$$

Setting the two expressions for H''_{z2}, Eqs. 5.47 and 5.52, equal to each other results in the condition

$$E''_{x2} = -\frac{\mu_3}{\mu_2}\frac{\kappa_{2m}}{\delta_{3m}} E'_{x2}. \tag{5.53}$$

This leaves only one free parameter, E'_{x2}, in the problem, which is determined by the power in the incident wave.[2]

Solving Eq. 5.45 for region 1 results in

$$-\delta_{1m}\left(-E'_{x2}\sin\kappa_{2m}d + E''_{x2}\cos\kappa_{2m}d\right)$$
$$= i\mu_1\omega\left(-H'_{z2}\sin\kappa_{2m}d + H''_{z2}\cos\kappa_{2m}d\right). \tag{5.54}$$

Plugging in Eqs. 5.50 and 5.52 leads to

$$\delta_{1m}\left(-E'_{x2}\sin\kappa_{2m}d + E''_{x2}\cos\kappa_{2m}d\right)$$
$$= \frac{\mu_1}{\mu_2}\kappa_{2m}\left(E''_{x2}\sin\kappa_{2m}d + E'_{x2}\cos\kappa_{2m}d\right) \tag{5.55}$$

---

[2] See p. 13 of [1] or p. 33 of [2].

or

$$E'_{x2}\left(\delta_{1m}\sin\kappa_{2m}d + \frac{\mu_1}{\mu_2}\kappa_{2m}\cos\kappa_{2m}d\right)$$

$$= E''_{x2}\left(-\frac{\mu_1}{\mu_2}\kappa_{2m}\sin\kappa_{2m}d + \delta_{1m}\cos\kappa_{2m}d\right). \quad (5.56)$$

Plugging in the relationship between E''_{x2} and E'_{x2}, Eq. 5.53, we find the condition

$$\left(\delta_{1m}\sin\kappa_{2m}d + \frac{\mu_1}{\mu_2}\kappa_{2m}\cos\kappa_{2m}d\right)$$

$$= -\frac{\mu_3}{\mu_2}\frac{\kappa_{2m}}{\delta_{3m}}\left(-\frac{\mu_1}{\mu_2}\kappa_{2m}\sin\kappa_{2m}d + \delta_{1m}\cos\kappa_{2m}d\right). \quad (5.57)$$

This condition can be shown to be the same condition that we previously found for the discrete angles θ_m of the guided TE modes, Eq. 5.11.[3]

Finally, we need expressions for H_{oy}. Because the y-component of the magnetic field is perpendicular to the interface, the boundary conditions are

$$B_{oy1}(-d) = B_{oy2}(-d) \quad (5.58)$$

or

$$\mu_1 H_{oy1}(-d) = \mu_2 H_{oy2}(-d) \quad (5.59)$$

and

$$B_{oy2}(0) = B_{oy3}(0) \quad (5.60)$$

or

$$\mu_2 H_{oy2}(0) = \mu_3 H_{oy3}(0). \quad (5.61)$$

Starting from the general solution of the same form as Eq. 5.36 for E_{ox},

$$H_{oy} = \begin{cases} H'_{y1}e^{\delta_{1m}(y+d)} & \text{for } y < -d \\ H'_{y2}\sin\kappa_{2m}y + H''_{y2}\cos\kappa_{2m}y & \text{for } -d \leq y \leq 0 \\ H''_{y3}e^{-\delta_{3m}y} & \text{for } 0 < y \end{cases}, \quad (5.62)$$

we find from the boundary conditions

$$H'_{y1} = \frac{\mu_2}{\mu_1}\left(-H'_{y2}\sin\kappa_{2m}d + H''_{y2}\cos\kappa_{2m}d\right) \quad (5.63)$$

and

$$H''_{y3} = \frac{\mu_2}{\mu_3}H''_{y2}. \quad (5.64)$$

This leads to

$$H_{oy} = \begin{cases} \frac{\mu_2}{\mu_1}\left(-H'_{y2}\sin\kappa_{2m}d + H''_{y2}\cos\kappa_{2m}d\right)e^{\delta_{1m}(y+d)} & \text{for } y < -d \\ H'_{y2}\sin\kappa_{2m}y + H''_{y2}\cos\kappa_{2m}y & \text{for } -d \leq y \leq 0 \\ \frac{\mu_2}{\mu_3}H''_{y2}e^{-\delta_{3m}y} & \text{for } 0 < y \end{cases}. \quad (5.65)$$

---

[3]See pp. 7 and 10 of [1].

The y-component of Eq. 2.19 is

$$-\left(\frac{\partial E_z}{\partial x} - \frac{\partial E_x}{\partial z}\right) = -\frac{\partial B_y}{\partial t} \tag{5.66}$$

or

$$-\left(\frac{\partial E_z}{\partial x} - \frac{\partial E_x}{\partial z}\right) = -\mu\frac{\partial H_y}{\partial t}. \tag{5.67}$$

Since the waveguide is infinite along x, the electric field cannot depend on x and the first term vanishes. Thus, this becomes

$$\frac{\partial E_x}{\partial z} = -\mu\frac{\partial H_y}{\partial t}. \tag{5.68}$$

and, taking the derivatives of 5.22 and the y-component of 5.21,

$$ik_m E_{ox} = i\mu\omega H_{oy} \tag{5.69}$$

or

$$H_{oy} = \frac{k_m}{\mu\omega}E_{ox}. \tag{5.70}$$

Thus, the coefficients for H_{oy} are

$$H''_{y2} = \frac{k_m}{\mu_2\omega}E''_{x2} \tag{5.71}$$

and

$$H'_{y2} = \frac{k_m}{\mu_2\omega}E'_{x2}. \tag{5.72}$$

In summary, the expressions for the field amplitudes for guided TE modes are

$$E_{ox} = \begin{cases} -E'_{x2}\left(\sin\kappa_{2m}d + \frac{\mu_3}{\mu_2}\frac{\kappa_{2m}}{\delta_{3m}}\cos\kappa_{2m}d\right)e^{\delta_{1m}(y+d)} & \text{for } y < -d \\ E'_{x2}\left(\sin\kappa_{2m}y - \frac{\mu_3}{\mu_2}\frac{\kappa_{2m}}{\delta_{3m}}\cos\kappa_{2m}y\right) & \text{for } -d \le y \le 0, \\ -\frac{\mu_3}{\mu_2}\frac{\kappa_{2m}}{\delta_{3m}}E'_{x2}e^{-\delta_{3m}y} & \text{for } 0 < y \end{cases} \tag{5.73}$$

$$H_{oz} = \begin{cases} i\frac{\kappa_{2m}}{\mu_2\omega}E'_{x2}\left(-\frac{\mu_3}{\mu_2}\frac{\kappa_{2m}}{\delta_{3m}}\sin\kappa_{2m}d + \cos\kappa_{2m}d\right)e^{\delta_{1m}(y+d)} & \text{for } y < -d \\ i\frac{\kappa_{2m}}{\mu_2\omega}E'_{x2}\left(\frac{\mu_3}{\mu_2}\frac{\kappa_{2m}}{\delta_{3m}}\sin\kappa_{2m}y + \cos\kappa_{2m}y\right) & \text{for } -d \le y \le 0, \\ i\frac{\kappa_{2m}}{\mu_2\omega}E'_{x2}e^{-\delta_{3m}y} & \text{for } 0 < y \end{cases} \tag{5.74}$$

and

$$H_{oy} = \begin{cases} -\frac{k_m}{\mu_1\omega}E'_{x2}\left(\sin\kappa_{2m}d + \frac{\mu_3}{\mu_2}\frac{\kappa_{2m}}{\delta_{3m}}\cos\kappa_{2m}d\right)e^{\delta_{1m}(y+d)} & \text{for } y < -d \\ \frac{k_m}{\mu_2\omega}E'_{x2}\left(\sin\kappa_{2m}y - \frac{\mu_3}{\mu_2}\frac{\kappa_{2m}}{\delta_{3m}}\cos\kappa_{2m}y\right) & \text{for } -d \le y \le 0. \\ -\frac{k_m}{\mu_2\omega}\frac{\kappa_{2m}}{\delta_{3m}}E'_{x2}e^{-\delta_{3m}y} & \text{for } 0 < y \end{cases} \tag{5.75}$$

Since commonly dielectric waveguide materials are non-magnetic, so

$$\mu_1 = \mu_2 = \mu_3 = \mu_o, \tag{5.76}$$

where μ_o is the permeability of free space, these simplify to

$$
E_{ox} = \begin{cases}
-E'_{x2}\left(\sin \kappa_{2m}d + \frac{\kappa_{2m}}{\delta_{3m}}\cos \kappa_{2m}d\right)e^{\delta_{1m}(y+d)} & \text{for } y < -d \\
E'_{x2}\left(\sin \kappa_{2m}y - \frac{\kappa_{2m}}{\delta_{3m}}\cos \kappa_{2m}y\right) & \text{for } -d \leq y \leq 0, \\
-\frac{\kappa_{2m}}{\delta_{3m}}E'_{x2}e^{-\delta_{3m}y} & \text{for } 0 < y
\end{cases} \tag{5.77}
$$

$$
H_{oz} = \begin{cases}
i\frac{\kappa_{2m}}{\mu_o\omega}E'_{x2}\left(-\frac{\kappa_{2m}}{\delta_{3m}}\sin \kappa_{2m}d + \cos \kappa_{2m}d\right)e^{\delta_{1m}(y+d)} & \text{for } y < -d \\
i\frac{\kappa_{2m}}{\mu_o\omega}E'_{x2}\left(\frac{\kappa_{2m}}{\delta_{3m}}\sin \kappa_{2m}y + \cos \kappa_{2m}y\right) & \text{for } -d \leq y \leq 0, \\
i\frac{\kappa_{2m}}{\mu_o\omega}E'_{x2}e^{-\delta_{3m}y} & \text{for } 0 < y
\end{cases} \tag{5.78}
$$

and

$$
H_{oy} = \begin{cases}
-\frac{k_m}{\mu_o\omega}E'_{x2}\left(\sin \kappa_{2m}d + \frac{\kappa_{2m}}{\delta_{3m}}\cos \kappa_{2m}d\right)e^{\delta_{1m}(y+d)} & \text{for } y < -d \\
\frac{k_m}{\mu_o\omega}E'_{x2}\left(\sin \kappa_{2m}y - \frac{\kappa_{2m}}{\delta_{3m}}\cos \kappa_{2m}y\right) & \text{for } -d \leq y \leq 0. \\
-\frac{k_m}{\mu_o\omega}\frac{\kappa_{2m}}{\delta_{3m}}E'_{x2}e^{-\delta_{3m}y} & \text{for } 0 < y
\end{cases} \tag{5.79}
$$

5.2.2   TM Modes

To find the expressions for the field amplitudes of guided TM modes, we follow the same chain of arguments as for TE modes. We begin with the magnetic field amplitude, which only has one component, pointed along the transverse direction, the x-direction in Figure 5.1. The x-component is parallel to the interface and must thus be continuous across each interface. Because the magnetic field wave, Eq. 5.21, has the same functional form as Eq. 5.20 for the electric field, and because the boundary conditions for the x-component of the magnetic field are identical to those for the x-component of the electric field, H_{ox} for TM modes has the same functional form as Eq. 5.41 for E_{ox} for TE modes, so

$$
H_{ox} = \begin{cases}
(-H'_{x2}\sin \kappa_{2m}d + H''_{x2}\cos \kappa_{2m}d)\,e^{\delta_{1m}(y+d)} & \text{for } y < -d \\
H'_{x2}\sin \kappa_{2m}y + H''_{x2}\cos \kappa_{2m}y & \text{for } -d \leq y \leq 0. \\
H''_{x2}e^{-\delta_{3m}y} & \text{for } 0 < y
\end{cases} \tag{5.80}
$$

Note that for ease of notation we are using the same symbols for κ_{2m}, δ_{1m}, δ_{3m}, and k_m as for TE modes, but their values are different for TM modes, because the angles θ_m satisfy a different condition, Eq. 5.12 instead of Eq. 5.11.

Since the z-component of the electric field is parallel to the interfaces and thus continuous across them, it has the same form as H_{ox}, or

$$
E_{oz} = \begin{cases} (-E'_{z2}\sin\kappa_{2m}d + E''_{z2}\cos\kappa_{2m}d)\, e^{\delta_{1m}(y+d)} & \text{for } y < -d \\ E'_{z2}\sin\kappa_{2m}y + E''_{z2}\cos\kappa_{2m}y & \text{for } -d \le y \le 0. \quad (5.81) \\ E''_{z2}e^{-\delta_{3m}y} & \text{for } 0 < y \end{cases}
$$

To relate the electric field coefficients to those of the magnetic field, we use Eq. 2.21. The z-component of this equation is

$$
\frac{\partial H_y}{\partial x} - \frac{\partial H_x}{\partial y} = \frac{\partial D_z}{\partial t} \tag{5.82}
$$

or

$$
\frac{\partial H_y}{\partial x} - \frac{\partial H_x}{\partial y} = \epsilon\frac{\partial E_z}{\partial t}. \tag{5.83}
$$

Again, because the slab waveguide is infinite along the x-direction, the magnetic field cannot depend on x and the first term is zero. Carrying out the derivative on the traveling wave expression for E_z (see Eq. 5.20) results in

$$
-\frac{\partial H_x}{\partial y} = -i\epsilon\omega E_z. \tag{5.84}
$$

In region 3, this yields the relationship

$$
\delta_{3m}H''_{x2} = -i\epsilon_3\omega E''_{z2} \tag{5.85}
$$

or

$$
E''_{z2} = i\frac{\delta_{3m}}{\epsilon_3\omega}H''_{x2}. \tag{5.86}
$$

In region 2, we find

$$
\begin{aligned} -\kappa_{2m}\left(H'_{x2}\cos\kappa_{2m}y - H''_{x2}\sin\kappa_{2m}y\right) \\ = -i\epsilon_2\omega\left(E'_{z2}\sin\kappa_{2m}y + E''_{z2}\cos\kappa_{2m}y\right). \end{aligned} \tag{5.87}
$$

The coefficients of the sine terms on either side of this equation must be equal for it to hold for all values of y in this region. This results in

$$
\kappa_{2m}H''_{x2} = -i\epsilon_2\omega E'_{z2} \tag{5.88}
$$

or

$$
E'_{z2} = i\frac{\kappa_{2m}}{\epsilon_2\omega}H''_{x2}. \tag{5.89}
$$

Similarly, comparing the coefficients of the cosine terms, we find

$$
-\kappa_{2m}H'_{x2} = -i\epsilon_2\omega E''_{z2} \tag{5.90}
$$

or

$$
E''_{z2} = -i\frac{\kappa_{2m}}{\epsilon_2\omega}H'_{x2}. \tag{5.91}
$$

Setting this equal to the expression we found for region 1, Eq. 5.86, leads to the condition

$$i\frac{\delta_{3m}}{\epsilon_3 \omega} H''_{x2} = -i\frac{\kappa_{2m}}{\epsilon_2 \omega} H'_{x2} \tag{5.92}$$

or

$$H''_{x2} = -\frac{\epsilon_3 \kappa_{2m}}{\epsilon_2 \delta_{3m}} H'_{x2}. \tag{5.93}$$

This leaves only one free parameter, H'_{x2}, which is determined by the power of the incident wave.[4]

Applying Eq. 5.84 to region 1 results in

$$\delta_{1m}(-H'_{x2}\sin\kappa_{2m}d + H''_{x2}\cos\kappa_{2m}d)$$
$$= i\epsilon_1\omega\left(-E'_{?2}\sin\kappa_{2m}d + E''_{?2}\cos\kappa_{2m}d\right). \tag{5.94}$$

Expressing the electric field coefficients in terms of the magnetic field coefficients using Eqs. 5.89 and 5.91 leads to

$$\delta_{1m}(-H'_{x2}\sin\kappa_{2m}d + H''_{x2}\cos\kappa_{2m}d)$$
$$= \frac{\epsilon_1\kappa_{2m}}{\epsilon_2}\left(H''_{x2}\sin\kappa_{2m}d + H'_{x2}\cos\kappa_{2m}d\right). \tag{5.95}$$

Replacing H''_{x2} with Eq. 5.93 and simplifying, this becomes

$$\delta_{1m}\left(\sin\kappa_{2m}d + \frac{\epsilon_3\kappa_{2m}}{\epsilon_2\delta_{3m}}\cos\kappa_{2m}d\right)$$
$$= \frac{\epsilon_1\kappa_{2m}}{\epsilon_2}\left(\frac{\epsilon_3}{\epsilon_2}\frac{\kappa_{2m}}{\delta_{3m}}\sin\kappa_{2m}d - \cos\kappa_{2m}d\right). \tag{5.96}$$

This condition can be shown to be equivalent to Eq. 5.12, the condition for the angles θ_m for guided TM modes.[5]

Finally, the solution for the y-component of the electric field amplitude can be found by applying the boundary conditions for electric field components perpendicular to the interfaces, so

$$\epsilon_1 E_{oy1}(-d) = \epsilon_2 E_{oy2}(-d) \tag{5.97}$$

and

$$\epsilon_2 E_{oy2}(0) = \epsilon_3 E_{oy3}(0), \tag{5.98}$$

to the general solution of the same functional form as E_{ox} for TE modes, Eq. 5.36,

$$E_{oy} = \begin{cases} E'_{y1}e^{\delta_{1m}(y+d)} & \text{for } y < -d \\ E'_{y2}\sin\kappa_{2m}y + E''_{y2}\cos\kappa_{2m}y & \text{for } -d \le y \le 0. \\ E''_{y3}e^{-\delta_{3m}y} & \text{for } 0 < y \end{cases} \tag{5.99}$$

---

[4]See pp. 16, 17 of [1] or p. 34 of [2].
[5]See pp. 7 and 15 of [1].

The boundary conditions yield the relationships

$$E'_{y1} = \frac{\epsilon_2}{\epsilon_1} \left(-E'_{y2} \sin \kappa_{2m}d + E''_{y2} \cos \kappa_{2m}d \right) \tag{5.100}$$

and

$$E''_{y2} = \frac{\epsilon_3}{\epsilon_2} E''_{y3} \tag{5.101}$$

or

$$E''_{y3} = \frac{\epsilon_2}{\epsilon_3} E''_{y2}. \tag{5.102}$$

Furthermore, we can relate the electric field coefficients to the magnetic field coefficients, using the y-component of Maxwell's equation, Eq. 2.21,

$$-\left(\frac{\partial H_z}{\partial x} - \frac{\partial H_x}{\partial z} \right) = \frac{\partial D_y}{\partial t} \tag{5.103}$$

or

$$-\left(\frac{\partial H_z}{\partial x} - \frac{\partial H_x}{\partial z} \right) = \epsilon \frac{\partial E_y}{\partial t}. \tag{5.104}$$

Because the slab waveguide is infinite along x, the first term vanishes. Plugging in the traveling wave expressions for H_x and E_y from Eqs. 5.21 and 5.20 and taking the derivatives yields

$$ik_m H_x = -i\epsilon\omega E_y \tag{5.105}$$

or

$$E_y = -\frac{k_m}{\epsilon\omega} H_x. \tag{5.106}$$

Thus,

$$E'_{y2} = -\frac{k_m}{\epsilon_2\omega} H'_{x2} \tag{5.107}$$

and

$$E''_{y2} = -\frac{k_m}{\epsilon_2\omega} H''_{x2}, \tag{5.108}$$

which in turn is

$$E''_{y2} = \frac{k_m \epsilon_3 \kappa_{2m}}{\epsilon_2^2 \omega \delta_{3m}} H'_{x2} \tag{5.109}$$

after plugging in Eq. 5.93.

In summary, the field amplitudes for guided TM modes are

$$H_{ox} = \begin{cases} -H'_{x2} \left(\sin \kappa_{2m}d + \frac{\epsilon_3}{\epsilon_2} \frac{\kappa_{2m}}{\delta_{3m}} \cos \kappa_{2m}d \right) e^{\delta_{1m}(y+d)} & \text{for } y < -d \\ H'_{x2} \left(\sin \kappa_{2m}y - \frac{\epsilon_3}{\epsilon_2} \frac{\kappa_{2m}}{\delta_{3m}} \cos \kappa_{2m}y \right) & \text{for } -d \le y \le 0, \\ -\frac{\epsilon_3}{\epsilon_2} \frac{\kappa_{2m}}{\delta_{3m}} H'_{x2} e^{-\delta_{3m}y} & \text{for } 0 < y \end{cases} \tag{5.110}$$

$$E_{oz} = \begin{cases} -i\frac{\kappa_{2m}}{\epsilon_2\omega} H'_{x2} \left(-\frac{\epsilon_3}{\epsilon_2} \frac{\kappa_{2m}}{\delta_{3m}} \sin \kappa_{2m}d + \cos \kappa_{2m}d \right) e^{\delta_{1m}(y+d)} & \text{for } y < -d \\ -i\frac{\kappa_{2m}}{\epsilon_2\omega} H'_{x2} \left(\frac{\epsilon_3}{\epsilon_2} \frac{\kappa_{2m}}{\delta_{3m}} \sin \kappa_{2m}y + \cos \kappa_{2m}y \right) & \text{for } -d \le y \le 0, \\ -i\frac{\kappa_{2m}}{\epsilon_2\omega} H'_{x2} e^{-\delta_{3m}y} & \text{for } 0 < y \end{cases}$$
$$\tag{5.111}$$

and

$$
E_{oy} = \begin{cases}
\frac{k_m}{\epsilon_2 \omega} H'_{x2} \left(\frac{\epsilon_2}{\epsilon_1} \sin \kappa_{2m} d + \frac{\epsilon_3}{\epsilon_1} \frac{\kappa_{2m}}{\delta_{3m}} \cos \kappa_{2m} d \right) e^{\delta_{1m}(y+d)} & \text{for } y < -d \\
-\frac{k_m}{\epsilon_2 \omega} H'_{x2} \left(\sin \kappa_{2m} y - \frac{\epsilon_3}{\epsilon_2} \frac{\kappa_{2m}}{\delta_{3m}} \cos \kappa_{2m} y \right) & \text{for } -d \leq y \leq 0 \\
\frac{k_m}{\epsilon_2 \omega} \frac{\kappa_{2m}}{\delta_{3m}} H'_{x2} e^{-\delta_{3m} y} & \text{for } 0 < y
\end{cases}
$$

$$(5.112)$$

Assuming non-magnetic materials and using Eq. 2.45, we substitute

$$
\epsilon_j = n_j^2 \epsilon_o, \tag{5.113}
$$

where $j = 1, 2, 3$ signifies materials 1, 2, 3, respectively, and where ϵ_o is the permittivity of free space. We can thus write the field amplitudes for guided TM modes as

$$
H_{ox} = \begin{cases}
-H'_{x2} \left(\sin \kappa_{2m} d + \frac{n_3^2}{n_2^2} \frac{\kappa_{2m}}{\delta_{3m}} \cos \kappa_{2m} d \right) e^{\delta_{1m}(y+d)} & \text{for } y < -d \\
H'_{x2} \left(\sin \kappa_{2m} y - \frac{n_3^2}{n_2^2} \frac{\kappa_{2m}}{\delta_{3m}} \cos \kappa_{2m} y \right) & \text{for } -d \leq y \leq 0 \\
-\frac{n_3^2}{n_2^2} \frac{\kappa_{2m}}{\delta_{3m}} H'_{x2} e^{-\delta_{3m} y} & \text{for } 0 < y
\end{cases}
$$

$$(5.114)$$

$$
E_{oz} = \begin{cases}
-i \frac{\kappa_{2m}}{n_2^2 \epsilon_o \omega} H'_{x2} \left(-\frac{n_3^2}{n_2^2} \frac{\kappa_{2m}}{\delta_{3m}} \sin \kappa_{2m} d + \cos \kappa_{2m} d \right) e^{\delta_{1m}(y+d)} & \text{for } y < -d \\
-i \frac{\kappa_{2m}}{n_2^2 \epsilon_o \omega} H'_{x2} \left(\frac{n_3^2}{n_2^2} \frac{\kappa_{2m}}{\delta_{3m}} \sin \kappa_{2m} y + \cos \kappa_{2m} y \right) & \text{for } -d \leq y \leq 0 \\
-i \frac{\kappa_{2m}}{n_2^2 \epsilon_o \omega} H'_{x2} e^{-\delta_{3m} y} & \text{for } 0 < y
\end{cases}
$$

$$(5.115)$$

and

$$
E_{oy} = \begin{cases}
\frac{k_m}{n_2^2 \epsilon_o \omega} H'_{x2} \left(\frac{n_2^2}{n_1^2} \sin \kappa_{2m} d + \frac{n_3^2}{n_1^2} \frac{\kappa_{2m}}{\delta_{3m}} \cos \kappa_{2m} d \right) e^{\delta_{1m}(y+d)} & \text{for } y < -d \\
-\frac{k_m}{n_2^2 \epsilon_o \omega} H'_{x2} \left(\sin \kappa_{2m} y - \frac{n_3^2}{n_2^2} \frac{\kappa_{2m}}{\delta_{3m}} \cos \kappa_{2m} y \right) & \text{for } -d \leq y \leq 0 \\
\frac{k_m}{n_2^2 \epsilon_o \omega} \frac{\kappa_{2m}}{\delta_{3m}} H'_{x2} e^{-\delta_{3m} y} & \text{for } 0 < y
\end{cases}
$$

$$(5.116)$$

5.2.3 Evanescent Waves

As shown above, the electric and magnetic field amplitudes of the guided modes in a slab waveguide do not stop abruptly at the interfaces, but rather some light exists in the outer materials. These portions of the guided modes are called evanescent waves. For both the TE and TM modes *all* field amplitudes in the regions $y > 0$ and $y < -d$ have the form of an exponential decay as a function of distance from the waveguide surface. The parameters δ_{1m} and δ_{3m} are the decay constants in materials 1 and 3. They depend on whether the mode is TE or TM, and on the mode number, since

$$
\delta_{1m} = \sqrt{k_m^2 - k_1^2} \tag{5.117}
$$

and

$$
\delta_{2m} = \sqrt{k_m^2 - k_3^2}, \tag{5.118}
$$

with

$$k_m = k_2 \sin \theta_m \qquad (5.119)$$

and the θ_m are different for TE and TM modes.

One application of evanescent waves is in the field of atomic physics. The light of the evanescent wave can be used to repel atoms. If region 3 is vacuum, and laser-cooled atoms are dropped on the waveguide from above, the evanescent wave acts as an atom mirror [5]. Furthermore, the difference in decay constants allows the creation of an evanescent wave interference pattern with localized dark spots away from the waveguide surface. These dark spots can be used as atom traps, as detailed in [6].

Bibliography

[1] D. Marcuse, *Theory of Dielectric Optical Waveguides*, Second Edition, Academic Press, Inc., San Diego, 1991.

[2] R.G. Hunsperger, *Integrated Optics: Theory and Technology*, Second Edition, Springer-Verlag, Berlin, 1984.

[3] F.L. Pedrotti, L.S. Pedrotti, and L.M. Pedrotti, *Introduction to Optics*, Third Edition, Pearson Prentice Hall, Upper Saddle River, New Jersey, 2007.

[4] D.J. Griffiths, *Introduction to Electrodynamics*, Third Edition, Prentice-Hall, Inc., Upper Saddle River, New Jersey, 1999.

[5] W. Seifert, R. Kaiser, A. Aspect, and J. Mlynek, "Reflection of atoms from a dielectric wave-guide," Opt. Comm. **111**, 566–576 (1994).

[6] K. Christandl, G.P. Lafyatis, S.-C. Lee, and J.-F. Lee, "One- and two-dimensional optical lattices on a chip for quantum computing," Phys. Rev. A **70**, 032302 (2004).

6

Paraxial Propagation of Gaussian Beams

> In this chapter we present mathematical models for the propagation of Gaussian beams. Topics covered are:
>
> - propagation, focal parameters, and focusability behaviors of diverging or converging TEM_{00} beams
>
> - ABCD matrix treatment of TEM_{00} beams propagating through single or multiple optical elements
>
> - Equations for the electric fields for higher-order Gaussian beams
>
> - historical evolution and current usage of the M^2 beam propagation parameter

---

6.1 Introduction

Description of light propagation usually reduces to answering the question: with the electric field distribution of light known in some region of space, what is the distribution at other points? The general answer to this question comes from the solution to the Helmholtz equation, the integral solution of which involves a double integral with a rapidly oscillating integrand, which is very challenging to evaluate. However, for the special case of paraxial propagation, in which the light beam is confined to regions near the axis along which it propagates, such as in the case of a propagating laser beam, the Helmholtz equation takes a much simpler form, called the paraxial wave equation. A solution to the paraxial wave equation is the fundamental (or TEM_{00}) Gaussian beam. Description of propagation of such a beam through simple optical systems, such as a transparent homogeneous medium, an interface of two transparent media, a thin lens, etc. can be obtained algebraically without having to solve a differential equation or to evaluate an integral at every step of the way. In this chapter we will first derive the paraxial wave equation, then obtain the fundamental solution to it and then describe the method used for paraxial propagation.

The paraxial wave equation has solutions other than the fundamental one, which can be analytically expressed in terms of polynomials in the spatial coordinates. These general solutions are referred to as the higher order modes, and the fundamental (TEM$_{00}$) solution is a special (lowest order) case of such polynomial solutions. The general solutions for the higher order modes will also be described in this chapter. Moreover, different states of polarization associated with these higher order modes will be described. Finally, a measure of the quality of a laser beam, in terms of its deviation from the fundamental (TEM$_{00}$) beam distribution, will be discussed in detail.

6.1.1   Paraxial Wave Equation

For a paraxial beam, the divergence angles are small and the minimum beam size (obtained at the beam waist) is much larger than the wavelength of light. Paraxial beam propagation usually takes place through optical components whose transverse dimensions are much larger than the width of the intensity profile of the beam. When the waist size is of the same order as the wavelength, the divergence angle becomes large and the beam is said to be non-paraxial. Chapter 10 presents vector diffraction theory for the propagation of non-paraxial Gaussian beams and Ref. [1] has a discussion on the regions of validity of the paraxial approximation in terms of the beam sizes.

As shown earlier, for light propagation through linear source-free regions of space, the electric field must obey the wave equation:

$$\left(\nabla^2 - \epsilon\mu\frac{\partial^2}{\partial t^2} \right) \mathbf{E} = 0. \tag{6.1}$$

It is assumed here that one component of the electric field is always much larger than the other components such that the vector equation becomes a scalar equation representing the dominant polarization dimension.[1]

Moreover, because we have already assumed that the energy flow is primarily along one direction (our chosen z-direction), we can say that the phase of the oscillating electric field $\widetilde{E}$ has a strong dependence upon z, which allows factoring out a phase term of e^{ikz} from the scalar electric field, or

$$\widetilde{E}\left(x, y, z, t\right) = \widetilde{U}\left(x, y, z, t\right) e^{ikz}, \tag{6.2}$$

where $\widetilde{U}\left(r, \theta, z, t\right)$ is a complex scalar function of space and time. We can also assume (as we have in previous chapters) that the time dependence is simply that of a single frequency, ω, creating a global time-dependent phase factor of $e^{-i\omega t}$. The general form of the scalar electric field becomes

$$\widetilde{E}\left(x, y, z, t\right) = U\left(x, y, z\right) e^{i(kz - \omega t)}; \tag{6.3}$$

---

[1]The limits of this approximation are further discussed in Sections 7.1 and 7.4. In short, the vector components are coupled together by the divergence equation, $\nabla \cdot \mathbf{E} = 0$, and the scalar approximation is valid as long as the spatial rate of change of the dominant polarization is not too great.

in other words, the light field has the mathematical form of a plane wave with a position-dependent complex amplitude.

Writing the Laplacian ∇^2 as

$$\nabla^2 = \nabla_T^2 + \frac{\partial^2}{\partial z^2}, \tag{6.4}$$

where ∇_T^2 represents the transverse terms of the Laplacian, using the scalar approximation, substituting Eq. 6.3 into Eq. 6.1, and canceling out the common phase term, $e^{i(kz-\omega t)}$, we get

$$\nabla_T^2 U + \left[\frac{\partial^2 U}{\partial z^2} + 2ik\frac{\partial U}{\partial z} - k^2 U\right] + k^2 U = 0, \tag{6.5}$$

or

$$\nabla_T^2 U + 2ik\frac{\partial U}{\partial z} = 0, \tag{6.6}$$

by canceling the k^2 terms and using a slowly varying approximation. The slowly varying approximation assumes that the longitudinal variations of the function U are such that

$$k\frac{\partial U}{\partial z} \gg \frac{\partial^2 U}{\partial z^2}. \tag{6.7}$$

Equation 6.6 is also commonly referred to as the **paraxial wave equation.**

The paraxial wave equation can be tackled in either the Cartesian (i.e., x, y, z) or cylindrical (r, ϕ, z) coordinate systems, depending on the symmetry (or lack thereof) of the optical components. For a radially symmetric system, the cylindrical coordinates are convenient because the dependence of U on the azimuthal angle ϕ drops off and U becomes a function of only two variables, r and z.

In cylindrical coordinates, (r, ϕ, z), the Laplacian takes the form

$$\nabla^2 = \frac{1}{r}\frac{\partial}{\partial r}\left(r\frac{\partial}{\partial r}\right) + \frac{1}{r^2}\frac{\partial^2}{\partial \phi^2} + \frac{\partial^2}{\partial z^2}. \tag{6.8}$$

6.2 TEM$_{00}$ Gaussian Beam Propagation and Parameters

Books by Marcuse [2], Siegman [3], or Yariv [4] and the early paper by Kogelnik [5] provide thorough discussions of Gaussian beams and their characteristics.[2] For completeness, we describe here some of these properties of Gaussian beams.

---

[2] For detailed discussions of TEM$_{00}$ see Chapter 6 of Ref. [2], Chapters 16 and 17 of Ref. [3], Chapter 6 of Ref. [4]

TEM$_{00}$ Beams

In either Cartesian or cylindrical coordinate systems, the paraxial wave equation can be transformed into a second order differential equation in one independent variable, and the general solution to it can be expressed either in terms of a series of Hermite polynomials (for Cartesian coordinates) or Laguerre polynomials (for cylindrical coordinates). For either coordinate system, solutions are expressed in terms of a pair of integers that take non-negative values, i.e., go from 0 to higher values. As the values of these integers increase, so does the size of the beams associated with them. The lowest value of the integers is the pair 0,0, and the beam associated with this pair is known as the TEM$_{00}$ beam, which has a Gaussian cross-section at any value z.

Although laser beams can be configured in many different shapes, the TEM$_{00}$ beam is perhaps the most widely used or sought after. Their characteristics can be obtained as special (lowest order) cases of either the Hermite Gaussian or the Laguerre Gaussian beams, which will be discussed here later. But because of their importance and unique properties, we start by deriving them directly from the paraxial wave equation using the method shown by Kogelnik and Li [5].

As in Ref [5], we assume the function U to have the general form of

$$U(r,z) = \exp\left[iP(z) + ik\frac{r^2}{2q(z)}\right],\qquad(6.9)$$

where P and q are complex functions. In other words, the complex amplitude function of Eq. 6.3 is expressed as a product of two complex phase terms, where the first phase term, $\exp[iP]$ is only a function of z, and the second has the general form of $\exp[fr^2]$ in order to yield the final expected radial intensity decay of the form of

$$I(r) \propto e^{\left(-\frac{r^2}{\omega^2}\right)}.\qquad(6.10)$$

Substitution of Eq. 6.9 into the paraxial wave equation and arranging the terms according to powers of r yields the differential equation

$$\left(\frac{k^2}{q^2}\frac{\partial q}{\partial z} - \frac{k^2}{q^2}\right)r^2 + \left(2k\frac{i}{q} - 2k\frac{\partial P}{\partial z}\right) = 0.\qquad(6.11)$$

The left side of Eq. 6.11 is equal to zero only if all terms of the same power of r sum to zero. Setting each of the parentheses equal to zero gives us two simpler first-order differential equations:

$$\frac{\partial q}{\partial z} = 1,\qquad(6.12)$$

and

$$\frac{\partial P}{\partial z} = \frac{i}{q},\qquad(6.13)$$

which have the straightforward solutions

$$q(z) = z + q_o, \tag{6.14}$$

and

$$P(z) = i \ln\left(1 + \frac{z}{q_o}\right), \tag{6.15}$$

where q_o is a complex integration constant. To determine the propagation laws for the beam, two real beam parameters, denoted by $R(z)$ and $w(z)$, related to the complex parameter q by the relation are *introduced* following Kogelnik and Li [5]:

$$\frac{1}{q(z)} \equiv \frac{1}{R(z)} + i\frac{\lambda}{n\pi w^2(z)}. \tag{6.16}$$

Inserting $q(z)$ from Eq. 6.16 in Eq. 6.9 it is seen that $R(z)$ represents the curvature of the wave front that intersects the axis at distance z and w represents the radial distance where the amplitude of the electric field has decreased by a factor of e^{-1} from the value at $r = 0$. The parameter w is usually referred to as the beam radius or beam width. If the origin of the coordinate system used is assumed to be on the plane on which wave front is plane, i.e., at $z = 0$, $R(0) = \infty$, Eq. 6.16 shows that $q(0)$ is purely imaginary, i.e., from Eq. 6.14, q_o is purely imaginary. Writing q_o as

$$q_o = -iz_o \tag{6.17}$$

where z_o is as yet unspecified, and by substituting in Eq. 6.16, the radius of curvature of the phase front at a distance z from the origin is obtained as

$$R(z) = z\left(1 + \frac{z_o^2}{z^2}\right) \tag{6.18}$$

along with the square of the beam width at the same distance

$$w(z)^2 = \frac{\lambda}{n\pi}\frac{z^2 + z_o^2}{z_o}. \tag{6.19}$$

Setting $z = 0$ in Eq. 6.19, the parameter z_o is obtained as

$$z_o = \frac{n\pi w(0)^2}{\lambda} = \frac{n\pi w_o^2}{\lambda} = \frac{1}{2}kw_o^2 \tag{6.20}$$

where $w_o = w(0)$ denotes the beam width at $z = 0$.

Eq. 6.19 can then be rewritten as

$$w(z) = w_o\sqrt{\left(1 + \frac{z^2}{z_o^2}\right)} \tag{6.21}$$

showing that the minimum beam width is obtained at $z = 0$ where the phase front is plane, i.e., where the radius of curvature R of the phase front is ∞. The position of the minimum beam width is called the beam waist.

The parameter z_o is known as the *Rayleigh range* and is the axial distance from the $z = 0$ plane to a plane where the beam width has increased by a factor of $\sqrt{2}$, i.e., the beam area increased by a factor of 2. The confocal parameter of a Gaussian beam, usually denoted by the symbol b, is defined to be twice the Rayleigh range

$$b = 2z_o. \tag{6.22}$$

The confocal parameter is the length over which a Gaussian beam remains somewhat collimated.

Figure 6.1 is an illustration of the behavior of the width of the beam and the surfaces of uniform phases for a TEM_{00} beam with beam waist at $z = 0$. As the beam approaches the plane of the waist (say from $z \leq -3z_o$) the beam width monotonically decreases, just as geometrical optics and ray optics would predict. Near the waist, the width of the beam deviates from expected geometrical convergence behavior and only converges to some finite minimum width. After the beam has passed the waist, it monotonically increases in width. Also, far from the waist, the phase fronts of a Gaussian beam approach those of a spherical wave that originated at the waist. For distances $z \gg z_o$

$$R(z) \approx z, \tag{6.23}$$

$$\omega(z) \approx \frac{z}{z_o}\omega_o, \tag{6.24}$$

and the half-angle of the far-field divergence, θ_o, is

$$\theta_o = \tan^{-1}\frac{\omega(z)}{z} \approx \frac{\omega_o}{z_o} \approx \frac{\lambda}{n\pi\omega_o}. \tag{6.25}$$

A commonly used, and rather handy, parameter to describe focused laser beams is the *beam parameter product* (also referred to as the *beam propagation product*), or BPP, which is the product of a beam's minimum width ω_o multiplied by the far-field divergence angle θ_o. For a TEM_{00} beam propagating through vacuum or air ($n \approx 1$) the BPP is

$$\text{BPP} = \omega_o\theta_o = \frac{\lambda}{\pi}. \tag{6.26}$$

BPP is usually expressed in units of mm mrad. Its usefulness comes from the fact that the product of these two parameters for a TEM_{00} beam of a given wavelength is constant. This direct relationship between the focal spot width and the convergence angle makes it straightforward to predict and manipulate the focal properties of a beam with a relatively high degree of accuracy. For example, if you desire a new focal spot to be twice as small as the original focal spot you will need twice the convergence angle. In other words, if one lens has half the focal length of another, then simply replacing the original

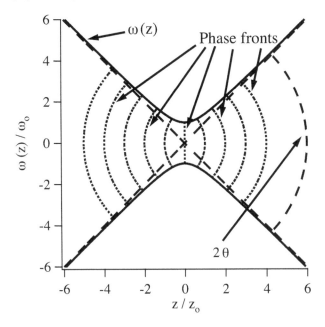

FIGURE 6.1
Beam width, phase fronts, and far-field divergence angle for a focused TEM$_{00}$ beam.

lens with the shorter focal length lens would result in the desired doubling of the convergence angle, and the new beam would have a focal spot with a width half that of the original.

The ratio of the measured value of BPP for an actual beam of the same wavelength to the ideal value of $\lambda/(n\pi)$ is known as M^2 and it provides a measure of the deviation of that beam from the a TEM$_{00}$ beam, i.e., it quantifies the 'quality' of a laser beam. Further discussion of BPP and M^2 is provided later in this chapter in Section 6.6.

By substituting Eqs. 6.15 and 6.16 into Eq. 6.9 it can be shown that the product of the complex phase terms results in a real amplitude and complex phase for each, or

$$U(r, z) = \frac{w_o}{w(z)} \exp\left[-i\tan^{-1}\left(\frac{z}{z_o}\right)\right] \exp\left[-\frac{r^2}{w^2(z)} + i\frac{k}{2}\frac{r^2}{R(z)}\right]. \quad (6.27)$$

Now, with substitution of Eq. 6.27 into Eq. 6.3 we can arrive at an expression for the scalar electric field of a TEM$_{00}$ Gaussian beam:

$$E(r, z, t) = \left[E_o\frac{w_o}{w(z)} \exp\left(-\frac{r^2}{w^2(z)}\right)\right]$$
$$\times \exp\left[i\left(\frac{k}{2}\frac{r^2}{R(z)} - \tan^{-1}\frac{z}{z_o}\right)\right] e^{i(kz - \omega t)}. \quad (6.28)$$

Equation 6.28 is written as the product of several terms. The first few terms, including the exponential term with the first set of square brackets in Eq. 6.28, represents the real amplitude as a function of r and z and the exponential term with the second set of square brackets in Eq. 6.28 represents the complex phase of the focused TEM$_{00}$ beam.

Using the definition of irradiance to be the time average of the Poynting vector, or

$$I = \langle |\mathbf{S}| \rangle \tag{6.29}$$

we can arrive at an expression for the intensity of a focused Gaussian beam as a function of the radial and axial position of the point of interest

$$I(r, z) = I_o \left(\frac{\omega_o}{\omega(z)} \right)^2 \exp\left[-2 \frac{r^2}{\omega(z)^2} \right], \tag{6.30}$$

where $\omega(z)$ is determined by Eq. 6.21, and I_o denotes peak (on-axis) value of the irradiance at the beam waist.

By taking a close look at Eqs. 6.21 and 6.30 we find that along the optical axis the irradiance follows a Lorentzian form, or

$$I(r = 0, z) = \frac{I_o}{1 + (z/z_o)^2}. \tag{6.31}$$

Also, for any axial position, the radial intensity follows the Gaussian function

$$I(r, z) = I_z \exp\left[-2 \frac{r^2}{\omega_z^2} \right], \tag{6.32}$$

where the peak intensity, I_z, and the beam width, ω_z, are determined by Eqs. 6.31 and 6.21, respectively.

Figure 6.2 is another illustration of the beam width behavior for a focused TEM$_{00}$ Gaussian beam with the radial intensity profiles for a few axial locations. All of the intensity profiles of Figure 6.2 are normalized to the same relative intensity scaling.

6.3 *ABCD* Matrix Treatment of Gaussian Beam Propagation

As described in the previous section, the complex Gaussian beam parameter $q(z)$ yields both the curvature of the wavefront and the width of a Gaussian beam for a particular axial location z. When combined with matrix methods used for ray optics, the q-parameter can be used to describe the propagation of Gaussian beams through optical systems. In this section we briefly review

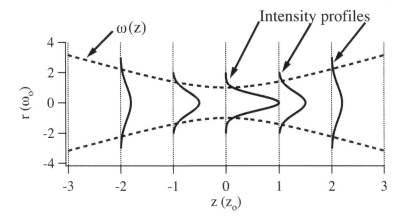

FIGURE 6.2
Beam width and radial intensity profiles as a function of axial distance for a focused TEM_{00} Gaussian beam.

matrix methods,[3] explain how matrix methods can be applied to the propagation of Gaussian beams, and illustrate their usefulness with some common examples.

6.3.1 ABCD Matrices

According to ray optics matrix methods, an optical system can be represented by a single 2×2 matrix, M, with the general form

$$M = \begin{bmatrix} A & B \\ C & D \end{bmatrix}. \tag{6.33}$$

For a particular axial location any ray can be completely described by two parameters: y and α, which represent the ray's tangential distance from the optical axis and the ray's angle with respect to the optical axis, respectively. The matrix M is used to determine the ray's parameters in the output plane, y' and α', through matrix multiplication of the known input parameters, y and α, and the system matrix, or

$$\begin{bmatrix} y' \\ \alpha' \end{bmatrix} = \begin{bmatrix} A & B \\ C & D \end{bmatrix} \begin{bmatrix} y \\ \alpha \end{bmatrix}. \tag{6.34}$$

The system matrix is calculated from the individual system elements. A system of multiple components is depicted in Figure 6.3. The particular optical system illustrated in Figure 6.3 is composed of four elements. Each element represents a single interaction of the light ray with the environment.

---

[3]Books by Siegman [3] or Yariv [4] provide a detailed review of ray optics matrix methods.

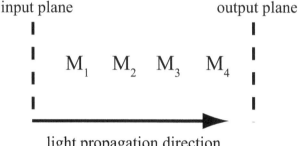

FIGURE 6.3
Optical system composed of four sequential matrix operations on the input plane (axial translations, lenses, refraction, reflection, etc.) where each is operation is represented by its own matrix.

Table 6.1 lists just a few of the common matrices used for ray optics matrix calculations (axial translation, refraction at a planar surface, and a thin lens). Other commonly encountered matrices are for refraction at a curved surface, reflection at a curved surface, reflection at a planar mirror, etc. More examples of ray matrices are available in Refs. [3] and [4].

The overall system matrix is computed by multiplying the matrices of the system in the *opposite* order in which the light propagates through the system; the system matrix for the four-component system illustrated in Figure 6.3 would be calculated from

$$M = M_4 \times M_3 \times M_2 \times M_1. \tag{6.35}$$

6.3.2 Propagation of a Gaussian Beam through Multiple Optical Elements

The propagation of Gaussian beams through optical systems (whether they are simple or rather complicated) reveals the convenient yet powerful method of using matrix methods in combination with the complex parameter q. Any optical system can be represented by a single system matrix, M, as discussed in the previous section and in more detail in Refs. [2], [3], and [4]. The system matrix is determined by matrix multiplication of each element or translation matrix as in Eq. 6.35, with the final form of

$$M = \begin{bmatrix} A & B \\ C & D \end{bmatrix}. \tag{6.36}$$

This is where the complex parameter q becomes very useful: q is defined in Eq. 6.16, which is repeated here for convenience:

$$\frac{1}{q(z)} = \frac{1}{R(z)} + i \frac{\lambda}{n\pi\omega^2(z)}. \tag{6.37}$$

Description	Matrix
Axial translation of distance L	$\begin{bmatrix} 1 & L \\ 0 & 1 \end{bmatrix}$
Refraction from medium n_1 into medium n_2	$\begin{bmatrix} 1 & 0 \\ 0 & \dfrac{n_1}{n_2} \end{bmatrix}$
Thin lens of focal length f	$\begin{bmatrix} 1 & 0 \\ -\dfrac{1}{f} & 1 \end{bmatrix}$
Reflection from a spherical mirror of radius of curvature R	$\begin{bmatrix} 1 & 0 \\ -\dfrac{2}{R} & 1 \end{bmatrix}$

TABLE 6.1
$ABCD$ matrix elements for an axial translation, ray refraction at a surface normal to the optical axis, a thin lens, and reflection off of a spherical mirror.

A paraxial Gaussian wave can be completely described at any axial location, z, by the parameter q. The real part of q yields the radius of curvature of the wave front. The sign of the real part indicates whether the wave is converging or diverging. The imaginary part of q reveals the radial width of the Gaussian function.

When combined with the matrix method of beam propagation, the parameter q of the output beam can be easily determined from the system matrix, M. In other words, if the properties of a TEM$_{00}$ Gaussian beam at the input plane and the system matrix for an optical system are known, the full beam properties at the output plane can be predicted. If q_1 represents the complex parameter of the Gaussian beam on the input plane, then the complex parameter in the output plane, q_2, is determined by the matrix elements of the system matrix by the relation

$$q_2 = \frac{Aq_1 + B}{Cq_1 + D}. \tag{6.38}$$

Equation 6.38 is known as the **$ABCD$ Propagation Law**.

A commonly used method for calculating the propagation of Gaussian beams using the $ABCD$ Propagation Law is by following these steps through a system for which the A, B, C, D elements have been determined:

• Substitute initial values of R and ω into Eq. 6.16 to find the complex q_1

• Substitute q_1 into the $ABCD$ Propagation Law, Eq. 6.38 to find q_2

Description	$q_2(q_1)$
Axial translation of distance L	$q_2 = q_1 + L$
Refraction from medium n_1 into medium n_2	$q_2 = \dfrac{n_2}{n_1} q_1$
Thin lens of focal length f	$\dfrac{1}{q_2} = \dfrac{1}{q_1} - \dfrac{1}{f}$
Reflection from a spherical mirror of radius of curvature R	$\dfrac{1}{q_2} = \dfrac{1}{q_1} - \dfrac{2}{R}$

TABLE 6.2
Relationship between input, q_1, and output, q_2, parameters for an axial translation, ray refraction at a surface normal to the optical axis, a thin lens, and reflection off of a spherical mirror.

- Take the inverse of q_2 and rewrite it as having a real and imaginary part

- Determine the new values for R and ω from the real and imaginary parts of $1/q_2$, respectively, using Eq. 6.16.

For a single optical element, such as any of those listed in Table 6.1, the new parameter q_2 can be expressed as a simple relationship. Table 6.2 is a summary of the relationship between the input q_1 and the output q_2.

Here, we will also present a more straightforward method of calculating the new beam parameters using a combined equation for determining the new radius of phase front curvature and beam width. Let us first assume that the known parameter q_1 can be written as

$$q_1 = q_R + i q_I, \tag{6.39}$$

where q_R and q_I are the real and imaginary parts of the input parameter q_1, respectively.

Substitution of Eq. 6.39 into the $ABCD$ Propagation Law, to find the new parameter, q_2, taking the inverse and rewriting as a real and an imaginary part yields the slightly cumbersome equation:

$$\frac{1}{q_2} = \left[\frac{(Aq_R + B)(Cq_R + D) + ACq_I^2}{(Aq_R + B)^2 + A^2 q_I^2} \right]$$
$$+ i \left[\frac{(BC - AD) q_I}{(Aq_R + B)^2 + A^2 q_I^2} \right], \tag{6.40}$$

so that the new phase front radius of curvature and beam width parameters can be determined from

$$R_2 = \frac{(Aq_R + B)^2 + A^2 q_I^2}{(Aq_R + B)(Cq_R + D) + ACq_I^2},$$ (6.41)

and

$$\omega_2^2 = \frac{n_2 \pi}{\lambda} \left[\frac{(Aq_R + B)^2 + A^2 q_I^2}{(BC - AD) q_I} \right],$$ (6.42)

where n_2 is the refractive index of the medium at the output plane, and A, B, C, and D are the matrix elements of the optical system matrix. The variables q_R and q_I of Eqs. 6.41 and 6.42 are related to the known input parameters R_1 and ω_1 by

$$q_R = \frac{\dfrac{1}{R_1}}{\dfrac{1}{R_1^2} + \dfrac{\lambda^2}{n_1^2 \pi^2 \omega_1^4}}, \quad \text{and} \quad q_I = -\frac{\dfrac{\lambda}{n_1 \pi \omega_1^2}}{\dfrac{1}{R_1^2} + \dfrac{\lambda^2}{n_1^2 \pi^2 \omega_1^4}}$$ (6.43)

or

$$q_R = \frac{\dfrac{1}{R_1}}{\dfrac{1}{R_1^2} + \dfrac{1}{s_1^2}}, \quad \text{and} \quad q_I = -\frac{\dfrac{1}{s_1}}{\dfrac{1}{R_1^2} + \dfrac{1}{s_1^2}},$$ (6.44)

where s_1 is a function of the width of the beam incident upon the input plane

$$s_1 = \frac{n_1 \pi \omega_1^2}{\lambda}.$$ (6.45)

Note that s_1, being a function of the width of the beam not necessarily in a focal plane, is not the same as the Rayleigh range, $z_0 = (n\pi\omega_o^2)/\lambda$, which is a function of the beam's minimum width.

6.3.3 Focusing a Gaussian Beam by a Thin Lens

General Case: Incident Beam is Converging, Collimated, or Diverging

In this subsection we present the general case of a converging or diverging TEM$_{00}$ beam incident upon a thin lens with focal length f. This general case is illustrated in Figure 6.4 where a diverging beam is incident upon a thin lens. There are three planes of interest: the input plane just before the lens, the output plane just after the lens, and the new focal plane some distance L after the lens. The system matrix for this example is a product of an axial propagation matrix and a thin lens matrix, or

$$M = M_{trans} \times M_{lens} = \begin{bmatrix} 1 & L_2 \\ 0 & 1 \end{bmatrix} \begin{bmatrix} 1 & 0 \\ -\dfrac{1}{f} & 1 \end{bmatrix},$$ (6.46)

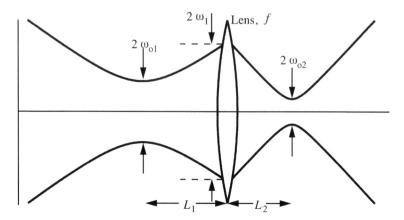

FIGURE 6.4
General case of a diverging TEM_{00} Gaussian beam having a minimum beam
waist of ω_{o1} and a spot width of ω_1 incident upon a lens of focal length f.
The new focal plane is located a distance L_2 away from the lens with a new
focal spot size of ω_{o2}.

or

$$M = \begin{bmatrix} \left(1 - \dfrac{L_2}{f}\right) & L_2 \\ -\dfrac{1}{f} & 1 \end{bmatrix}. \tag{6.47}$$

For the general case of a Gaussian laser beam incident upon a lens, there
are three parameters of usual interest characterizing the beam after the lens:

- the location of the beam waist (or axial distance from the lens to the new minimum beam waist),

- the radius of curvature of the wave front just after the lens, and

- the new minimum beam waist at the waist.

The location of the beam waist after the lens is determined from the output plane's parameter q_2 after applying the $ABCD$ Propagation Law, where
matrix elements of Eq. 6.47 are substituted into Eq. 6.40 to find the inverse
of the output plane parameter q_2. At the waist, the real part of q_2 is equal
to zero; i.e., the radius of curvature of the wavefront is infinite. Setting the
numerator of the real part of Eq. 6.40 to zero we get

$$(Aq_R + B)(Cq_R + D) + ACq_I^2 = 0, \tag{6.48}$$

where A, B, C, and D are the system matrix elements.
After substitution of the matrix elements of Eq. 6.47 into Eq. 6.48, solving

for L_2, substitution of Eq. 6.44 for q_{1R} and q_{1I}, and finally some simplification, the distance from the lens to the focal plane can be found to be

$$L_2 = \frac{f}{1 - \dfrac{f}{R_1} + \dfrac{f^2}{s_1^2}\left(\dfrac{1}{1 - f/R_1}\right)}, \qquad (6.49)$$

where f is the focal length of the lens, R_1 is the radius of curvature of the wave front incident upon the lens, and z_o is the Rayleigh range of the beam incident upon the lens. q_{1R} and q_{1I} denote the real end imaginary parts of the parameter q just before the lens in Figure 6.4. Equation 6.49 can also be written in a geometrical optics thin-lens-style as

$$\frac{1}{R_1} + \frac{1}{L_2} = \frac{1}{f} + \frac{\dfrac{f}{s_1^2}}{1 - \dfrac{f}{R_1}}. \qquad (6.50)$$

Similar to the derivation of Eq. 6.49, the new minimum beam waist is determined from the imaginary part of Eq. 6.40 with the real part equal to zero. Defining

$$\omega_{o2} \equiv \omega_2\left(z = L_2\right), \qquad (6.51)$$

solving for q_I and converting q_I to ω_2 we can arrive at an expression for the new minimum beam waist, ω_{o2} as a function of the beam width incident upon the lens, ω_1, the radius of curvature of the phase front on the input plane, R_1, the focal length of the lens, f, and the input beam width parameter s_1:

$$\omega_{o2} = \omega_1 \frac{\dfrac{f}{s_1}}{\sqrt{\left(1 - \dfrac{f}{R_1}\right)^2 + \dfrac{f^2}{s_1^2}}}. \qquad (6.52)$$

An Alternative Derivation of the Properties of a Gaussian Beam Focused by a Lens

The expressions for the beam waist ω_{o2} and the waist location L_2 were presented above in terms of the parameter s_1, which is related to the beam width ω_1 at the position of the lens. It is sometimes more convenient to have these expressions in terms of the beam width ω_{o1} at the focus prior to the lens. We present here the derivation of such expressions, mainly following the treatment by Kogelnik [5].

The planes located at the first beam waist, the incident face of the lens, the exit face of the lens, and the second beam waist are designated the planes 0, 1, 1_+, and 2, respectively, as shown in Figure 6.4. The complex beam parameter

q at these four locations can be written as

$$
\begin{aligned}
q_o &= -iz_{o1} \\
q_1 &= q_o + L_1 \\
q_{1_+} &= \frac{f q_1}{f - q_1} \\
q_2 &= q_{1_+} + L_2 = -iz_{o2}
\end{aligned}
\tag{6.53}
$$

where

$$
z_{o1} = \frac{\pi \omega_{o1}^2}{\lambda} \quad \text{and} \quad z_{o2} = \frac{\pi \omega_{o2}^2}{\lambda}
\tag{6.54}
$$

where we have assumed the refractive index of the medium on the two sides of the lens to be 1.

Eq. 6.53 can be rewritten as

$$
q_2 = \frac{A q_0 + B}{C q_0 + D} \quad \text{i.e.,} \quad -i z_{o2} = \frac{-A i z_{o1} + B}{-C i z_{o1} + D}
\tag{6.55}
$$

where

$$
\begin{aligned}
A &\equiv 1 - \frac{L_2}{f} \\
B &\equiv L_1 + L_2 - \frac{L_1 L_2}{f} \\
C &\equiv -\frac{1}{f} \\
D &\equiv 1 - \frac{L_1}{f}.
\end{aligned}
\tag{6.56}
$$

From Eq. 6.55 two relations are obtained:

$$
B + C z_{o1} z_{o2} = 0 \quad \text{and} \quad z_{o1} A = z_{o2} D.
\tag{6.57}
$$

Defining a parameter f_o by the relation

$$
f_o \equiv \frac{\pi \omega_{o1} \omega_{o2}}{\lambda}
\tag{6.58}
$$

and two new variables

$$
u_1 \equiv \frac{L_1}{f} \quad \text{and} \quad u_2 \equiv \frac{L_2}{f}
\tag{6.59}
$$

we obtain the equations relating u_1 and u_2

$$
\begin{aligned}
u_1 + u_2 - u_1 u_2 &= \left(\frac{f_o}{f} \right)^2 \\
\frac{1 - u_2}{1 - u_1} &= \frac{z_{o1}}{z_{o2}},
\end{aligned}
\tag{6.60}
$$

which can be reduced to quadratic equations in u_1 and u_2, the solutions to which are

$$u_1 = 1 \pm \frac{\omega_{o1}}{f\omega_{o2}} \sqrt{f^2 - f_o^2}$$

$$u_2 = 1 \pm \frac{\omega_{o2}}{f\omega_{o1}} \sqrt{f^2 - f_o^2} \qquad (6.61)$$

so that

$$L_1 = f \pm \frac{\omega_{o1}}{\omega_{o2}} \sqrt{f^2 - f_o^2}$$

$$L_2 = f \pm \frac{\omega_{o2}}{\omega_{o1}} \sqrt{f^2 - f_o^2}. \qquad (6.62)$$

The plus and minus signs in Eqs. 6.62 occur together, i.e., both equations have either the plus sign or the minus sign. These equations also show that $f \geq f_o$, so that the focal length f and the waist sizes ω_{o1} and ω_{o1} must obey the relationship

$$f \geq \frac{\pi}{\lambda} \omega_{o1} \omega_{o2}. \qquad (6.63)$$

Relations between ω_{o1}, ω_{o2}, L_1, L_2 and f

For a given wavelength λ, the five parameters ω_{o1}, ω_{o2}, L_1, L_2, and f are related by the two relations given in Eqs. 6.62. Thus with three of the parameters known, the other two can be obtained by solving the two equations. Which three of the parameters are known depends on the particular experimental situation. In a situation in which ω_{o1}, L_1, and f are known, the value of ω_{o2} can be obtained as

$$\omega_{o2} = \frac{f\omega_{o1}}{\sqrt{(L_1 - f)^2 + z_{o1}^2}} \qquad (6.64)$$

where as usual, $z_{o1} = (\pi\omega_{o1}^2)/\lambda$. With ω_{o2} known, the value of L_2 can be easily obtained as

$$L_2 = f + (L_1 - f)\frac{\omega_{o2}^2}{\omega_{o1}^2}$$

$$= f + (L_1 - f)\frac{f^2}{(L_1 - f)^2 + z_{o1}^2}. \qquad (6.65)$$

These expressions for ω_{o2} and L_2 are identical with those given in Eqs. 6.52 and 6.49 when s_1 is substituted by $z_{o1}(1 + L_1^2/z_{o1}^2)$.

Special Case: Input Beam is Collimated, or Focused onto the Lens

In this subsection we investigate the special case where the wavefront of the incident light is planar; i.e., the incident beam is either collimated or is focused

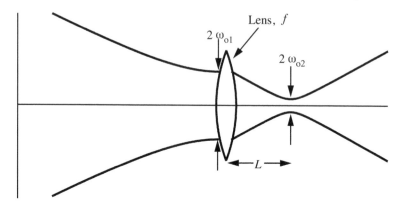

FIGURE 6.5
Special case of a collimated, or focused, TEM_{00} Gaussian beam incident upon a lens of focal length f. The width of the beam incident upon the lens *is* the minimum beam waist, ω_{o1}. The new focal plane is located a distance $L(= L_2)$ away from the lens with a new focal spot size of ω_{o2}.

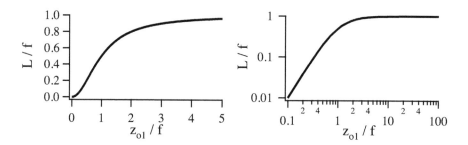

FIGURE 6.6
Ratio of the location of the new focal plane to the lens focal length as a function of the ratio of the Rayleigh range of the incident beam to the focal length of the lens for a focused TEM_{00} Gaussian beam incident upon a lens.

onto the lens as illustrated in Figure 6.5. When a beam is incident on the lens as a collimated beam, the expressions for L_2 and ω_{o2} are further simplified. Setting L_1 equal to 0 in Eqs. 6.64 and 6.65 gives

$$\omega_{o2} = \omega_{o1} \frac{f}{\sqrt{z_{o1}^2 + f^2}}. \quad \text{and} \quad L_2 = \frac{f z_{o1}^2}{z_{o1}^2 + f^2} = \frac{f}{1 + (f/z_{o1})^2} \quad (6.66)$$

Figure 6.6 is an illustration of the effects of the Rayleigh range of the incident beam on the location of the new focal plane. A "well-collimated" beam would be one where the Rayleigh range of the incident beam is very long with respect to the focal plane; i.e., the width of the beam does not change appreciably over a significant distance before the lens. As we can see in the figure, when the Rayleigh range of the incident beam becomes large with respect to the focal length of the lens, the location of the focal plane is a focal length, f, away from the lens.

A "tightly focused" beam would be one where the Rayleigh range of the incident beam is small compared to the focal length of the lens; i.e., the width of the incident beam changes dramatically before encountering the lens. The more "tightly focused" the incident beam the closer the new focal plane moves towards the lens, and moves away from the geometrical location of the focal plane for a planar wave front.

Figure 6.7 is an illustration of the effects of the Rayleigh range on the spot size of the new beam waist after the lens. For tightly focused beams we see that, in addition to the new focal spot moving closer to the lens, the lens has a decreasing effect on further focusing of the beam. As the Rayleigh range of the incident beam gets very small compared to the focal length of the lens, the spot size of the beam after the lens approaches that of the incident beam. In other words, the more tightly focused the incident beam, the less effect the lens has on the beam. The limit, of course, is that as $z_{o1} \ll f$ the lens has no effect on the location of the beam waist or the size of the minimum beam waist.

At the other extreme, for collimated incident beams, the ratio of the new spot size to the incident spot size continues to decrease as the ratio of the Rayleigh range to the focal length continues to decrease. Illustrated another way in Figure 6.8, for collimated beams the ratio of the spot sizes equals that of the ratio of the focal length to the incident Rayleigh range.

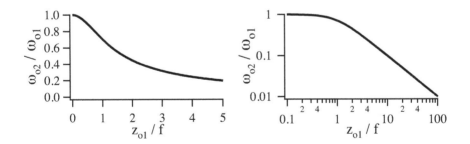

FIGURE 6.7

Ratio of the new focal spot width to the incident spot width as a function of the ratio of the Rayleigh range of the incident beam to the focal length of the lens for a focused TEM_{00} Gaussian beam incident upon a lens.

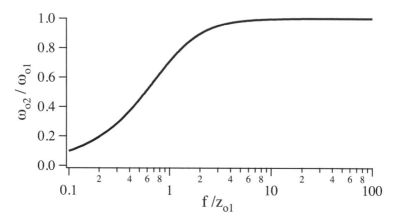

FIGURE 6.8
Ratio of the new focal spot width to the incident spot width as a function of
the ratio of the focal length of the lens to the Rayleigh range of the incident
beam for a focused TEM_{00} Gaussian beam incident upon a lens.

Another Special Case: The Waist of the Incident Beam is a Focal Length Away from the Lens

Eq. 6.62 shows that if $L_1 = f$, then $f = f_o$ and then $L_2 = f$. Thus if the waist
of a beam is a focal length away on one side of a lens, the waist on the other
side will also be a focal length away. Since $f = f_o$, the waist sizes on the two
sides are related:

$$\omega_{o2} = \frac{\lambda f}{\pi \omega_{o1}} \tag{6.67}$$

Focusing through a Slab

In many optical systems, for example in a camera, one or more optical windows
may be placed between the focusing lens and the plane of the beam waist, also
called the focal plane. The distance between the exit plane of the lens and the
focal plane in such a case can be easily determined using the ABCD matrix
formalism.

As shown in Figure 6.9, suppose light is incident on a lens of focal length
f from the left and a material of thickness d and refractive index n is placed
in air between the lens and the focal plane on the right, at a distance L_a from
the lens, and suppose the focus is located a distance L_b from the right surface
of the material. The exit plane of the lens, the two surfaces of the material,
and the focal plane are designated the planes 1_+, 2, 3, and 4, and the planes
just inside the material to the right of plane 2 and just outside the material
to the right of plane 3 are designated the planes 2_+ and 3_+. The parameter q

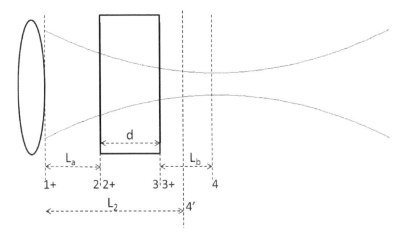

FIGURE 6.9
A material of thickness d and refractive index n placed between the lens and the focal plane. The refraction of the beam in the material is not shown. The focal plane in the presence of the material is at position 4 and the focal plane in the absence of the material is at position $4'$.

at these planes is designated by the corresponding subscripts and is given by

$$q_2 = q_{1_+} + L_a$$

$$q_{2_+} = q_2 n$$

$$q_3 = q_{2_+} + d = nq_2 + d$$

$$q_{3_+} = \frac{q_3}{n} = q_2 + \frac{d}{n}$$

$$q_4 = q_{3_+} + L_b = q_2 + \frac{d}{n} + L_b$$

$$= q_{1_+} + L_a + \frac{d}{n} + L_b = -iz_{o2}. \tag{6.68}$$

Comparing the last equations in Eqs. 6.53 and 6.68, we see that the distance L_{air} (equal to L_2 defined earlier) for focusing in air is to be replaced by the distance $L_a + \dfrac{d}{n} + L_b$ for focusing through a slab of thickness d and refractive index n. If the physical distance between the lens exit (plane 1_+) and the focal plane (plane 4) is L_4, then

$$L_4 = L_a + d + L_b \tag{6.69}$$

Since $n > 1$, L_4 is greater than L_{air}, i.e., the focus shifts to the right when the slab is present. The amount of the shift is given by

$$L_4 - L_{air} = d - \frac{d}{n} = d\frac{n-1}{n}. \tag{6.70}$$

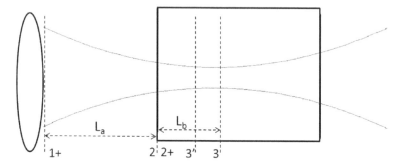

FIGURE 6.10
A material of refractive index n placed enclosing the focal plane. The refraction of the beam in the material is not shown. The focal plane in the material is at position 3 and the focal plane in the absence of the material is at the position denoted by $3'$.

The amount of shift is independent of the location of the slab, i.e., of the value of L_a. Moreover, Eq. 6.64 shows that the beam waist size w_{o2} on the right is independent of L_2 (or L_4) and is therefore the same with or without the slab in place—it is only the location of the focus that shifts when the slab is present.

Focusing in a Slab

A similar situation shown in Figure 6.10 is also encountered often in laser laboratories when a laser beam needs to be focused inside a material, say having refractive index n, for example for nonlinear frequency conversion.

It is assumed that the slab is placed a distance L_a from the lens exit and that the beam is focused a distance L_b inside the slab. Designating the planes at the exit of the lens, at the entrance face of the slab, just inside the slab, and at the focal plane as planes 1_+, 2, 2_+, and 3, respectively, and the corresponding q parameters by the appropriate subscripts, we have

$$q_2 = q_{1_+} + L_a$$
$$q_{2_+} = q_2 n$$
$$q_3 = q_{2_+} + L_b = nq_2 + L_b = -inz_{o2}. \tag{6.71}$$

In this case the distance L_{air} (equal to L_2 defined earlier) for focusing in air is to be replaced by the distance

$$L_a + \frac{L_b}{n} \tag{6.72}$$

for focusing inside the slab of thickness d and refractive index n. If the physical distance between the lens exit (plane 1_+) and the focal plane (plane 3) is L_3, then

$$L_3 = L_a + L_b. \tag{6.73}$$

Since $n > 1$, L_3 is greater than L_{air}, i.e., the focus shifts to the right when the slab is present. The amount of the shift is given by

$$L_3 - L_{\text{air}} = L_b - \frac{L_b}{n} = L_b \frac{n-1}{n}. \tag{6.74}$$

Before placing the slab in the beam, the distance (L_{air}) of the focus from the lens exit in air can be measured. If the focus is desired a distance L_b inside the material of refractive index n then the face of the material needs to be placed a distance $L_a = L_{\text{air}} - L_b/n$ from the lens exit.

For example, suppose the focus of a lens has been determined to be 10 cm from the exit plane, i.e., $L_{\text{air}} = 10$ cm. Suppose it is desired that the beam is focused at the center of a 4-cm-long crystal with refractive index 3. Then the incident face of the crystal will need to be 10 cm $-\frac{2 \text{ cm}}{3} = 9.33$ cm from the lens exit. The physical location of the focus is then 11.33 cm from the lens exit, i.e., the focal plane shift is 1.33 cm. This is equal to $L_b(n-1)/n$ with $L_b = 2$ cm and $n = 3$.

6.4 Higher-Order Gaussian Beams

Similar to the discussion at the beginning of this chapter, we will continue to use the scalar paraxial approximations to the general vector wave equation, or

$$\nabla_t^2 U + 2ik \frac{\partial U}{\partial z} = 0. \tag{6.75}$$

The partial differential equation of Eq. 6.75 can have many solutions other than the TEM_{00} Gaussian solution. The specific type of wave (or superposition of multiple types) that exit a particular laser cavity is determined by many parameters, including the shape, symmetry, and boundary conditions of the

- gain medium,

- laser cavity,

- optical components.

In this section we will briefly discuss two more types of solutions to the paraxial wave equation and the coordinate system used for each:

- Hermite-Gaussian beams $\longrightarrow$ Cartesian coordinates,

- Laguerre-Gaussian beams $\longrightarrow$ cylindrical coordinates.

6.4.1 Hermite-Gaussian Beams

To infer a solution to the paraxial wave equation in Cartesian coordinates, we need to first identify boundary conditions and expected axial field and intensity characteristics:

- the beam intensity profile decays for large axial distances,

- the beam has symmetry along perpendicular directions in the axial plane; i.e., use Cartesian coordinates,

- the beam can have different behaviors in each axial direction; i.e., x and y-dependent functions.

The first of these conditions allows us to start with the general form of U used for our TEM$_{00}$ Gaussian function, Eqs. 6.9. The second of these conditions informs us of which coordinate system to use for mathematical convenience. The last of these conditions can be applied to the general form as a product of independent functions for each axial coordinate, x and y. Our new function U in Cartesian coordinates for the general form of the solution can be expressed as

$$U(x, y, z) = f(x)\, g(y) \exp\left[ip(z) + ik \frac{x^2 + y^2}{2q(z)}\right]. \tag{6.76}$$

Substitution of Eq. 6.76 into the wave equation and solving (eventually) yields the electric field[4]

$$
\begin{aligned}
E_{\ell,m}(x, y, z, t) =\ & \left[H_\ell\left(\frac{\sqrt{2}x}{\omega(z)}\right) H_m\left(\frac{\sqrt{2}y}{\omega(z)}\right)\right] \\
& \times \left[E_o \frac{\omega_o}{\omega(z)} \exp\left[-\frac{x^2 + y^2}{\omega^2(z)}\right]\right] \\
& \times \left[\exp i\left(\frac{k}{2}\frac{x^2 + y^2}{R(z)} - (\ell + m + 1)\tan^{-1}\frac{z}{z_o}\right)\right] \\
& \times e^{i(kz - \omega t)},
\end{aligned} \tag{6.77}
$$

where H_ℓ and H_m are Hermite polynomial functions of the degree (or "order") ℓ and m for the x- and y-coordinates, respectively. All of the other variables are the same as those defined for TEM$_{00}$ Gaussian beams in Section 6.2. The generating function for the Hermite polynomials is Rodrigue's formula:

$$H_n(\alpha) = (-1)^n e^{\alpha^2} \frac{d^n}{d\alpha^n}\left(e^{-\alpha^2}\right). \tag{6.78}$$

The first few functions for the Hermite-polynomial functions found in Eq. 6.77 are summarized in Table 6.3.

---

[4]For a more descriptive presentation of the derivation of electric field functions for Hermite-Gaussian beams see Refs. [3] or [4].

Order of $H_n\left(\dfrac{\sqrt{2}x}{\omega^2}\right)$	Function
$n = 0$	1
$n = 1$	$\dfrac{2\sqrt{2}x}{\omega}$
$n = 2$	$\dfrac{8x^2}{\omega^2} - 2$
$n = 3$	$\dfrac{16\sqrt{2}x^3}{\omega^3} - \dfrac{12\sqrt{2}x}{\omega}$

TABLE 6.3
The first few Hermite polynomial functions for the Hermite functions of Eq. 6.77.

Equation 6.77 is written as a product of four terms. The first two terms comprise the real amplitude of the field and the last two terms contain the phase of the wave. The first term by itself is the effect of the higher-order Hermite functions. Note that if $\ell = m = 0$, the electric field is exactly that of Eq. 6.28; i.e., the TEM$_{00}$ Gaussian beam *is* the lowest order Hermite-Gaussian beam.

6.4.2 Laguerre-Gaussian (LG) Beams

The paraxial wave equation, Eq. 6.6, can be expressed in either Cartesian (x, y, z) or cylindrical (r, ϕ, z) coordinates. General solutions that are functions of the cylindrical coordinates, expressed in terms of Laguerre polynomials, were obtained in the early work by Kogelnik and Li, [5] and were presented by Siegman [3] with the normalization constants worked out. Most of the later work refers back to these two publications, and since the details of the derivation leading to the solution are not easy to find we present here some steps leading to the LG beam expression using the procedure outlined in Ref. [5].

A Laguerre Gaussian beam with angular mode number ℓ (which will be defined below) has been shown to possess orbital angular momentum of amount $\ell\hbar$ per photon [6]. Although they were initially postulated as possible laser resonator modes, they have had little application in practical lasers. Although laser mirrors and cavities are often cylindrically symmetric, some slight asymmetry in the cavity, a Brewster's window, etalon, or any misaligned surface usually forces the higher order modes of a laser to be the HG rather than LG beams. However, there has been continued interest in generating the LG beams outside a laser cavity, for applications in trapping small particles [7], atoms [8], or even Bose Einstein condensates [9].

Simpson et al. [10] modeled the axial trapping forces within optical tweezers arising from Laguerre-Gaussian laser modes and state that for an 8 μm diameter sphere suspended in water, the higher-order modes produce an axial trapping force several times larger than that of the fundamental. Transfer of the orbital angular momentum from the Laguerre-Gaussian mode to the trapped particle resulted in the rotation of the particle by what they called an optical spanner (wrench).

The steps leading to the expressions for the LG beams are outlined here. Following Ref [5] we assume that the solution for Eq. 6.6 can be written as

$$U(r, \phi, z) = v^{\ell/2} G(v) \exp\left[i\left(P(z) + \frac{kr^2}{2q(z)} + \ell\phi\right)\right], \tag{6.79}$$

where

$$v \equiv 2\frac{r^2}{\omega(z)^2}, \tag{6.80}$$

and $P(z)$, $q(z)$ and $\omega(z)$ are to be determined later and will be referred to as P, q, and ω for short. Defining a parameter m_1 as

$$m_1(z) \equiv \frac{k\omega^2}{4q}, \tag{6.81}$$

and using the expression for ∇_T^2 in cylindrical coordinates

$$\nabla_T^2 = \frac{1}{r}\frac{\partial}{\partial r}\left(r\frac{\partial}{\partial r}\right) + \frac{1}{r^2}\frac{\partial^2}{\partial \phi^2} \tag{6.82}$$

we obtain (after lengthy but straightforward calculations)

$$\nabla_T^2 U = \frac{8}{\omega^2}\frac{U}{G}$$
$$\times \left(v\frac{\partial^2 G}{\partial v^2} + (1 + \ell + 2im_1 v)\frac{\partial G}{\partial v} + [-vm_1^2 + im_1(\ell+1)]G\right). \tag{6.83}$$

Similarly, denoting the differentiation with respect to z by the primed symbol, it can be shown in a few lines that

$$\frac{\partial U}{\partial z} = U'$$
$$= U\left(iP' - i\frac{kr^2}{2q^2}q' - \ell\frac{\omega'}{\omega} + \frac{1}{G}\frac{\partial G}{\partial v}\right). \tag{6.84}$$

The paraxial wave equation, i.e., Eq. 6.6, then can be expressed as

$$v\frac{\partial^2 G}{\partial v^2} + (1 + \ell + 2im_1 v)\frac{\partial G}{\partial v} + [-m_1^2 v + im_1(1 + \ell)]G$$

$$= -2ik\frac{\omega^2}{8}\left(iP'G - i\frac{kr^2}{2q^2}q'G - \ell G\frac{\omega'}{\omega} + \frac{\partial G}{\partial v}\right). \tag{6.85}$$

Ignoring the $\partial^2 G/\partial v^2$ and the $\partial G/\partial v$ terms as being small near the axis, and equating the terms with same power of r, we obtain

$$-m_1^2 vG = 2ik\frac{\omega^2}{8}\frac{ikr^2}{2q^2}q'G, \tag{6.86}$$

which shows that as for the case of the TEM$_{00}$ beam, Eq. 6.12 holds for the higher order beams and we have again

$$\frac{\partial q}{\partial z} = 1, \tag{6.87}$$

the solution of which is $q(z) = z - iz_o$ as given in Eq. 6.17. The function $\omega(z)$ used in Eq. 6.80 has so far been undetermined. *Defining* the relation between $q(z)$ and $\omega(z)$ to be (as earlier)

$$\frac{1}{q(z)} \equiv \frac{1}{R} + \frac{2i}{k\omega^2}, \tag{6.88}$$

we find that again

$$R(z) = \frac{z^2 + z_0^2}{z} \quad \text{and} \quad \frac{k\omega^2}{2} = z_0\left(1 + \frac{z^2}{z_0^2}\right) \tag{6.89}$$

and

$$im_1 = \frac{ik\omega^2}{4q} = -\frac{1}{2} + i\frac{k\omega^2}{4R}. \tag{6.90}$$

The solutions for $P(z)$ and $G(v)$ are still to be found. To do that, the expression Eq. 6.90 is inserted in Eq. 6.85 and the terms are rearranged so that

$$v\frac{\partial^2 G}{\partial v^2} + (1 + \ell - v)\frac{\partial G}{\partial v}$$

$$= \frac{k\omega^2}{4}\left(P'G - i\frac{\partial G}{\partial v} - 2iv\frac{\partial G}{\partial v}\right)$$

$$+ \frac{G(\ell + 1)}{2} - iG\frac{k\omega^2}{4R} \tag{6.91}$$

where we have used the relation

$$\frac{\omega'}{\omega} = \frac{1}{R}, \tag{6.92}$$

which can be derived from Eq. 6.89. We now assume that the expression on the left side of Eq. 6.91 is proportional to G and the proportionality constant is $-p$

$$v\frac{\partial^2 G}{\partial v^2} \; + \; (1+\ell-v)\frac{\partial G}{\partial v} = -pG;$$

i.e.,

$$v\frac{\partial^2 G}{\partial v^2} \; + \; (1+\ell-v)\frac{\partial G}{\partial v} + pG = 0, \tag{6.93}$$

and also

$$\frac{k\omega^2}{4}\left(P'G - i\frac{\partial G}{\partial v} - 2iv\frac{\partial G}{\partial v}\right) + \frac{G(\ell+1)}{2} - iG\frac{k\omega^2}{4R} + pG = 0. \tag{6.94}$$

Ignoring the $\partial G/\partial v$ terms as being small near the axis, Eq. 6.94 can be solved with the explicit expressions for $\omega(z)$ and $R(z)$ obtained from Eq. 6.89, and it can be shown that

$$e^{iP(z)} = \frac{\omega_0}{\omega(z)}\exp\left[i(\ell+1+2p)\tan^{-1}\frac{z}{z_0}\right], \tag{6.95}$$

where $\omega_0 = \omega(0)$. So far the variables ℓ and p have been left unspecified except for the assumption that they are real numbers. Eq. 6.93 is identical with the differential equation satisfied by the Laguerre polynomials [11] L_p^ℓ if ℓ and p are restricted to nonnegative integer values. Thus the full solution to the paraxial wave equation in cylindrical coordinates can be written as

$$E_{p,\ell}(r,\phi,z,t) = E_0\frac{\omega_o}{\omega(z)}\left(\frac{\sqrt{2}r}{\omega(z)}\right)^\ell L_p^\ell\left(\frac{2r^2}{\omega^2(z)}\right)e^{i(kz-\omega t)}$$

$$\times \exp\left[-\frac{r^2}{\omega^2(z)} - ik\frac{r^2}{2R(z)} - i\ell\phi + (2p+\ell+1)\tan^{-1}\frac{z}{z_o}\right] \tag{6.96}$$

where ℓ and p are integers 0, 1, 2, etc. This matches the expression found in Ref. [4] for a Laguerre-Gaussian beam. ℓ and p are referred to as the angular and radial mode numbers of the beam [5].

The generating function for the Laguerre polynomials is Rodrigue's formula:

$$L_p^\ell(x) = \frac{1}{p!}e^x x^{-\ell}\frac{d^p}{dx^p}\left(e^{-x}x^{p+\ell}\right). \tag{6.97}$$

The first few Laguerre-polynomial functions found in Eq. 6.96 are

$$
\begin{aligned}
L_0^\ell(x) &= 1 \\
L_1^\ell(x) &= -x+\ell+1 \\
L_2^\ell(x) &= \frac{1}{2}[x^2 - 2(\ell+2)x + (\ell+1)(\ell+2)] \\
L_3^\ell(x) &= -\frac{1}{6}[x^3 - 3(\ell+3)x^2 + 3(\ell+2)(\ell+3)x \\
&\quad -(\ell+1)(\ell+2)(\ell+3)]
\end{aligned}
\tag{6.98}
$$

As for the Hermite Gaussian beams, if $\ell = p = 0$ in Eq. 6.96, the electric field is exactly that of Eq. 6.28; i.e., the TEM_{00} Gaussian beam is also the lowest order Laguerre-Gaussian beam.

6.5 Azimuthal and Radial Polarization

Natural light is typically unpolarized or partially polarized through specular reflection. Use of polarizers and waveplates enables light to be linearly or circularly polarized and such polarized light has many common everyday applications, such as in display technology. Beams of light can have other types of polarization distributions—for example, in a cross section of a beam the directions of polarization of the electric field can be along radial lines. Such a beam is called "radially polarized." If the polarization directions are along the perpendiculars to the radial lines, the beam is said to be "azimuthally polarized." Such radially or azimuthally polarized beams also have important applications, for example in laser acceleration and focusing of electrons using the inverse Cerenkov effect [12] or in laser micromachining [13]. Different approaches are taken to produce such beams, such as the use of liquid crystal based devices, a combination of waveplates, or the modification of a laser resonator.

The discussion on LG beams in the last section provides a way to describe radially and azimuthally polarized beams. The $e^{i\ell\phi}$ dependence of electric field of such beams on the azimuthal angle ϕ implies that such beams can be expressed as the linear combinations of beams with $\cos\ell\phi$ and $\sin\ell\phi$ dependence on ϕ. When $\ell = 1$, the electric field distribution is proportional to either $\cos\phi$ or $\sin\phi$, where ϕ is measured with respect to a direction designated the x axis. If the x and y axes are so chosen that the polarization direction of the electric field in the first quadrant is along the x direction, Figure 6.11 (a) shows the polarization directions for the fields with $\cos\phi$ and $\sin\phi$ dependence in the other three quadrants. Similarly, if the polarization direction of the electric field in the first quadrant is along the y direction, Figure 6.11 (b) shows the polarization directions for the fields with $\cos\phi$ and $\sin\phi$ dependence in the other three quadrants. Reference [5] shows how different polarization configurations can be synthesized from the linearly polarized TEM_{01} mode. Two of the possible configurations are shown in Figure 6.12. If two TEM_{01} beams with equal value ($\mathbf{E}_0$, say) for the maximum field amplitudes are combined as in Figure 6.12 (a), the resulting field is given by

$$\mathbf{E} = \mathbf{E}_0(\hat{x}\cos\phi + \hat{y}\sin\phi). \tag{6.99}$$

If the resultant vector $\mathbf{E}$ makes an angle ψ with the x axis, we have

$$\tan\psi = \tan\phi; \tag{6.100}$$

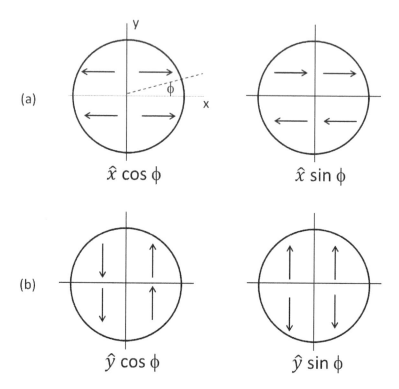

FIGURE 6.11
Polarization directions, shown by the arrows, in the four quadrants of TEM_{01} beams with azimuthal angle (ϕ) dependence of $\cos\phi$ and $\sin\phi$: (a) polarization in the first quadrant along $\hat{x}$, (b) polarization in the first quadrant along $\hat{y}$.

i.e., the angle between the polarization direction and the x axis is the same as the azimuthal angle ϕ, and the polarization directions are along the radial lines.

Similarly, combining two TEM_{01} beams with equal values $\mathbf{E}_0$ for the maximum field amplitudes as in Figure 6.12 (b)

$$\mathbf{E} = \mathbf{E}_0(\hat{x}\sin\phi - \hat{y}\cos\phi) \tag{6.101}$$

so that if the resultant $\mathbf{E}$ makes an angle ψ with the x axis, we have

$$\tan\psi = -\cot\phi; \tag{6.102}$$

i.e., $\psi = 90° - \phi$. The polarization direction is therefore perpendicular to the radius vectors, and the azimuthal polarization configuration is obtained.

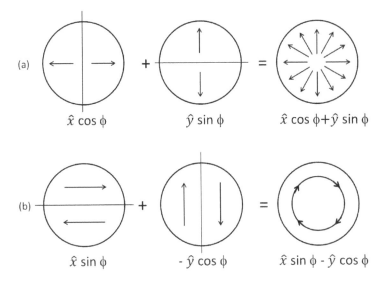

FIGURE 6.12
Combinations of TEM_{01} modes that give rise to beams having (a) radial and (b) azimuthal polarization.

6.6 M^2 Parameter

A frequently used parameter for describing real-world laser beams is the M^2 parameter (pronounced "M-squared"). The common usages of M^2 are usually:

- to describe the "quality" of the laser beam in comparison with an ideal TEM_{00} beam

- to describe the focusability of experimental real-world laser beams, or

- to serve as a fitting parameter to describe the propagation of multi-mode, or partially coherent, laser beams.

M^2 is also commonly referred to as the *beam quality factor*, *propagation constant*, or the *propagation factor*. In this section, we briefly describe the historical evolution of this parameter, and how it came to be used as both a fitting parameter and a beam propagation descriptor.

However, before we begin our discussion of the evolution and usage of M^2, we must first revisit our BPP (or beam propagation product) from Section 6.2. The BPP for a focused laser beam is the product of the minimum spot size half-width and the far-field divergence half-angle. For a pure TEM_{00} Gaussian beam the BPP is (Eq. 6.26)

$$\omega_o \theta_o = \frac{\lambda}{\pi}, \tag{6.103}$$

where the o subscript is used to clarify that this is for the lowest (zeroth) order Gaussian beam.

The useful property of the BPP is that for a given wavelength the product is equal to a constant. Thus, we have a fairly rigid relationship between the minimum beam waist and the far-field divergence; i.e., in order to decrease the spot size by a factor of 2 one would have to increase the convergence angle by a factor of 2. Additionally, if you know the spot size incident upon a lens, ω_1, and the focal length of the lens, f, you can fairly accurately estimate what the theoretical minimum focal spot size should be by calculating the convergence half-angle and knowing the wavelength of the laser light, or

$$\omega_o \approx \frac{\lambda}{\pi \tan^{-1}\left(\dfrac{\omega_1}{f}\right)}. \tag{6.104}$$

Equations 6.103 and 6.104 are very useful for TEM$_{00}$ laser beams. However, there is a disconnect between this model and reality due to the fact that *very few* real-world laser beams are "perfect Gaussian" beams (single-mode TEM$_{00}$ laser beams). This is where scientists and engineers turn to the M^2 model for real-world laser beams.

6.6.1   Historical Evolution of M^2

Theoretically, Gaussian beams extend radially out to infinity. However, all laser cavities have finite radial dimensions. Additionally, the geometrical shapes and features of components within the cavity determine which higher-order modes could be found to oscillate within the cavity. The components that determine what modes will be supported can be the mirrors, gain medium cross-section, gain medium surfaces, apertures, or windows, among many others. Frequently, a "spatial filter" will be inserted in the laser cavity to control which transverse laser modes can oscillate without loss due to the limited transverse size of the filter. Transverse intensity distributions of higher order Hermite-Gaussian or Laguerre-Gaussian modes (where $n > 0$) will have n nodes and $n + 1$ maxima. The widths of the beam waists of higher-order modes are always bigger than the diffraction limited zero-order TEM$_{00}$ Gaussian beam waist. In Siegman's 1986 text [3], *Lasers*, he describes the half-width of the n-th mode to be

$$x_n \approx \sqrt{n}\omega_o, \tag{6.105}$$

where ω_o is the half-width of the zero-order mode. Siegman goes on to describe how higher-order transverse modes will propagate through a limiting aperture. Only modes with widths less than that of the limiting aperture will propagate within the laser cavity without significant loss. The highest order that could be sustained is

$$N \equiv \left(\frac{a}{\omega_o}\right)^2, \tag{6.106}$$

where a is the radius of the limiting aperture. In other words, "N" is the highest order mode that will "fit" within that limiting aperture.

For the higher-mode of order N the far-field half-angle divergence, θ, can also be written as a function of the far-field angle for the lowest order, θ_o, or

$$\theta = \sqrt{N}\theta_o = \sqrt{N}\frac{\lambda}{\pi\omega_o}, \tag{6.107}$$

where we have used the relationship between the half-angle far-field divergence and the minimum beam waist half-width, Eq. 6.25, for light propagating through a refractive index of $n \approx 1$.

In Siegman's 1986 text he continues on to calculate the product of the source aperture area, A, and the far-field solid angular spread, Ω, to be

$$A \times \Omega = \left(\pi a^2\right)\left(\pi\theta^2\right) = N^2\lambda^2. \tag{6.108}$$

The important result here is that Siegman found the spot area and far-field solid angular spread product for multi-mode beams, or non-diffraction limited beams, to be N^2 times that of a lowest-order diffraction limited beam.

A few years later, N^2 morphed into M^2 with a few key differences:

- N mathematically describes what modes will "fit" within the aperture

- M^2 is used as a fitting parameter for experimental multi-mode and/or incoherent beams,

- N^2 is a result of the product of the focal spot area and the far-field solid angular spread,

- M^2 is a result of the product of the focal spot half-width and the far-field half-angle.

It is difficult to attribute the adaptation of a mathematical model describing possible transverse laser cavity modes to a fitting parameter for the output beams of real-world lasers to a single person. During the years 1988 to 1990, in addition to Siegman's work [14], similar models are also observed in the writings of Lavi [15], and Sasnett [16].

6.6.2   Current Usage of M^2

Most current usage of the M^2 parameter follows the approach of Sasnett [16] and Siegman [14], which is also found in some more recent texts such as Silfvast's first [17] or second edition [18] textbook. The mathematical models and theories presented earlier in this chapter in sections 6.2 and 6.3 are for pure TEM$_{00}$ laser beams. In other words, the laser beams are "perfect" in the sense that they are entirely composed of a single lowest order mode. However, real-world laser beams are usually composed of multiple transverse modes, or have other components or properties that adversely effect the beam's focusability and propagation characteristics. These real-world laser beams will always have beam waists or divergence angles larger than that of a pure TEM_{00} beam.

Having its roots in the mathematical model for higher-order beams, where both the spot size and far-field divergence are the same multiple of the zeroth-order beam, the M^2 propagation model predicts the beam waist half-width, W_o, and far-field divergence half-angle, Θ_o, for real-world beams to be

$$W_o = M\omega_o, \tag{6.109}$$

and

$$\Theta_o = M\theta_o, \tag{6.110}$$

where $M > 1$ for non-diffraction-limited beams, and $M = 1$ for diffraction-limited pure TEM_{00} beams.

If we now follow the model presented in Section 6.2 for the BPP we easily find the BPP for a real-world beam to be

$$\text{BPP} \equiv W_o \Theta_o = M^2 \omega_o \theta_o = M^2 \frac{\lambda}{\pi}. \tag{6.111}$$

In other words, the beam propagation product for a real-world laser beam is a factor of M^2 larger than that for a diffraction-limited TEM_{00} laser beam. Another variation for interpreting what an M^2 value for an experimental beam means is that it represents the *quality* of the beam. Here, "quality" is defined as the ratio of the experimental beam's propagation product to that of an ideal, or "perfect," beam; i.e.,

$$M^2 = \frac{\text{BPP}_{actual}}{\text{BPP}_{ideal}} = \frac{W_o \Theta_o}{\omega_o \theta_o}. \tag{6.112}$$

If we allow for the complex phase functions p and q of Eq. 6.9 to be modified to represent the beam width and divergence angle of Eqs. 6.109 and 6.110 and we follow the same procedure as Eqs. 6.11 to 6.16 we arrive at an expression for the half-width of the beam as a function of axial position, z, the diffraction-limited beam width, ω_o, and the diffraction-limited Rayleigh range, z_o, as

$$W^2(z) = M^2 \omega^2(z) = M^2 \omega_o^2 \left(1 + \frac{z^2}{z_o^2}\right). \tag{6.113}$$

However, for real-world laser beams, it is the non-diffraction-limited beam width that can be measured. Therefore, Eq. 6.113 can be rewritten in a much more convenient form where it is only a function of the measured minimum half-width, W_o, and the axial distance from the beam waist, z, as

$$W^2(z) = W_o^2 + M^4 \frac{\lambda^2}{\pi W_o^2} z^2. \tag{6.114}$$

A commonly used method for determining the M^2 value for an experimental beam is to first focus the beam with a lens, and then measure the beam profile for multiple axial positions before, within, and after the focal plane. Finally, one would fit the experimentally measured beam width data to Eq. 6.114 using M as a fitting parameter. Several commercial products are available that do this with a simple push of a button.

The M^2 model of beam propagation is based upon an easily measured parameter that yields a series of equations for predicting the focusability and propagation behavior for real-world laser beams. The model works fairly well for evaluating the output of lasers, or for quantifying the effects of minor aberrations introduced into the laser beam by optical components. However, the model does have its limits if significant diffraction or clipping of the laser beam occurs, as discussed in Section 10.3.2.

Bibliography

[1] S. Nemoto, "Nonparaxial Gaussian beams," Applied Optics, 29, 1940-1946, 1990.

[2] D. Marcuse, *Light Transmission Optics*, Second Edition, Van Nostrand Reinhold Company, New York, 1982.

[3] A. Siegman, *Lasers*, University Science Books, Mill Valley, California, 1986.

[4] A. Yariv, *Quantum Electronics*, Third Edition, John Wiley & Sons, New York, 1989.

[5] H. Kogelnik and T. Li, "Laser beam resonators," Proc. IEEE **54**, 1312–1329, 1966.

[6] L. Allen, M.W. Beijersbergen, R.J.C. Spreeuw, and J.P. Woerdman, "Orbital angular momentum of light and the transformation of Laguerre-Gaussian laser modes," Phys. Rev. A 45, 8185–8189, 1992.

[7] M.E.J. Friese, J. Enger, H. Rubinsztein-Dunlap, and N.R. Heckenberg, "Optical angular momentum transfer to trapped absorbing particles," Phys. Rev. A 54, 1593–1596, 1996.

[8] T. Kuga, Y. Tori, N. Shiokawa, T. Hirano, Y. Shimizu, and H. Sasada, "Novel optical trap of atoms with a doughnut beam," Phys. Rev. Lett., 78, 4713-4716, 1997.

[9] E.M. Wright, J. Arlt, and K. Dholakia, "Toroidal optical dipole traps for atomic Bose-Eistein condensates using Laguerre-Gaussian beams," Phys. Rev. A 63, 013608–013614, 2000.

[10] N. B. Simpson, L. Allen, and M. J. Padgett, "Optical tweezers and optical spanners with Laguerre-Gaussian modes," Journal of Modern Optics, Volume 43, Issue 12, pages 2485–2491, 1996.

[11] I.S. Gradshteyn and I.M. Ryzhik, *Table of Integrals, Series and Products*, Academic Press, San Diego, 2000.

[12] J. A. Edighoffer, W. D. Kimura, R. H. Pantell, M. A. Piestrup, and D. Y. Wang, "Observation of inverse Cerenkov interaction between free electrons and laser light," Phys. Rev. A 23, 1848–1854, 1981.

[13] M. Meier, V. Romano and T. Fuerer, "Materials Processing with Pulsed Radially and Azimuthally Polarized Laser Beams," Applied Physics A, 86, 329–334, 2007.

[14] A. E. Siegman, "New developments in laser resonators," Proc. SPIE 1224, 2–12, 1990.

[15] S. Lavi, R. Prochaska, and E. Keren, "Generalized beam parameters and transformation laws for partially coherent light," Appl. Opt. 27, 3696–3703, 1988.

[16] M. W. Sasnett, *Physics and Technology of Laser Resonators*, Taylor & Francis Group, New York, 1989.

[17] W. T. Sifvast, *Laser Fundamentals*, Cambridge University Press, New York, 1996.

[18] W. T. Sifvast, *Laser Fundamentals*, Second Edition, Cambridge University Press, New York, 2004.

7

Scalar and Vector Diffraction Theories

In this chapter we address the situation where light is incident upon a spatially limiting aperture. We present mathematical models for the following scalar and vector diffraction theories:

- Rayleigh-Sommerfeld scalar diffraction

- Fresnel-Kirchhoff scalar diffraction

- Hertz vector diffraction theory (HVDT)

- Kirchhoff vector diffraction theory (KVDT)

A thorough mathematical model using surface integrals is presented for each of these diffraction theories, as well as a discussion of the limitations of each and guidelines for when the reader should use which diffraction model. For the special case where the point of interest is along the optical axis, we present a simple analytical model for both HVDT and KVDT. The power transmission function of light passing through an aperture is also investigated.

By definition, diffraction is the phenomenon of the spreading of light (and its deviation from a rectilinear path) brought about when an aperture is illuminated by the light. In spite of the historical importance of this phenomenon to optics (the understanding of which first established the wave nature of light), a complete description of light distributions at points immediately beyond the illuminated aperture is still not available in the commonly used textbooks of optics [1, 2, 3, 4, 5, 6]. Treatment of the subject of near-field diffraction in undergraduate textbooks ranges from simply stating the equations to be used for near-field and far field diffraction regions [1], to briefly discussing the physical bases for Fresnel and Fraunhofer diffraction models [2], to deriving Fresnel and Fraunhofer diffraction using scalar diffraction theory [3, 4]. For a more detailed discussion of scalar diffraction theory and derivations of models other than Fresnel and Fraunhofer, readers are usually referred to advanced graduate-level textbooks [5, 6].

The simplest case of diffraction of light is that of plane waves traveling beyond a hard planar aperture. The complexity of accurately modeling and predicting the electric field and intensity distributions for every point beyond the aperture can be computationally quite intensive. Frequently, various approximations are implemented to decrease the level of complexity and decrease computational times required to model the diffracted beam propagation. The various approximations used are either based upon the distance between the region of interest and the diffracting plane, or based upon the relationship between the dimensions of the diffracting aperture to the wavelength of the light.

Over the past few decades, interest has slowly grown in reexamining the field of near-field diffraction theory, driven by advancements in laser technology, the continual shrinking of technological components, and advancements in computer technology. One of the many recent advances in laser technology is the generation of terahertz radiation [7]. The conversion of light wavelengths from the nanometer regime to the centimeter regime also results in a proportionate growth of the scale of near-field diffraction patterns, from the microscopic (sub-micron) scale to the macroscopic (millimeter and centimeter) scale. The presence of "near-field" optics on this macroscopic scale has fueled efforts to accurately characterize [7, 8] and model near-field propagation of terahertz electromagnetic radiation [9]. In addition to laser technology expanding the scales of near-field beam propagation, manufacturing technology continues to shrink the scale of optical components. Micro-electrical-mechanical systems (MEMs) and optical MEMs components are now reaching the dimensions of tens [10] to hundreds [11] of microns. Modeling beam propagation within complex systems on this scale presented new challenges to find accurate theoretical models to predict optical activity within these systems [12, 13].

In recent years the computational times required for calculating diffraction patterns, using various models, have decreased dramatically with the advancements of computer technology. The relatively recent advancements of computer technology have made calculations using models employing fewer and fewer mathematical approximations attainable with an average desktop computer.

7.1   Scalar Diffraction Theories

Studies investigating propagation of light beyond an aperture have been extensively performed as early as the late nineteenth century by Kirchhoff [14], Sommerfeld [15], and Rayleigh [16], among many others. Further progress in the field to find closed complete solutions that accurately describe the electric field distributions for all points beyond the aperture were hampered by the complexity of the integrals to be performed. Over the years, several different mathematical models have been developed to predict the electric field and intensity distributions for points beyond the aperture. The difference between the models depends upon the assumptions and approximations made in order to simplify the mathematics and reduce computational time. Table 7.1 lists

Model	Approximations Used
Full wave equations	None
Complete Rayleigh-Sommerfeld	Boundary conditions on surfaces
Approximated Rayleigh-Sommerfeld	$\rho \gg \lambda$
Fresnel-Kirchhoff	$\rho \gg \lambda$
Fresnel	$z_1^3 \gg \dfrac{\pi}{4\lambda}\left((x_1-x_0)^2+(y_1-y_0)^2\right)^2$
Fraunhofer	$z_1 \gg \dfrac{\pi}{\lambda}\left(x_0^2+y_0^2\right)$

TABLE 7.1
Various diffraction theory models and the mathematical approximations used by each model [17].

some of the models and the approximations used for each. As a result of the speed of modern computer technology, numerical results for even the full wave equations can now be obtained using a desktop computer.

We begin with the situation depicted in Figure 7.1. Monochromatic plane waves of light are traveling in the positive z direction and are incident upon an aperture in the $z = 0$ plane. A volume of space after the $z = 0$ plane, ν, is enclosed by two surfaces, S_0 and S_2. The surface S_0 is planar in shape, and lies in the $z = 0$ plane. The surface S_2 is an arbitrary surface in the $z > 0$ region enclosing some volume ν. Note that the volume of space enclosed by S_0 and S_2 has a small "hollow" sphere with a surface of S_1. This sphere surrounds the point of interest, P_1.

Using the divergence form of Gauss' theorem, vector identities and a particular choice for a vector function, we arrive at Green's scalar theorem [6],

$$\iint_S (U\boldsymbol{\nabla}V - V\boldsymbol{\nabla}U)\cdot\hat{n}\,ds = \iiint_\nu (U\nabla^2 V - V\nabla^2 U)\,dv, \qquad (7.1)$$

where U and V are arbitrary scalar fields that are functions of position. At this point the only restriction on the functions U and V are that they are smooth and continuous at all points in space. The surface integration is over all surfaces; S_0, S_1, and S_2. The volume integration is over the volume enclosed by S_0, and S_2, with the exception of the volume enclosed by S_1. If U and V are also restricted to be solutions of the Helmholtz wave equation,

$$(\nabla^2 + k^2)U = 0 \text{ and } (\nabla^2 + k^2)V = 0, \qquad (7.2)$$

then the volume integral is equal to zero, and Eq. 7.1 becomes

$$\iint_{S_0} (U\boldsymbol{\nabla}V - V\boldsymbol{\nabla}U)\cdot\hat{n}\,ds_0 + \iint_{S_2} (U\boldsymbol{\nabla}V - V\boldsymbol{\nabla}U)\cdot\hat{n}\,ds_2$$
$$+ \iint_{S_1} (U\boldsymbol{\nabla}V - V\boldsymbol{\nabla}U)\cdot\hat{n}\,ds_1 = 0. \qquad (7.3)$$

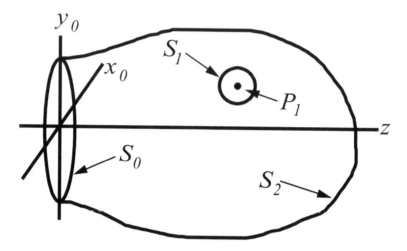

FIGURE 7.1
Representation of volumes and surfaces used in establishing a diffraction theory [17].

Letting the function V represent each component of the electric field, and the radius of S_1 go to zero, it can be shown (see Ref. [6] p. 42) that the integral over S_1 becomes

$$\iint_{S_1} (U\boldsymbol{\nabla}E_i - E_i\boldsymbol{\nabla}U) \cdot \hat{n} \, ds_1 = 4\pi E_i(P_1). \tag{7.4}$$

where E_i represents either the x, y, or z component of the electric field for each point on the surface, and $E_i(P_1)$ is the ith component at position P_1. Using the Sommerfeld radiation condition [6], that the radiation across surface S_2 is composed of only outward waves, and letting the distance from P_1 to the surface of S_2 become very large, the integral over S_2 also vanishes. Equation 7.3 now becomes

$$E_i(P_1) = -\frac{1}{4\pi} \iint_{S_0} (U\boldsymbol{\nabla}E_i - E_i\boldsymbol{\nabla}U) \cdot \hat{n} \, ds_0. \tag{7.5}$$

7.1.1 Rayleigh-Sommerfeld Diffraction Integral

Rayleigh-Sommerfeld boundary conditions on the aperture plane state that:

1. in the aperture region of the aperture plane the field distribution U and its normal derivative are exactly as they would be in the absence of the screen, and

2. everywhere else in the aperture plane either the field distribution U or its normal derivative is exactly zero.

Both of these boundary conditions on U in the aperture plane are satisfied by the following Green's function:

$$U = \frac{e^{ik|\mathbf{r}_2 - \mathbf{r}_0|}}{|\mathbf{r}_2 - \mathbf{r}_0|} - \frac{e^{ik|\mathbf{r}_1 - \mathbf{r}_0|}}{|\mathbf{r}_1 - \mathbf{r}_0|}, \tag{7.6}$$

where $\mathbf{r}_0$ denotes the vector from the origin to a point on the $z = 0$ plane, $\mathbf{r}_1$ denotes the vector from the origin to point P_1, and $\mathbf{r}_2$ is to a mirror image of point P_1 on the negative side of the $z = 0$ plane. This scalar function represents a point source located at $\mathbf{r}_1$ and a mirror image point source located at $\mathbf{r}_2$ oscillating exactly 180 degrees out of phase with each other, and satisfies the Rayleigh-Sommerfeld boundary conditions. Substituting Eq. 7.6 into Eq. 7.5 it can be shown that the field at P_1 is

$$\mathbf{E}(P_1) = \frac{kz_1}{i2\pi} \iint_{So} E_{z=0} \frac{e^{ik\rho}}{\rho^2} \left(1 - \frac{1}{ik\rho}\right) dx_0 \, dy_0, \tag{7.7}$$

where

$$\rho = \sqrt{(x_1 - x_0)^2 + (y_1 - y_0)^2 + z_1^2}. \tag{7.8}$$

Equation 7.7 is the full Rayleigh-Sommerfeld solution. The only assumptions made to this point are the boundary conditions on surfaces S_0 and S_2. It should also be noted here that the assumed electric field, $E_{z=0}$, is a solution to the wave equation *only*, which does not necessarily mean that it is also a solution to Maxwell's equations. The dangers and limitations of this oversight of the model are discussed further in Sections 7.4 and 7.5.

7.1.2 Fresnel-Kirchhoff Diffraction Integral

The Rayleigh-Sommerfeld diffraction integral is more commonly expressed in its simplified form of

$$\mathbf{E}(P_1) = \frac{kz_1}{i2\pi} \iint_{So} \mathbf{E}_{z=0} \frac{e^{ik\rho}}{\rho^2} dx_0 \, dy_0, \tag{7.9}$$

where it has been assumed that $\rho \gg \lambda$ [6]. Equation 7.9 is also known as the Fresnel-Kirchhoff diffraction integral.

If Eq. 7.8 is expanded using a binomial expansion and it is assumed that

$$z_1^3 \gg \frac{\pi}{4\lambda} \left((x_1 - x_0)^2 + (y_1 - y_0)^2\right)^2 \tag{7.10}$$

and

$$\rho^2 \approx z^2 \tag{7.11}$$

is assumed for the denominator of Eq. 7.9, we arrive at the Fresnel (paraxial) near field diffraction integral,

$$\mathbf{E}_1(P_1) = \frac{ke^{ikz_1}}{i2\pi z_1} \iint \mathbf{E}_{z=0} e^{\frac{ik}{2z_1}\left((x_1 - x_0)^2 + (y_1 - y_0)^2\right)} dx_0 \, dy_0. \tag{7.12}$$

For a cylindrically symmetric situation, Eq. 7.12 can be expressed in cylindrical coordinates for a circular aperture of radius a, as

$$\mathbf{E}(P_1) = \frac{ke^{ikz_1}}{iz_1} e^{\frac{ikr_1^2}{2z_1}} \int_0^a \mathbf{E}_{z=0} e^{\frac{ikr_0^2}{2z_1}} J_0\left(\frac{kr_0r_1}{z_1}\right) r_0 \, dr_0, \qquad (7.13)$$

where r_0 and r_1 are the radial coordinates in the planes $z = 0$ and $z = z$, respectively, and point P_1 is (x_1, y_1, z).

7.2   Comparisons of Scalar Diffraction Model Calculations

Figure 7.2 illustrates the on-axis calculated intensities for the three models discussed thus far by integrating the diffraction integrals derived in the previous section for each model. Figure 7.2(a) is the calculated on-axis intensity using the complete Rayleigh-Sommerfeld model, Eq. 7.7. Figure 7.2(b) is calculated using the approximated Rayleigh-Sommerfeld model, Eq. 7.9, also known as the Fresnel-Kirchhoff model, and Figure 7.2(c) is calculated using Fresnel's paraxial approximation, Eq. 7.12. Each graph of Figure 7.2 is calculated as a function of position along the z-axis with $x_1 = y_1 = 0$, for incident plane waves, with a wavelength of 10 microns, and a round aperture with a radius (denoted by a) of 100 microns in the aperture plane. The amplitude of the electric field of the incident light is assumed to be unity.

According to Figure 7.2(a), if we were to place a detector at a position $(0, 0, z)$, where z is on the order of $1000a$, or 1000 aperture radii, and observe the central intensity as we reduce z, we would first observe an increasing intensity as a function of smaller distances to the aperture; what would be expected intuitively. But, the increasing axial intensity does not simply continue to increase in intensity all the way up to the location of the aperture, as would be expected for a z^{-2} intensity dependence for a point source. Instead, a primary maximum is reached at a location of $10a$ and an intensity oscillation is observed for smaller values of z, with an overall amplitude decrease. It would also be expected that as z asymptotically approaches the position of the aperture the intensity should approach that of the incident plane waves, or unity for this example. Closer inspection of Figure 7.2(a) reveals that the number of on-axis maxima observed follows a/λ (which is also observed for other calculations performed using a variety of $a : \lambda$ ratios). For optically dense materials, $n > 1$, beyond the aperture, the longitudinal location of the diffraction pattern also scales proportional to n, and will be discussed further in the next section.

The vertical arrows in Figure 7.2(b) and Figure 7.2(c) represent the lower region of validity of each model, as defined in Table 7.1. For Figure 7.2(b), the vertical arrow is positioned at a value of $z = \lambda$, while the arrow in Figure 7.2(c)

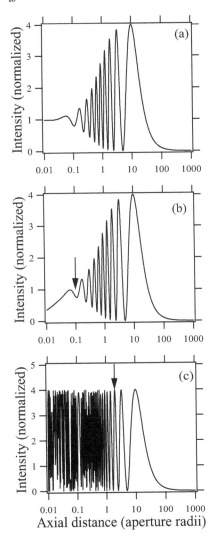

FIGURE 7.2
Calculations of on-axis intensity distributions for various scalar theory approximations [17]. Part (a) is calculated using the complete Rayleigh-Sommerfeld theory, (b) uses the approximate Rayleigh-Sommerfeld theory, and (c) uses the Fresnel diffraction model. The arrow in parts (b) and (c) represents the approximate lower limit of on-axis distances allowed for the mathematical validity of each model according to Table 7.1. All intensities are normalized to the uniform intensity incident upon the diffracting aperture.

is positioned at a value of $\frac{z}{a} \simeq \left(\frac{\pi a}{4\lambda}\right)^{1/3} \simeq 2$, using Eq. 7.10 and assuming x_o or $y_o \approx a$. Figure 7.3 is a closer view of the longitudinal region where the approximated models begin to deviate from the axial intensity distributions calculated

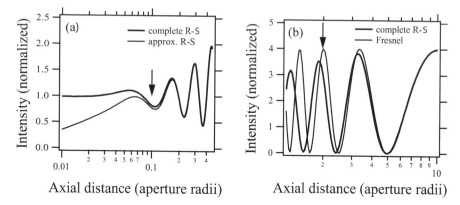

FIGURE 7.3

A close-up view of the deviation of (a) the approximate Rayleigh-Sommerfeld model, and (b) the Fresnel model, from the complete Rayleigh-Sommerfeld model [17]. The vertical arrows represent the approximate lower limit of on-axis distances allowed for the mathematical validity of each model according to Table 7.1.

using the complete Rayleigh-Sommerfeld model. The vertical arrows for Figure 7.3(a) and Figure 7.3(b) are located at the same longitudinal distance as those described for Figure 7.2(b) and Figure 7.2(c). As would be expected, the deviation of each approximated model is gradual; starting slightly beyond the defined cutoff distance and deviating farther and farther from the complete model as you get closer to the aperture. The difference between the complete Rayleigh-Sommerfeld model, Eq. 7.7, and the approximated model, Eq. 7.9, is primarily an amplitude inaccuracy, whereas the deviation of the paraxial approximation, Eq. 7.13, has both amplitude and phase inaccuracies.

Figure 7.4 is a calculated image plot as a function of the distance from the aperture, z, and the radial distance from the z-axis, using the complete Rayleigh-Sommerfeld model and the same aperture and laser field as in Figure 7.2. The integration of Eq. 7.7 is repeated for every point in a grid in the $y - z$ plane. In Figure 7.4, white represents the maximum intensity, black the minimum intensity, and gray the in-between. If a screen were to be placed at a particular distance from the aperture then the image observed would be a radially symmetric image with a radial intensity profile equal to a vertical stripe through Figure 7.4. To help illustrate this, Figure 7.5 is a collection of radial intensity distributions for various distances from the aperture, calculated directly from Eq. 7.7, for points along a $(0, y, z)$ line where z is held constant.

The four radial beam profiles of Figure 7.5 are a demonstration of the variety of radial intensity distributions that can be observed as a function of distance from the aperture. Figure 7.5(a) is a radial intensity distribution

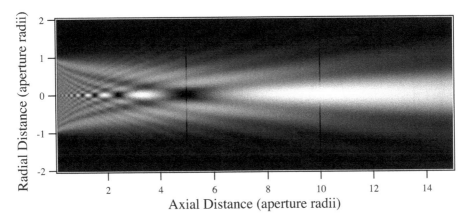

FIGURE 7.4
Calculated radial and axial intensity distribution for light diffracted by a circular aperture using the complete Rayleigh-Sommerfeld model [17].

calculated for a distance of 10 aperture radii, 1 mm, or the location of the primary maxima. A single spot would be observed on an image screen placed at this location with intensity at the center four times greater than that of the incident light on the aperture plane. Figure 7.5(b) is calculated for a distance of 4.9 aperture radii, 490 μm, or the location of the first minima. If an image screen were to be placed here, a single "doughnut" shape would be observed. Figure 7.5(c) is a calculated radial intensity distribution for a distance of 1.86 aperture radii, or the location of the third maxima. At this location, a bright central spot would be observed circumscribed by 2 blurry circles of light. In general, as the image screen is moved closer to the aperture the central region oscillates between a maximum and a minimum and the number of rings surrounding it increase in both number and size, eventually fading together (as the contrast between the maxima and minima decreases) to form a near uniform spot with a radius equal to that of the aperture and an intensity of unity, as observed in Figure 7.5(d). Figure 7.5(d) is a calculation for a distance of only 5 microns from a 200-micron diameter aperture.

7.3 Verification of Snell's Laws Using Diffraction

In this section we apply the complete Rayleigh Sommerfeld diffraction model to the phenomenon of the refraction of light at the interface of two different media. The results are qualitatively and quantitatively compared to those predicted using Snell's law of refraction and geometrical optics. For the case

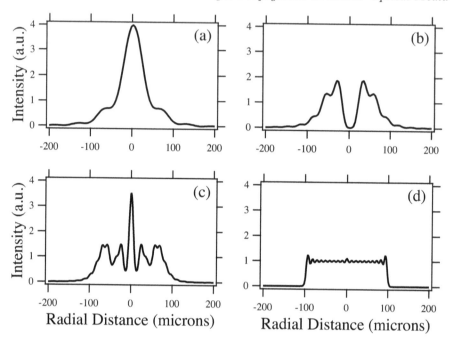

FIGURE 7.5
Calculated radial and intensity profiles for the diffraction pattern illustrated in Figure 7.4 for axial distances of (a) 10, (b) 4.9, (c) 1.86, and (d) 0.05 multiples of the aperture radius. All intensities are normalized to the intensity incident upon the aperture [17].

of plane waves incident upon the aperture at an angle other than normal, a position-dependent phase term must be included at the aperture plane. The electric field of the incident light can be written as

$$\mathbf{E}\left(x_0, y_0, 0\right) = e^{i2\pi n_o x_0 \sin(\theta)/\lambda}\hat{x}, \tag{7.14}$$

where n_o is the refractive index of in the medium before the aperture, x_o is the radial distance from the center of the aperture in the x direction, θ is the angle of incidence, and $\hat{x}$ is the polarization direction of the laser light. Once again, the amplitude of the electric field is assumed to be unity. Substituting Eq. 7.14 into Eq. 7.7 and using n as the refractive index of the medium after the aperture, we get

$$\mathbf{E}(P_1) = \frac{nz_1}{i\lambda} \iint e^{i2\pi n_o x_0 \sin(\theta)/\lambda} \frac{e^{i2\pi n\rho/\lambda}}{\rho^2} \left(1 - \frac{\lambda}{i2\pi n\rho}\right) dx_0\, dy_0\hat{x}. \tag{7.15}$$

Equation 7.15 is integrated over all points in the aperture plane for each point on the image plane. The integral is then repeated for points along the line

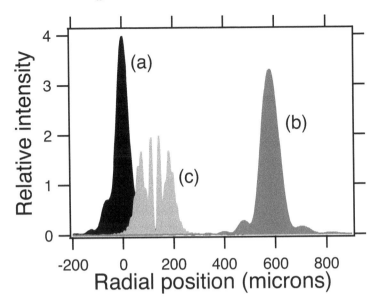

FIGURE 7.6
Radial intensity profiles for plane waves, with a wavelength of 10 microns, incident upon a 200 micron diameter aperture, and an aperture-image plane distance of 1 mm for: (a) $n = n_o = 1$, $\theta = 0$ degrees, (b) $n = n_o = 1$, $\theta = 30$ degrees, and (c) $n = 4$, $n_o = 1$, $\theta = 30$ degrees [17].

$(x, 0, z_p)$, where z_p is constant, and is the distance from the aperture to the longitudinal position of the image plane of interest. Using this method, beam intensity profiles as a function of x are calculated and displayed in Figure 7.6. The experimental conditions modeled for Figure 7.6 are those of a collimated CO_2 laser, with a wavelength of 10 microns, incident upon an aperture with a radius (denoted by a) of 100 microns. The beam waist is assumed to be much larger than the aperture (so a constant field amplitude across the aperture can be assumed). The distance from the aperture to the image plane is chosen to coincide with the primary maxima of the interference pattern for beam propagation through air or vacuum after the aperture.

   First, a calculation using Eqs. 7.15 is performed for light waves normally incident upon the aperture, and for a refractive index of the medium after the aperture equal to that before the aperture plane, $n = n_o = 1$. The resulting beam profile is shown in Figure 7.6(a) and is centered at $x = 0$. This beam intensity profile is the same as that of Figure 7.4(a) which was calculated using Eqs. 7.7 and the same experimental conditions. Using geometrical optics and Snell's Law, the expected intensity profile would be a perfect copy of the uniform "flat-top" intensity profile incident upon the aperture and centered around $x = 0$, i.e., a beam profile having a constant intensity with amplitude of unity would be expected from -100 μm to +100 μm with sharp edges. Figure 7.6(a) shows that incorporating the wave nature of light and a mathematically rigorous model of diffraction theory we observe a much different intensity profile than that predicted using geometrical optics.

If we perform a similar calculation as above, except allowing for an angle of incidence of 30 degrees, we would initially expect a similar beam profile offset from $x = 0$ degrees on the image plane by $z \tan \theta$, or 577 μm at $z = 1$ mm. The resulting intensity profile is illustrated in Figure 7.6(b) and centered on $x = 578$ μm. The difference between the transverse locations of the two is within a step size of the Rayleigh-Sommerfeld calculation. The geometrical shadow of the 200-μm diameter pinhole for light incident at an angle of 30 degrees would be an ellipse with a minor diameter of 200 μm, in the y-direction, and a major diameter of 231 μm, in the x-direction. The uniform intensity would have a value of 0.866 for an incident electric field amplitude of unity. This is slightly higher than the ratio of the peak heights of Figure 7.6(a) and Figure 7.6(b) of 0.825. But, neither of these intensity profiles of Figure 7.6 are close to uniform, as predicted using geometrical optics. Closer inspection of Figure 7.6(b) reveals that the intensity distribution is not just simply wider and lower than that of Figure 7.6(a), as would be expected using geometrical optics, but actually has some asymmetries in the outer-lying rings; another unique result of the wave nature of light using diffraction theory.

Finally, to truly compare the results of Rayleigh-Sommerfeld calculations to Snell's Law we include an optically dense medium in the region after the aperture. The refractive index chosen for this calculation was that of germanium, a common optical material for long-wave infrared applications, with a linear refractive index of $n = 4$. For each point on the x-axis, and a distance $z = 1$ μm away, Eq. 7.15 is integrated using $n_o = 1$, $n = 4$, and $\theta = 30$ degrees. The resulting intensity profile is illustrated in Figure 7.6(c). Using Snell's Law of refraction, the image should be centered on $x = 126\mu$m, which is the precise location of the minimum in the middle of Figure 7.6(c).

The most dominant difference between Figure 7.6(b) and Figure 7.6(c), other than their locations on the x-axis, is the difference in their intensity profiles. As the refractive index of the medium after the aperture changes, so does the relative distance to a particular interference pattern. For example, the intensity profile calculated for Figure 7.6(b) is very similar to the beam profile calculated for $n = 1$, $\theta = 0$ degrees and a longitudinal distance closer to the aperture by a factor of 4 (the n value of germanium). The net result of beam propagation through a non-absorbing, optically dense medium after the aperture is a diffraction pattern similar to that of Figure 7.4, except the longitudinal axis is increased by a factor of n.

7.4 Vector Diffraction Theories

Even though the diffraction of light by an aperture is a fundamental phenomenon in optics and has been studied for a long time [14, 15, 16], it continues to be of modern interest [18, 19, 20, 17, 21]. It has long been recognized that *vector* diffraction theory needs to be used to describe propagation of light in and around structures that are of the same length scale or smaller than the wavelength of light [22]. Not to mention that light itself is a *vector* electromagnetic wave. So, it is only natural to model it using a *vector*-based theory. Study of diffraction of light through an aperture with a radius (a) comparable to the wavelength of light (λ) is a challenging phenomenon to model completely, especially for the case of a/λ ranging between 0.1 and 10, which falls in between the theory of transmission of light through very small apertures $(a \ll \lambda)$ [18, 23] and vector diffraction theory using Kirchhoff boundary conditions and high-frequency approximations [19, 20, 22, 24]. The high-frequency approximations and Kirchhoff boundary conditions assume that the light field in the aperture plane is known and is unperturbed by the presence of the aperture. While mathematically consistent and correctly predicting light distributions beyond the aperture plane, these assumptions fail to represent physical waves in the vicinity of the aperture plane and violate Maxwell's equations.

Paraxial approximations have historically been used to reduce the mathematical complexity and the computational time required for numerical integration, albeit at the cost of limiting the regions of their validity. The finite difference time domain (FDTD) method provides an accurate description of light distributions in the vicinity of the aperture without Kirchhoff or paraxial approximations, but can become computationally time-intensive at larger distances of propagation.

In sections 7.5 and 7.6 we present two different vector diffraction theories for calculating the vector light fields for points within, or beyond, the aperture. The expressions for the field components are first obtained here as double integrals. Then the field components are expressed as two computationally efficient single integrals for two mutually exclusive and all-encompassing volumes of space beyond the aperture plane. For the specific case of points along the optical axis, analytical forms of the electromagnetic field components are derived and presented in Sec. 7.7. The total power transmitted through a plane parallel to the aperture plane is calculated as a function of the aperture size in Sec. 7.8.

7.5 Hertz Vector Diffraction Theory (HVDT)

The electric ($\mathbf{E}$) and magnetic ($\mathbf{H}$) fields of an electromagnetic wave of wavelength λ and traveling in vacuum can be determined from the polarization potential [5], or Hertz vector ($\mathbf{\Pi}$), through the relations

$$\mathbf{E} = k^2\mathbf{\Pi} + \nabla\left(\nabla \cdot \mathbf{\Pi}\right), \qquad (7.16)$$

and

$$\mathbf{H} = -\frac{k^2}{i\omega\mu_o}\nabla \times \mathbf{\Pi} = ik\sqrt{\frac{\epsilon_o}{\mu_o}}\nabla \times \mathbf{\Pi}, \qquad (7.17)$$

where the Hertz vector is a smooth and continuous function everywhere and satisfies *both* the wave equation *and* Maxwell's equations. The variable k is the wave number, $k = \dfrac{2\pi}{\lambda}$, and ϵ_o and μ_o denote the vacuum permeability and permittivity constants. For the case of a plane wave that is linearly polarized (say, along the x-direction) and propagating in the $+z$ direction, all of the $\mathbf{E}$ and $\mathbf{H}$ field components can be calculated from Π_x, the x-component of the Hertz vector [25], and have the following forms:

$$
\begin{aligned}
E_x &= k^2\Pi_x + \frac{\partial^2\Pi_x}{\partial x^2}, \\
E_y &= \frac{\partial^2\Pi_x}{\partial y\partial x}, \\
E_z &= \frac{\partial^2\Pi_x}{\partial z\partial x},
\end{aligned}
\qquad (7.18)
$$

and

$$
\begin{aligned}
H_x &= 0, \\
H_y &= -\frac{k^2}{i\omega\mu_o}\frac{\partial\Pi_x}{\partial z} = ik\sqrt{\frac{\epsilon_o}{\mu_o}}\frac{\partial\Pi_x}{\partial z}, \\
H_z &= \frac{k^2}{i\omega\mu_o}\frac{\partial\Pi_x}{\partial y} = -ik\sqrt{\frac{\epsilon_o}{\mu_o}}\frac{\partial\Pi_x}{\partial y}.
\end{aligned}
\qquad (7.19)
$$

From the fields given in Eqs. (7.18) and (7.19) it is straightforward to obtain the Poynting vector, $\mathbf{S}$, where

$$
\begin{aligned}
\mathbf{S} &= Re\left(\mathbf{E} \times \mathbf{H}^*\right) \\
&= Re\left(E_yH_z^* - E_zH_y^*\right)\hat{i} + Re\left(-E_xH_z^*\right)\hat{j} + Re\left(E_xH_y^*\right)\hat{k}, \quad (7.20)
\end{aligned}
$$

and $\hat{i}, \hat{j}, \hat{k}$ denote the unit vectors in the x, y, and z directions.

7.5.1 Double Integral Forms for the Field Components

We assume the plane wave described above is incident on an aperture that is perfectly conducting, of negligible thickness, and located in the x-y plane at $z = 0$. Because the aperture is assumed to be conducting, it is required that the tangential components of the electric field and the normal components of the magnetic field vanish in the plane of the aperture. Applying these boundary conditions on the electromagnetic fields and requiring that the Hertz vector and the electromagnetic fields satisfy Maxwell's equations for all space, the Hertz vector component at the point of interest, (x, y, z), is given by Bekefi [25] to be

$$\Pi_x(x, y, z) = \frac{iE_o}{2\pi k} \int \int \frac{e^{-ik\rho}}{\rho} dx_o dy_0, \qquad (7.21)$$

where E_o is the electric field amplitude of the incident plane wave and ρ is the distance from a point in the aperture plane, $(x_o, y_o, 0)$, to the point of interest, or

$$\rho = \sqrt{(x - x_0)^2 + (y - y_0)^2 + z^2}, \qquad (7.22)$$

and the integration is performed over the two-dimensional open aperture area.

To express the results in dimensionless parametric forms, a length scale a is first chosen, and a quantity $z_0 \equiv ka^2$ and a dimensionless parameter $p_1 \equiv 2\pi a/\lambda$ are defined, along with the dimensionless coordinates $\mathbf{r_1}(x_1, y_1, z_1)$, $\mathbf{r_{01}}(x_{01}, y_{01}, 0)$, and a dimensionless variable ρ_1:

$$x_1 \equiv \frac{x}{a}, \quad y_1 \equiv \frac{y}{a}, \quad z_1 \equiv \frac{z}{z_0}, \quad x_{01} \equiv \frac{x_o}{a}, \quad y_{01} \equiv \frac{y_o}{a}, \qquad (7.23)$$

$$\rho_1 \equiv \sqrt{(x_1 - x_{01})^2 + (y_1 - y_{01})^2 + p_1{}^2 z_1{}^2}. \qquad (7.24)$$

Expressing Eq. (7.21) in the dimensionless form and substituting into Eqs. (7.18) and (7.19), we obtain

$$
\begin{aligned}
E_x(\mathbf{r_1}) &= \frac{iE_o}{2\pi} \left[p_1 A_1(x_1, y_1, z_1) + \frac{1}{p_1} \frac{\partial^2 A_1(x_1, y_1, z_1)}{\partial x_1^2} \right], \\
E_y(\mathbf{r_1}) &= \frac{iE_o}{2\pi p_1} \frac{\partial^2 A_1(x_1, y_1, z_1)}{\partial y_1 \partial x_1}, \\
E_z(\mathbf{r_1}) &= \frac{iE_o}{2\pi p_1^2} \frac{\partial^2 A_1(x_1, y_1, z_1)}{\partial z_1 \partial x_1},
\end{aligned}
\qquad (7.25)
$$

and

$$
\begin{aligned}
H_x(\mathbf{r_1}) &= 0, \\
H_y(\mathbf{r_1}) &= -\frac{H_o}{2\pi p_1} \frac{\partial A_1(x_1, y_1, z_1)}{\partial z_1}, \\
H_z(\mathbf{r_1}) &= \frac{H_o}{2\pi} \frac{\partial A_1(x_1, y_1, z_1)}{\partial y_1},
\end{aligned}
\qquad (7.26)
$$

where

$$A_1(\mathbf{r_1}) \equiv \int\int \frac{e^{-ip_1\rho_1}}{\rho_1} dx_{01} dy_{01} \tag{7.27}$$

and $H_o \equiv E_o\sqrt{\dfrac{\epsilon_o}{\mu_o}}$.

Carrying out the differentiations of A_1 in Eqs. (7.25) and (7.26) inside the integrals in Eq. (7.27), the electric and magnetic field components are obtained as:

$$E_x(x_1, y_1, z_1) = \frac{iE_o p_1}{2\pi}$$
$$\times \int\int f_1 \left[(1+s_1) - (1+3s_1)\frac{(x_1-x_{01})^2}{\rho_1^2}\right] dx_{01} dy_{01}, \tag{7.28}$$

$$E_y(x_1, y_1, z_1) = -\frac{iE_o p_1}{2\pi}$$
$$\times \int\int f_1 (1+3s_1) \frac{(x_1-x_{01})(y_1-y_{01})}{\rho_1^2} dx_{01} dy_{01}, \tag{7.29}$$

$$E_z(x_1, y_1, z_1) = -\frac{iE_o p_1^2 z_1}{2\pi} \int\int f_1 (1+3s_1) \frac{(x_1-x_{01})}{\rho_1^2} dx_{01} dy_{01}, \tag{7.30}$$

$$H_x(x_1, y_1, z_1) = 0, \tag{7.31}$$

$$H_y(x_1, y_1, z_1) = -\frac{H_o p_1^3 z_1}{2\pi} \int\int f_1 s_1 dx_{01} dy_{01}, \tag{7.32}$$

$$H_z(x_1, y_1, z_1) = \frac{H_o p_1^2}{2\pi} \int\int f_1 s_1 (y_1 - y_{01}) dx_{01} dy_{01}, \tag{7.33}$$

where

$$f_1 \equiv \frac{e^{-ip_1\rho_1}}{\rho_1} \tag{7.34}$$

and

$$s_1 \equiv \frac{1}{ip_1\rho_1}\left(1 + \frac{1}{ip_1\rho_1}\right). \tag{7.35}$$

For a circular aperture of radius a, the limits of the integrals in Eqs. (7.28)–(7.33) are:

$$\int_{-1}^{1} \int_{-\sqrt{1-y_{01}^2}}^{\sqrt{1-y_{01}^2}} dx_{01} dy_{01}. \tag{7.36}$$

7.5.2   Single Integral Forms for the Field Components

Within the Geometrically Illuminated Region $(r_1 < 1)$

Although the double integral forms for the electric and magnetic fields presented in the previous section give the complete solution to the Hertz vector diffraction theory for any point in space with $z_1 \geq 0$, and can be numerically evaluated in minutes on currently available desktop computers, to obtain a detailed description of the light diffracted by the aperture over a large number of points in space, it is desirable to reduce the computation time further. Schoch [26] demonstrated that the surface integral of the Hertz vector in Eq. (7.21) can be expressed as a line integral around the edge of the aperture for field points within the geometrically illuminated region, or $r_1 < 1$ where r_1 is defined below in Eq. (7.40). Expressing Schoch's line integral form in terms of the dimensionless parameters of Eqs. (7.23) and (7.24), the Hertz vector component for a circular aperture of radius a becomes

$$\Pi_x (x_1, y_1, z_1) = \frac{E_o a^2}{p_1^2} \left[e^{-ip_1^2 z_1} - \frac{1}{2\pi} \int_0^{2\pi} \frac{e^{-ip_1 q}}{L^2} \left(1 - r_1 \cos \phi\right) d\phi \right], \quad (7.37)$$

where

$$q^2 (x_1, y_1, z_1, \phi) = L^2 + p_1^2 z_1^2, \qquad (7.38)$$

$$L^2 (x_1, y_1, \phi) = 1 + r_1^2 - 2r_1 \cos \phi \qquad (7.39)$$

and

$$r_1 (x_1, y_1) = \sqrt{x_1^2 + y_1^2}. \qquad (7.40)$$

Substituting Eq. (7.37) into Eqs. (7.18) and (7.19), the electric and magnetic field components can be expressed as

$$E_x (x_1, y_1, z_1) = E_o \left(e^{-ip_1^2 z_1} - \frac{1}{2\pi} \int_0^{2\pi} f_{2a} d\phi \right),$$

$$E_y (x_1, y_1, z_1) = -\frac{E_o}{2\pi p_1^2} \int_0^{2\pi} f_{2b} d\phi,$$

$$E_z (x_1, y_1, z_1) = -\frac{E_o}{2\pi p_1^3} \int_0^{2\pi} f_{2c} d\phi, \qquad (7.41)$$

and

$$H_x (x_1, y_1, z_1) = 0,$$

$$H_y (x_1, y_1, z_1) = H_o \left(e^{-ip_1^2 z_1} - \frac{i}{2\pi p_1^2} \int_0^{2\pi} f_{2d} d\phi \right),$$

$$H_z (x_1, y_1, z_1) = -\frac{iH_o}{2\pi p_1} \int_0^{2\pi} f_{2e} d\phi, \qquad (7.42)$$

where

$$
\begin{aligned}
f_{2a} &\equiv \alpha\beta\gamma + \frac{1}{p_1^2}\left(\alpha_{11}\beta\gamma + \beta_{11}\alpha\gamma + \gamma_{11}\alpha\beta\right) \\
&\quad + \frac{2}{p_1^2}\left(\alpha_1\beta_1\gamma + \beta_1\gamma_1\alpha + \gamma_1\alpha_1\beta\right), \\
f_{2b} &\equiv \alpha_{12}\beta\gamma + \beta_{12}\alpha\gamma + \gamma_{12}\alpha\beta \\
&\quad + \alpha\left(\beta_1\gamma_2 + \gamma_1\beta_2\right) + \beta\left(\alpha_1\gamma_2 + \gamma_1\alpha_2\right) + \gamma\left(\alpha_1\beta_2 + \beta_1\alpha_2\right), \\
f_{2c} &\equiv \alpha_{13}\beta\gamma + \alpha_3\left(\beta\gamma_1 + \gamma\beta_1\right), \\
f_{2d} &\equiv \alpha_3\beta\gamma, \\
f_{2e} &\equiv \alpha_2\beta\gamma + \beta_2\alpha\gamma + \gamma_2\alpha\beta.
\end{aligned}
\tag{7.43}
$$

The variables α, β, and γ are defined as

$$
\begin{aligned}
\alpha &\equiv e^{-ip_1 q}, \\
\beta &\equiv \frac{1}{L^2}, \\
\gamma &\equiv 1 - r_1\cos\phi.
\end{aligned}
\tag{7.44}
$$

The subscripts of α, β, and γ in Eq. (7.43) correspond to partial derivatives; for example,

$$
\alpha_1 = \frac{\partial\alpha}{\partial x_1}, \quad \alpha_2 = \frac{\partial\alpha}{\partial y_1}, \quad \alpha_3 = \frac{\partial\alpha}{\partial z_1},
$$
$$
\alpha_{11} = \frac{\partial^2\alpha}{\partial x_1^2}, \quad \alpha_{12} = \frac{\partial^2\alpha}{\partial y_1\partial x_1}, \text{ etc.}
\tag{7.45}
$$

The partial derivative terms of α are

$$
\alpha_1 \equiv -ip_1\alpha q_1, \quad \alpha_2 \equiv -ip_1\alpha q_2, \quad \alpha_3 \equiv -ip_1\alpha q_3,
\tag{7.46}
$$

$$
\alpha_{11} \equiv -ip_1\left(\alpha_1 q_1 + \alpha q_{11}\right), \quad \alpha_{12} \equiv -ip_1\left(\alpha_2 q_1 + \alpha q_{12}\right),
\tag{7.47}
$$

and

$$
\alpha_{13} \equiv -ip_1\left(\alpha_3 q_1 + \alpha q_{13}\right),
\tag{7.48}
$$

where

$$
q_1 \equiv \frac{x_1\delta}{q}, \quad q_2 \equiv \frac{y_1\delta}{q}, \quad q_3 \equiv \frac{p_1^2 z_1}{q},
\tag{7.49}
$$

$$
q_{11} \equiv \frac{\delta}{q} + \frac{x_1^2}{q}\left(\frac{\cos\phi}{r_1^3} - \frac{\delta^2}{q^2}\right),
\tag{7.50}
$$

$$q_{12} \equiv \frac{x_1 y_1}{q}\left(\frac{\cos\phi}{r_1^3} - \frac{\delta^2}{q^2}\right), \qquad q_{13} \equiv -p_1^2 x_1 z_1 \frac{\delta}{q^3} \qquad (7.51)$$

and

$$\delta \equiv \frac{r_1 - \cos\phi}{r_1}. \qquad (7.52)$$

The partial derivative terms of β are

$$\beta_1 \equiv -\frac{2L_1}{L^3}, \qquad \beta_2 \equiv -\frac{2L_2}{L^3}, \qquad (7.53)$$

$$\beta_{11} \equiv \frac{6L_1^2}{L^4} - \frac{2L_{11}}{L^3}, \qquad \beta_{12} \equiv \frac{6L_1 L_2}{L^4} - \frac{2L_{12}}{L^3}, \qquad (7.54)$$

where

$$L_1 \equiv \frac{x_1 \delta}{L}, \qquad L_2 \equiv \frac{y_1 \delta}{L}, \qquad (7.55)$$

$$L_{11} \equiv \frac{\delta}{L} + \frac{x_1^2}{L}\left(\frac{\cos\phi}{r_1^3} - \frac{\delta^2}{L^2}\right), \qquad L_{12} \equiv \frac{x_1 y_1}{L}\left(\frac{\cos\phi}{r_1^3} - \frac{\delta^2}{L^2}\right). \qquad (7.56)$$

The partial derivative terms of γ are

$$\gamma_1 \equiv -\frac{x_1 \cos\phi}{r_1}, \qquad \gamma_2 \equiv -\frac{y_1 \cos\phi}{r_1}, \qquad (7.57)$$

$$\gamma_{11} \equiv -\frac{\cos\phi}{r_1}\left(1 - \frac{x_1^2}{r_1^2}\right), \qquad \gamma_{12} \equiv \frac{x_1 y_1 \cos\phi}{r_1^3}. \qquad (7.58)$$

Within the Geometrical Shadow Region $(r_1 > 1)$

For points of interest in the geometrical shadow region, or where $z_1 \geq 0$ and $r_1 > 1$, Carter and Williams [27] demonstrated that the double integral of the Hertz vector component in Eq. (7.21) can also be expressed as a single integral. Using the Carter and Williams' integral form (Eq. 12 of Ref. [27]) for the double integral in Eq. (7.21), the Hertz vector component can be expressed as

$$\Pi_x(x_1, y_1, z_1) = \frac{a^2 E_o}{\pi p_1^2} \int_0^{\pi/2} \frac{u}{v}\cos\psi\,d\psi, \qquad (7.59)$$

where

$$u \equiv e^{-ip_1 g} - e^{-ip_1 h}, \qquad v \equiv \sqrt{r_1^2 - \sin^2\psi}, \qquad (7.60)$$

and

$$g \equiv \sqrt{p_1^2 z_1^2 + (v + \cos\psi)^2}, \quad h \equiv \sqrt{p_1^2 z_1^2 + (v - \cos\psi)^2}. \tag{7.61}$$

Substituting Eq. (7.59) into Eqs. (7.18) and (7.19), the electric and magnetic field components become

$$E_x(x_1, y_1, z_1) = \frac{E_o}{\pi} \int_0^{\pi/2} f_{3a} d\psi,$$

$$E_y(x_1, y_1, z_1) = \frac{E_o}{\pi p_1^2} \int_0^{\pi/2} f_{3b} d\psi,$$

$$E_z(x_1, y_1, z_1) = \frac{E_o}{\pi p_1^3} \int_0^{\pi/2} f_{3c} d\psi, \tag{7.62}$$

and

$$H_x(x_1, y_1, z_1) = 0,$$

$$H_y(x_1, y_1, z_1) = \frac{iH_o}{\pi p_1^2} \int_0^{\pi/2} f_{3d} d\psi,$$

$$H_z(x_1, y_1, z_1) = -\frac{iH_o}{\pi p_1} \int_0^{\pi/2} f_{3e} d\psi, \tag{7.63}$$

where

$$f_{3a} \equiv \frac{\cos\psi}{v} \left[u + \frac{1}{p_1^2} \left(u_{11} - \frac{2x_1 u_1}{v^2} + \frac{3ux_1^2}{v^4} - \frac{u}{v^2} \right) \right],$$

$$f_{3b} \equiv \frac{\cos\psi}{v} \left(u_{12} - \frac{y_1 u_1}{v^2} + \frac{3x_1 y_1 u}{v^4} - \frac{x_1 u_2}{v^2} \right),$$

$$f_{3c} \equiv \frac{\cos\psi}{v} \left(u_{13} - \frac{x_1 u_3}{v^2} \right),$$

$$f_{3d} \equiv \frac{\cos\psi}{v} u_3,$$

$$f_{3e} \equiv \frac{\cos\psi}{v} \left(u_2 - \frac{y_1 u}{v^2} \right). \tag{7.64}$$

The subscripts of u correspond to partial derivatives (as in Eq. (7.45)):

$$u_1 \equiv -ip_1 \left(g_1 e^{-ip_1 g} - h_1 e^{-ip_1 h} \right), \quad u_2 \equiv -ip_1 \left(g_2 e^{-ip_1 g} - h_2 e^{-ip_1 h} \right),$$

$$u_3 \equiv -ip_1^3 z_1 \left(\frac{e^{-ip_1 g}}{g} - \frac{e^{-ip_1 h}}{h} \right), \tag{7.65}$$

$$u_{11} \equiv -ip_1\left(g_{11}e^{-ip_1g} - h_{11}e^{-ip_1h}\right)$$
$$-p_1^2\left(g_1^2 e^{-ip_1g} - h_1^2 e^{-ip_1h}\right),$$

$$u_{12} \equiv -ip_1\left(g_{12}e^{-ip_1g} - h_{12}e^{-ip_1h}\right)$$
$$-p_1^2\left(g_1g_2 e^{-ip_1g} - h_1h_2 e^{-ip_1h}\right),$$

$$u_{13} \equiv -ip_1\left(g_{13}e^{-ip_1g} - h_{13}e^{-ip_1h}\right)$$
$$-p_1^2\left(g_1g_3 e^{-ip_1g} - h_1h_3 e^{-ip_1h}\right), \tag{7.66}$$

where

$$g_1 \equiv x_1 g_a, \quad g_2 \equiv y_1 g_a, \quad g_3 \equiv \frac{z_1 p_1^2}{g}, \tag{7.67}$$

$$g_{11} \equiv g_a - \frac{x_1^2}{g}\left(g_a^2 + \frac{\cos\psi}{v^3}\right),$$

$$g_{12} \equiv -\frac{x_1 y_1}{g}\left(g_a^2 + \frac{\cos\psi}{v^3}\right),$$

$$g_{13} \equiv -\frac{x_1 z_1 p_1^2}{g^2}g_a, \tag{7.68}$$

$$h_1 \equiv x_1 h_a, \quad h_2 \equiv y_1 h_a, \quad h_3 \equiv \frac{z_1 p_1^2}{h}, \tag{7.69}$$

$$h_{11} \equiv h_a - \frac{x_1^2}{h}\left(h_a^2 - \frac{\cos\psi}{v^3}\right),$$

$$h_{12} \equiv -\frac{x_1 y_1}{h}\left(h_a^2 - \frac{\cos\psi}{v^3}\right),$$

$$h_{13} \equiv -\frac{x_1 z_1 p_1^2 h_a}{g^2}, \tag{7.70}$$

and

$$g_a \equiv \frac{v + \cos\psi}{vg}, \quad h_a \equiv \frac{v - \cos\psi}{vh}. \tag{7.71}$$

Even though the single integral expressions for **E** and **H** (Eqs. 7.41, 7.42, 7.62, and 7.63) involve a large number of terms, they can be evaluated significantly (10 to 15 times) faster than the double integral expressions (Eqs. 7.28–7.33).

7.6 Kirchhoff Vector Diffraction Theory (KVDT)

The Hertz vector diffraction theory (HVDT) described above provides the values of the electromagnetic fields in the aperture plane and beyond. The

Kirchhoff boundary conditions, on the other hand, *specify* the values of the fields in the aperture plane. Starting from these specified values at $z = 0$ the fields for $z > 0$ can be obtained using the Green's function method of solution of the wave equation. Luneberg [24] has shown how the Green's function method can be used to derive the longitudinal component (E_z) of the fields from the known transverse components in the aperture plane.

Luneberg's method has been used by Lü and Duan [19, 20] to derive analytical expressions for the electric field components of diffracted waves beyond the aperture. However, despite the claim in Refs. [19, 20] that the treatment of the problem is nonparaxial, the analytical expressions derived there are based upon the assumption that the on-axis distance to the point of interest is large compared to the radial distance (Eq. 6 in Lü and Duan [19] and Eq. 7 in Duan and Lü [20]). Here we derive the vector components of the electric and magnetic fields in terms of double integrals in dimensionless parameters for incident planar beam distributions of arbitrary shape, using the Kirchhoff boundary conditions, with the aim of determining the region of their validity. Neither the paraxial (Fresnel) approximation nor the approximations used by Lü and Duan [19, 20] are invoked here.

Given a beam of light whose electric field distribution in the plane $z = 0$ is known in terms of the coordinates x_0 and y_0 in the plane as $\mathbf{E} = E_x(\mathbf{r_0})\hat{i} + E_y(\mathbf{r_0})\hat{j}$, the field at a point $\mathbf{r} = x\hat{i} + y\hat{j} + z\hat{k}$ is given by [24]

$$E_x(\mathbf{r}) = -\frac{1}{2\pi} \int\int_{-\infty}^{\infty} E_x(\mathbf{r_0}) \frac{\partial G(\mathbf{r}, \mathbf{r_0})}{\partial z} dx_0 dy_0, \tag{7.72}$$

$$E_y(\mathbf{r}) = -\frac{1}{2\pi} \int\int_{-\infty}^{\infty} E_y(\mathbf{r_0}) \frac{\partial G(\mathbf{r}, \mathbf{r_0})}{\partial z} dx_0 dy_0, \tag{7.73}$$

where $\mathbf{r_0} = x_0\hat{i} + y_0\hat{j}$. The Green's function G used in Eqs. (7.72) and (7.73) is given by

$$G(\mathbf{r}, \mathbf{r_0}) = \frac{e^{-ik\rho}}{\rho}, \tag{7.74}$$

and the distance ρ is defined in Eq. (7.22).

Using Eqs. (7.72) and (7.73) in the Maxwell equation $\nabla \cdot \mathbf{E} = 0$ in charge-free space, i.e.,

$$\frac{\partial E_z}{\partial z} = -\left(\frac{\partial E_x}{\partial x} + \frac{\partial E_y}{\partial y}\right), \tag{7.75}$$

and interchanging the orders of the partial derivatives of G, we obtain

$$\begin{aligned} 2\pi \frac{\partial E_z}{\partial z} &= \int\int_{-\infty}^{\infty} \left(E_x(\mathbf{r_0}) \frac{\partial^2 G}{\partial x \partial z} + E_y(\mathbf{r_0}) \frac{\partial^2 G}{\partial y \partial z}\right) dx_0 dy_0 \\ &= \frac{\partial}{\partial z} \left[\int\int_{-\infty}^{\infty} \left(E_x(\mathbf{r_0}) \frac{\partial G}{\partial x} + E_y(\mathbf{r_0}) \frac{\partial G}{\partial y}\right) dx_0 dy_0\right]. \end{aligned} \tag{7.76}$$

From Eq. (7.76), E_z is obtained as

$$E_z(\mathbf{r}) = \frac{1}{2\pi} \int\!\!\int_{-\infty}^{\infty} \left(E_x(\mathbf{r_0})\frac{\partial G}{\partial x} + E_y(\mathbf{r_0})\frac{\partial G}{\partial y} \right) dx_0 dy_0$$
$$+ F_1(x,y), \tag{7.77}$$

where $F_1(x,y)$ is a function to be determined from the boundary conditions. The term $F_1(x,y)$, was ignored in previous treatments [19, 20, 24], and is necessary for E_z to approach the imposed boundary condition as z approaches 0.

Using the length scale a, the quantity $z_0 \equiv ka^2$, the dimensionless parameter $p_1 \equiv 2\pi a/\lambda$, and the normalized units of Eqs. (7.23) and (7.24) as before, and the expression for the Green's function from Eq. (7.74), the three components of the electric field vector can now be rewritten as

$$E_x(\mathbf{r_1}) = -\frac{p_1^3 z_1}{2\pi} \int\!\!\int_{-\infty}^{\infty} E_x(\mathbf{r_{01}})f_1 s_1 dx_{01} dy_{01},$$

$$E_y(\mathbf{r_1}) = -\frac{p_1^3 z_1}{2\pi} \int\!\!\int_{-\infty}^{\infty} E_y(\mathbf{r_{01}})f_1 s_1 dx_{01} dy_{01},$$

$$E_z(\mathbf{r_1}) = \frac{p_1^2}{2\pi} \left[\int\!\!\int_{-\infty}^{\infty} \Big(E_x(\mathbf{r_{01}})(x_1 - x_{01}) \right.$$
$$\left. + E_y(\mathbf{r_{01}})(y_1 - y_{01}) \Big) f_1 s_1 dx_{01} dy_{01} \right] + F_1(x_1, y_1), \tag{7.78}$$

where f_1 and s_1 are defined in Eqs. (7.34) and (7.35).

The magnetic field components can be derived similarly to the derivation of the electric field components using Green's theorem in Eqs. (7.72) – (7.77), yielding

$$H_x(\mathbf{r_1}) = -\frac{p_1^3 z_1}{2\pi} \int\!\!\int_{-\infty}^{\infty} H_x(\mathbf{r_{01}})f_1 s_1 dx_{01} dy_{01},$$

$$H_y(\mathbf{r_1}) = -\frac{p_1^3 z_1}{2\pi} \int\!\!\int_{-\infty}^{\infty} H_y(\mathbf{r_{01}})f_1 s_1 dx_{01} dy_{01},$$

$$H_z(\mathbf{r_1}) = \frac{p_1^2}{2\pi} \left[\int\!\!\int_{-\infty}^{\infty} \Big(H_x(\mathbf{r_{01}})(x_1 - x_{01}) \right.$$
$$\left. + H_y(\mathbf{r_{01}})(y_1 - y_{01}) \Big) f_1 s_1 dx_{01} dy_{01} \right] + F_2(x_1, y_1), \tag{7.79}$$

where $F_2(x_1, y_1)$ is a function to be determined from the boundary conditions. Using Eqs. (7.78) and (7.79) the Poynting vector components can be calculated from Eq. (7.20).

In the Kirchhoff formalism [22] the field components on the dark side of the screen are assumed to be zero except at the opening, where they have their undisturbed values that they would have had in the absence of the screen. For a linearly polarized plane wave light beam (say with $\mathbf{E}$ in the x-direction, and

H in the y-direction) incident on a circular aperture of radius a in an opaque screen, the electric field components just at the exit of the aperture (at the $z = 0$ plane) are then

$$E_x(x_{01}, y_{01}, 0) = \begin{cases} E_0 & \text{if } x_{01}^2 + y_{01}^2 < 1 \\ 0 & \text{otherwise} \end{cases}$$

$$E_y(x_{01}, y_{01}, 0) = 0$$

$$E_z(x_{01}, y_{01}, 0) = 0, \tag{7.80}$$

and the magnetic field components are

$$H_x(x_{01}, y_{01}, 0) = 0$$

$$H_y(x_{01}, y_{01}, 0) = \begin{cases} H_0 & \text{if } x_{01}^2 + y_{01}^2 < 1 \\ 0 & \text{otherwise} \end{cases}$$

$$H_z(x_{01}, y_{01}, 0) = 0. \tag{7.81}$$

Inserting Eqs. (7.80) and (7.81) in Eqs. (7.78) and (7.79), the expressions for the electric and magnetic fields are obtained:

$$E_x(x_1, y_1, z_1) = -E_0 \frac{p_1^3 z_1}{2\pi} B_1(x_1, y_1, z_1), \tag{7.82}$$

$$E_y(x_1, y_1, z_1) = 0, \tag{7.83}$$

$$E_z(x_1, y_1, z_1) = E_0 \frac{p_1^2}{2\pi} \left[B_2(x_1, y_1, z_1) - B_2(x_1, y_1, 0) \right], \tag{7.84}$$

and

$$H_x(x_1, y_1, z_1) = 0, \tag{7.85}$$

$$H_y(x_1, y_1, z_1) = -H_0 \frac{p_1^3 z_1}{2\pi} B_1(x_1, y_1, z_1), \tag{7.86}$$

$$H_z(x_1, y_1, z_1) = H_0 \frac{p_1^2}{2\pi} \left[B_3(x_1, y_1, z_1) - B_3(x_1, y_1, 0) \right], \tag{7.87}$$

where

$$B_1(x_1, y_1, z_1) = \int_{-1}^{1} \int_{-\sqrt{1-y_{01}^2}}^{\sqrt{1-y_{01}^2}} f_1 s_1 dx_{01} dy_{01}, \tag{7.88}$$

$$B_2(x_1, y_1, z_1) = \int_{-1}^{1} \int_{-\sqrt{1-y_{01}^2}}^{\sqrt{1-y_{01}^2}} f_1 s_1 (x_1 - x_{01}) dx_{01} dy_{01}, \tag{7.89}$$

$$B_3(x_1, y_1, z_1) = \int_{-1}^{1} \int_{-\sqrt{1-y_{01}^2}}^{\sqrt{1-y_{01}^2}} f_1 s_1 (y_1 - y_{01}) dx_{01} dy_{01}. \tag{7.90}$$

The terms $B_2(x_1, y_1, 0)$ and $B_3(x_1, y_1, 0)$ in Eqs. (7.84) and (7.87) arise from the presence of the terms $F_1(x_1, y_1)$ and $F_2(x_1, y_1)$ in Eqs. (7.77)–(7.79). They ensure that the boundary conditions $E_z = H_z = 0$ at $z_1 = 0$

(Eqs. (7.80) and (7.81)) are valid. As mentioned earlier, they have been ignored in Refs. [19, 20, 24], so that the calculated longitudinal field components E_z and H_z obtained there do not vanish at $z_1 = 0$, contradicting the boundary condition.

It should be noted that both the E_x and H_y KVDT integrals (Eqs. (7.82), (7.86), and (7.88)) have the same form as the H_y HVDT double integral (Eq. (7.32)) and one of the four integral terms of the HVDT E_x component (Eq. (7.28)), and are radially symmetric with respect to x_1 and y_1. Therefore the calculated fields of E_x and H_y using KVDT will always have radial symmetry and are independent of the orientation of the polarization for normally incident light. In contrast, the HVDT method shows (Eqs. (7.28)–(7.33)) that the components of all the electromagnetic fields except H_x and H_y are dependent on the direction of the incident polarization.

7.7   Analytical On-Axis Expressions and Calculations

7.7.1   Analytical On-Axis Expressions Using HVDT

Starting with the Hertz vector diffraction theory, analytical expressions for the components of **E** and **H** for on-axis positions and $z_1 \geq 0$ can be obtained by direct integration. Using Eqs. (7.18) and (7.21) the on-axis integral expression for E_x can be found to be

$$E_x(0,0,z_1) = E_o \frac{i}{2\pi} \left[\iint \left(p_1 \frac{e^{-ip_1 r_2}}{r_2} - i \frac{e^{-ip_1 r_2}}{r_2^2} - \frac{e^{-ip_1 r_2}}{p_1 r_2^3} - p_1 x_{01}^2 \frac{e^{-ip_1 r_2}}{r_2^2} \right. \right.$$
$$\left. \left. + 3i x_{01}^2 \frac{e^{-ip_1 r_2}}{r_2^4} + 3x_{01}^2 \frac{e^{-ip_1 r_2}}{p_1 r_2^5} \right) dx_{01} dy_{01} \right], \tag{7.91}$$

where

$$r_2 = \sqrt{x_{01}^2 + y_{01}^2 + p_1^2 z_1^2}. \tag{7.92}$$

After converting to polar coordinates and integrating the angular terms, the radial integral is integrated by parts where the definite integral from each term in Eq. (7.91) cancels with that of another term's definite integral. The resulting analytical expression for the on-axis x-component of the electric field becomes

$$E_x(0,0,z_1) = E_o \frac{i}{2} \left[i\,(e_1 - e_2) - \frac{1}{p_1} \left(\frac{e_1}{d_1} - \frac{e_2}{d_2} \right) \right.$$
$$\left. + i p_1^2 z_1^2 \left(\frac{e_1}{d_1^2} - \frac{e_2}{d_2^2} \right) + p_1 z_1^2 \left(\frac{e_1}{d_1^3} - \frac{e_2}{d_2^3} \right) \right], \tag{7.93}$$

where

$$
\begin{aligned}
e_1 &= e^{-ip_1\sqrt{1+p_1^2 z_1^2}}, \\
e_2 &= e^{-ip_1^2 z_1}, \\
d_1 &= \sqrt{1+p_1^2 z_1^2}, \\
d_2 &= p_1 z_1.
\end{aligned}
\tag{7.94}
$$

Similarly, for on-axis points the double integral expression for the y-component of the magnetic field becomes

$$
H_y(0,0,z_1) = H_o \frac{p_1 z_1}{2\pi} \int\int \left(ip_1 \frac{e^{-ip_1 r_2}}{r_2^2} + \frac{e^{-ip_1 r_2}}{r_2^3} \right) dx_{01} dy_{01}, \tag{7.95}
$$

which can be integrated to

$$
H_y(0,0,z_1) = H_o \left(e^{-ip_1^2 z_1} - p_1 z_1 \frac{e^{-ip_1\sqrt{1+p_1^2 z_1^2}}}{\sqrt{1+p_1^2 z_1^2}} \right). \tag{7.96}
$$

Using Eqs. (7.93) and (7.96) the z-component of the Poynting vector can be evaluated for on-axis points.

At the center of the circular aperture the real and imaginary components of E_x reduce to

$$
\text{Real}\,[E_x(0,0,0)] = E_o \left[1 - \frac{1}{2}\left(\cos p_1 + \frac{\sin p_1}{p_1} \right) \right] \tag{7.97}
$$

and

$$
\text{Imaginary}\,[E_x(0,0,0)] = -\frac{E_o}{2}\left(\frac{\cos p_1}{p_1} - \sin p_1 \right), \tag{7.98}
$$

while H_y/H_o approaches unity for all values of p_1.

7.7.2 Analytical On-Axis Expressions Using KVDT

For on-axis points ($x_1 = y_1 = 0$) the non-zero components of the fields evaluated from the KVDT integral term B_1 (Eq. (7.88)) also reduce to the analytical forms:

$$
E_x(0,0,z_1) = E_o \left(e^{-ip_1^2 z_1} - p_1 z_1 \frac{e^{-ip_1\sqrt{1+p_1^2 z_1^2}}}{\sqrt{1+p_1^2 z_1^2}} \right), \tag{7.99}
$$

and

$$
H_y(0,0,z_1) = H_o \left(e^{-ip_1^2 z_1} - p_1 z_1 \frac{e^{-ip_1\sqrt{1+p_1^2 z_1^2}}}{\sqrt{1+p_1^2 z_1^2}} \right). \tag{7.100}
$$

As z_1 approaches the aperture plane, both E_x/E_o and H_y/H_o approach unity and are independent of the value of p_1 and the orientation of the incident polarization.

The KVDT integrals B_2 and B_3 are identically equal to zero for $x_1 = y_1 = 0$, so that the longitudinal components E_z and H_z vanish for on-axis points, and the Poynting vector only has a z-component, or

$$\mathbf{S}(0, 0, z_1) = S_0 \left(\frac{1 + 2p_1^2 z_1^2}{1 + p_1^2 z_1^2} \right.$$

$$\left. \times -\frac{2p_1 z_1}{\sqrt{1 + p_1^2 z_1^2}} \cos \left[p_1^2 z_1 \left(\sqrt{1 + \frac{1}{p_1^2 z_1^2}} - 1 \right) \right] \right) \hat{k}, \qquad (7.101)$$

where

$$S_o \equiv E_o H_o^* = \sqrt{\frac{\epsilon_o}{\mu_o}} |E_o|^2 \qquad (7.102)$$

denotes the intensity of the incident undisturbed beam.

Due to the cosine term in Eq. (7.101), the on-axis value of the Poynting vector oscillates with z_1 for $p_1 > \pi$, with locations of the maxima and the minima given by

$$z_{1m} = \frac{p_1^2 - m^2 \pi^2}{2m\pi p_1^2}, \qquad (7.103)$$

where m has integer values from 1 to p_1/π. Odd values of m correspond to maxima and even values correspond to minima. The value of $m = 1$ gives the location of the on-axis maximum farthest from the aperture, and is roughly $1/(2\pi)$ for large values of p_1.

7.8   Power Transmission Function

One important parameter for the transmission of light through an aperture is the *power transmission function*. There are two broad interpretations of what this term implies:

- fraction of the the light energy transmitted by the aperture or mask to the incident light's total energy, and

- fraction of the light transmitted by the aperture to the light incident upon the open aperture area.

There is a subtle, yet very important, difference between these two general definitions.

The former definition is usually used by experimentalists who deal with laser beams with finite beam widths where a power meter can be placed before the aperture, and then after the aperture. It is assumed that the power meter's active area is large enough that it can collect all significant portions of the light field. This definition implicitly includes the portions of the light incident upon the opaque sections of the aperture plane (or mask) in the denominator of the ratio. An example of this definition using classical ray optics would be if an incident laser beam can be represented using 100 rays, and only 50 of those rays are incident upon the open aperture area. Using ray optics, any ray incident upon an open aperture area would be transmitted. Thus for this example, the power transmission function would be 50/100, or 0.5.

The latter definition is usually employed by theorists who are concerned with the fraction of the electromagnetic fields incident upon the open aperture area that actually get through the open aperture area. This definition does not include the light energy incident upon the opaque sections of the aperture. Using the previous classical ray optics example, all 50 of the rays incident upon the open aperture area are transmitted. Thus, the power transmission function would be 50/50, or 1.

In this book, we will be using the latter definition of "power transmission function," or the ratio of the power transmitted through the open aperture area to the power incident upon the open aperture area.

The radiative power transmitted through any surface can be written as

$$P = \int \mathbf{S} \cdot \hat{n} \mathrm{d}a, \tag{7.104}$$

where $\mathbf{S}$ is the Poynting vector, $\hat{n}$ is the normal to the surface, and the integral is calculated over the area of the surface of interest.

For simplicity, we will assume that the light is propagating in the z-direction and that the aperture plane is perpendicular to the z-axis, and located at the position, z_0. We can now further define the power transmission function, T, to be the fraction of the integrals on either side of $z = z_0$,

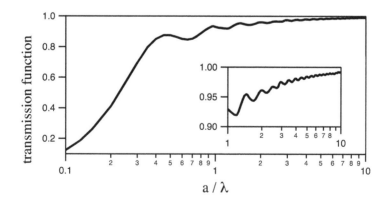

FIGURE 7.7
Calculated transmission function as a function of the aperture radius to wavelength of light ratio for a circular aperture, calculated using HVDT [28].

or

$$T \equiv \frac{P_{z+}}{P_{z-}} = \frac{\iint S_z\left(x, y, z_{0+}\right) \mathrm{d}x \mathrm{d}y}{\iint S_z\left(x, y, z_{0-}\right) \mathrm{d}x \mathrm{d}y}. \tag{7.105}$$

If the output power is to be calculated anywhere beyond $z = z_{0+}$ then the integration should be performed over the entire $x - y$ plane. Throughout this chapter we consider the input electromagnetic fields to be a plane wave, and we can define $P_o \equiv \pi a^2 S_o$ as the amount of unperturbed incident power that would be intercepted by the aperture. Using these assumptions, the power transmission function of the aperture can be expressed as

$$T \equiv \frac{P_z}{P_o} = \frac{1}{P_o} \int_{-\infty}^{\infty} \int_{-\infty}^{\infty} S_z\left(x, y, z_{0+}\right) \mathrm{d}x \mathrm{d}y, \tag{7.106}$$

where S_z is determined from the calculated values of **E** and **H** for a given value of a/λ.

The transmission function can also be used to test the regions of validity of various diffraction models. In theory, the transmission function should be independent of the value of z_{0+}. In other words, once the light passes through the aperture plane and into an unrestricted volume the power through any plane perpendicular to the net propagation direction should be the same.

It has previously been discussed that the results obtained using KVDT cannot be trusted for locations very near the aperture plane. If the fields calculated using KVDT are used to calculate P_z, the value of T is found to depend upon the value of z_+ for regions close to the aperture plane (and thus violates the conservation of energy principle).

If HVDT is used to calculate P_z the resulting total power through planes with locations of various z_+ values *is* found to be the same, and validates the HVDT model using conservation of energy.

Figure 7.7 shows a plot of P_z/P_o calculated using Eq. (7.106) as a function of a/λ for a circular aperture. For $a/\lambda < 0.5$ the power transmitted by the aperture decreases monotonically with decreasing a/λ. For $a/\lambda > 0.5$, the transmitted power oscillates weakly and in general increases with increasing a/λ and asymptotically approaches unity. However, as shown in the insert of Figure 7.7, even for $a/\lambda = 10$, the value of T is less than 1 (in fact it is equal to 99.2%).

Bibliography

[1] Frank L. Pedrotti and Leno S. Pedrotti, *Introduction to Optics* (Prentice-Hall, Englewood Cliffs, NJ, 1987), 1st. ed.

[2] Francis A. Jenkins and Harvey E. White, *Fundamentals of Optics* (McGraw-Hill, New York, NY, 1976), 4th. ed.

[3] Grant R. Fowles, *Introduction to Modern Optics* (Dover Publications, Inc., New York, NY, 1975), 2nd. ed.

[4] Eugene Hecht, *Optics* (Addison Wesley, San Francisco, CA, 2002), 4th. ed.

[5] M. Born and E. Wolf, *Principles of Optics*, Seventh Edition (Cambridge University Press, Cambridge, UK, 1999.)

[6] Joseph W. Goodman, *Introduction to Fourier Optics* (McGraw-Hill, New York, NY, 1996), 2nd. ed.

[7] B. I. Greene, J. F. Federici, D. R. Dykaar, R. R. Jones, and P. H. Bucksbaum, "Interferometric characterization of 160 fs far-infrared light pulses," Appl. Phys. Lett. **59**, 893–895 (1991).

[8] R. R. Jones, D. You, and P. H. Bucksbaum, "Ionization of Rydberg atoms by subpicosecond half-cycle electromagnetic pulses," Phys. Rev. Lett. **70**, 1236–1239 (1993).

[9] E. Budiarto, N.-W. Pu, S. Jeong, and J. Bokor, "Near-field propagation of terahertz pulses from a large-aperture antenna," Opt. Lett. **23**, 213–215 (1997).

[10] D. M. Bloom, "Grating light valve: revolutionizing display technology," Proc. of SPIE **3013**, 165–171 (1997).

[11] L. Y. Lin, E. L. Goldstein, and R. W. Tkach, "Free-space micro machined optical switches with sub millisecond switching time for large-scale optical cross connects," IEEE Phot. Technology Letters **10**, 525–527 (1998).

[12] T. P. Kurzweg, "Optical propagation methods for system-level modeling of optical MEM systems," Ph.D. dissertation, Dept. Elect. Eng., University of Pittsburgh, PA 2002.

[13] S. P. Levitan, J. A. Martinez, T. P. Kurzweg, A. J. Davare, M. Kahrs, M. Bails, and D. M. Chiarulli, "System simulation of mixed-signal multi-domain microsystems with piecewise linear models," IEEE Trans. Computer-Aided Design **22**, 139–154 (2003).

[14] G. R. Kirchhoff, "Zur Theorie der Lichtstrahlen," Ann. Phys. (Leipzig) **18**, 663-695 (1883).

[15] A. Sommerfeld, "Zur mathematischen Theorie der Beugungserscheinungen," Nachr. Kgl. Wiss. Göttingen **4**, 338-342 (1894).

[16] Lord Rayleigh, "On the passage of waves through apertures in plane screens, and allied problems," Philos. Mag. **43**, 259–272 (1897).

[17] G. D. Gillen and S. Guha, "Modeling and propagation of near-field diffraction patterns: A more complete approach," Am. J. Phys. **72**, 1195–1201 (2004).

[18] O. Mitrofanov, M. Lee, J. W. P. Hsu, L. N. Pfeifer, K. W. West, J. D. Wynn, and J. F. Federici, "Terahertz pulse propagation through small apertures," Appl. Phys. Lett. **79**, 907–909 (2001).

[19] B. Lü and K. Duan, "Nonparaxial propagation of vectorial Gaussian beams diffracted at a circular aperture," Opt. Lett **28**, 2440–2442 (2003).

[20] K. Duan and B. Lü, "Vectorial nonparaxial propagation equation of elliptical Gaussian beams in the presence of a rectangular aperture," J. Opt. Sec. Am. A **21**, 1613–1620 (2004).

[21] W. Freude and G. K. Grau, "Rayleigh-Sommerfeld and Helmholtz-Kirchhoff integrals: application to the scalar and vectorial theory of wave propagation and diffraction," J. Lightwave Technol. **13**, 24–32 (1995).

[22] J. A. Stratton and L.J. Chu, "Diffraction theory of electromagnetic waves," Phys. Rev. **56**, 99–107 (1939).

[23] H. A. Bethe, "Theory of diffraction by small holes," Phys. Rev. **66**, 163–182 (1944).

[24] R. K. Luneberg, *Mathematical Theory of Optics* (U. California Press, Berkeley, CA, 1964).

[25] G. Bekefi, "Diffraction of electromagnetic waves by an aperture in a large screen," J. App. Phys. **24**, 1123–1130 (1953).

[26] A. Schoch, "Betrachtungen über das Schallfeld einer Kolbenmembran,"Akust. Z. **6**, 318–326 (1941).

[27] A. H. Carter and A. O. Williams, Jr., "A New Expansion for the velocity potential of a piston source," J. Accoust. Soc. Am. **23**, 179–184 (1951).

[28] S. Guha and G. D. Gillen, "Description of light propagation through a circular aperture using nonparaxial vector diffraction theory," Opt. Express **13**, 1424–1447 (2005).

8

Calculations for Plane Waves Incident upon Various Apertures

> In this chapter we present calculations of electromagnetic field components, Poynting vectors, and laser beam intensity profiles and patterns for plane waves incident upon apertures of various shapes using the vector diffraction models presented in Chapter 7.

8.1 Beam Distributions in the Aperture Plane, Circular Aperture

In this section we present computational results of the electric field and light patterns for locations *within* the aperture plane. Because the points of interest here are located in the plane containing the diffracting element, Hertz Vector Diffraction theory (HVDT) is employed as it is the only model valid in that region whose results will inherently satisfy both the wave equation and Maxwell's equations. For a detailed discussion of the HVDT model see Section 7.5. Figures 8.1–8.9 are a collection of calculations of the electric field and the Poynting vector for circular apertures of a variety of radial sizes, showing the distributions of the normalized fields and normalized axial Poynting vector component, S_z/S_o, in the aperture plane. Each of these results were obtained from the previously derived integral expressions by setting $z_1 = 10^{-5}$ and making sure that the values of the integrals were unchanged when a smaller value of z_1 was chosen.

Figure 8.1 shows the calculation of the x-component of the electric field in the aperture plane as a function of x_1 and y_1, for $a/\lambda = 5$ [1]. The calculated E_x field is seen to exhibit radial asymmetry meaning that oscillations in the calculated net electric field are observed to be strongest in the direction perpendicular to that of the incident light field, the x-axis. This calculated behavior is in agreement with experimental measurements by Andrews [2] and Ehrlich et al. [3] in the aperture plane for diffraction of microwave radia-

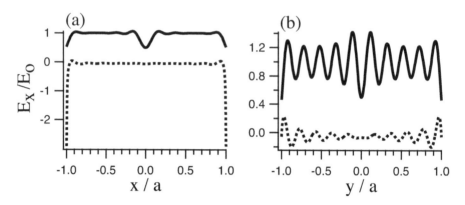

FIGURE 8.1
The real part (solid lines) and the imaginary part (dashed lines) of the x-component of the normalized electric field versus (a) the x-position and (b) the y-position in the aperture plane, calculated using the single integral HVDT for $a/\lambda = 5$ [1].

tion. The number of oscillations of E_x along the y-axis from the center of the aperture to the edge is equal to the aperture to wavelength ratio, a/λ.

Along with the strong asymmetry in the x and y-directions, it is seen that the value of the real part of E_x at the center of the aperture is $E_o/2$, whereas according to Kirchhoff approximations, E_x should be equal to E_o everywhere in the aperture. The real and imaginary parts of $E_x(0,0,0)$ obtained from Eqs. (7.97) and (7.98), are plotted as a function of a/λ in Figures 8.2(a) and (b), respectively. Figure 8.2 shows that at the aperture plane, the Kirchhoff approximation is not valid even for large values of a/λ, with the real part of E_x oscillating between 0.5 and 1.5, for all $a/\lambda > 0.5$. The oscillations of Figure 8.2, which continue indefinitely as p_1 increases, with values of 1.5 occurring for a/λ values having half-integer values, and 0.5 for integer values, have previously been experimentally demonstrated [3]. Eq. (7.98) also shows that as p_1, i.e. a/λ, goes to zero, the value of the imaginary part of $E_x(0,0,0)$ diverges as $1/p_1$, thereby implying that the HVDT theory described here is invalid for small p_1 (< 0.1.) The results presented here match the experimental measurements for $a/\lambda \geq 0.5$ shown in Ref. [2, 3, 4].

Figure 8.1 also shows that near the rim of the aperture the imaginary component of E_x increases rapidly in magnitude, causing the modulus square of the electric field to diverge as $x_1 \to \pm 1$. At the same positions in space, however, H_y rapidly approaches zero. The resulting z-component of the Poynting vector, S_z, has a smooth and continuous transition across the rim of the aperture, unlike $|E_x|^2$, as illustrated in Figure 8.3(a). There are also significant differences between $|E_x|^2$ and S_z in the aperture plane along the y-axis, as

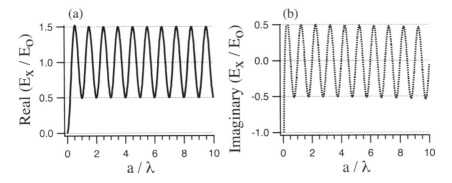

FIGURE 8.2
(a) The real part, and (b) the imaginary part of the normalized x-component of the electric field at the center of the aperture plane, (0,0,0), versus the aperture to wavelength ratio, a/λ, using HVDT [1].

shown in Figure 8.3(b). The values of E_x and H_y displayed in Figure 8.3 were calculated using Eqs. (7.41) and (7.42) for $r_1 < 1$ and Eqs. (7.62) and (7.63) for $r_1 > 1$. Other electromagnetic field components (E_y, E_z and H_z) also possess some structure in the aperture plane, but with negligible ($< 10^{-3}$) magnitudes.

To emphasize the difference between the squared modulus of the elec-

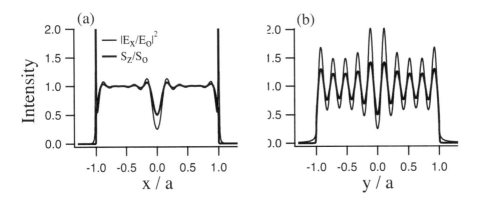

FIGURE 8.3
The modulus square of the x-component of the electric field (thin lines) and the z-component of the Poynting vector (thick lines) versus (a) the x-position and (b) the y-position in the aperture plane, calculated using the single integral HVDT for $a/\lambda = 5$ [1].

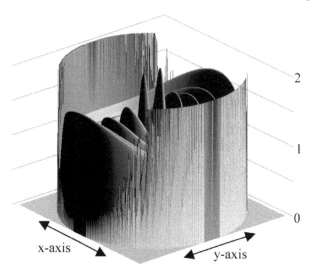

FIGURE 8.4
Calculated modulus square of the x-component of the electric field ($|E_x/E_o|^2$) versus x and y in the aperture plane using single integral HVDT, $\sqrt{(x/a)^2 + (y/a)^2} < 1$ and for $a/\lambda = 5$ [1].

tric field, $|E_x|^2$, and the Poynting vector component, S_z, the two-dimensional distributions of $|E_x/E_o|^2$ and S_z/S_o are plotted in Figures 8.4 and 8.5, respectively, both for $a/\lambda = 5$. As mentioned before, even in the aperture plane, the transition of the Poynting vector component S_z between the illuminated and the dark region is smooth, whereas according to the Kirchhoff boundary conditions, there is a sharp discontinuity of the electric field (as well as S_z) at the boundary.

In Figures 8.6–8.9, the x-y distribution of S_z in the aperture plane is plotted for $a/\lambda = 2.5$, 2, 1, and 0.5, respectively. From the center of the aperture out to the rim along the y-axis, the number of oscillations of S_z equals the value of a/λ. Figure 8.9 illustrates that for $a/\lambda < 1$ the oscillatory behavior of E_x, and subsequently S_z, in the aperture plane have been eliminated, and an ellipticity in the beam profile has become the dominant characteristic.

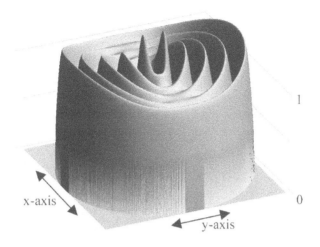

FIGURE 8.5
Calculated z-component of the Poynting vector (S_z/S_o) versus x and y in the aperture plane using single integral HVDT, $\sqrt{(x/a)^2 + (y/a)^2} < 1$ and for $a/\lambda = 5$ [1].

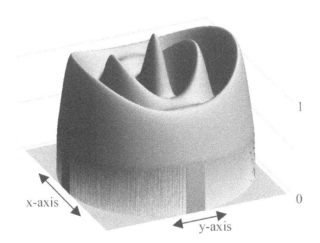

FIGURE 8.6
Calculated z-component of the Poynting vector (S_z/S_o) versus x and y in the aperture plane using single integral HVDT, $\sqrt{(x/a)^2 + (y/a)^2} < 1$ and for $a/\lambda = 2.5$ [1].

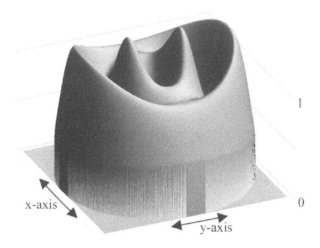

FIGURE 8.7
Calculated z-component of the Poynting vector (S_z/S_o) versus x and y in the aperture plane using single integral HVDT, $\sqrt{(x/a)^2 + (y/a)^2} < 1$ and for $a/\lambda = 2$ [1].

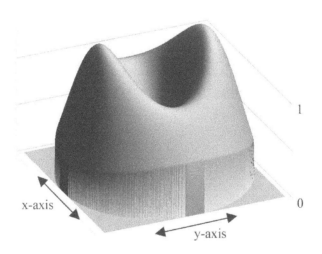

FIGURE 8.8
Calculated z-component of the Poynting vector (S_z/S_o) versus x and y in the aperture plane using single integral HVDT, $\sqrt{(x/a)^2 + (y/a)^2} < 1$ and for $a/\lambda = 1$ [1].

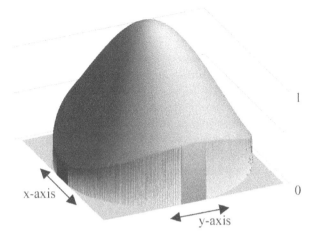

FIGURE 8.9
Calculated z-component of the Poynting vector (S_z/S_o) versus x and y in the aperture plane using single integral HVDT, $\sqrt{(x/a)^2 + (y/a)^2} < 1$ and for $a/\lambda = 0.5$ [1].

8.2 Beam Distributions beyond the Aperture Plane for a Circular Aperture

8.2.1 On-Axis Calculations

The analytical expressions obtained for the fields for on-axis points, ($x = y = 0$), in Eqs. (7.93), (7.96), (7.99), and (7.100) provide an understanding of the behavior of the beam away from the aperture plane. In Figure 8.10 the on-axis values of S_z/S_o calculated using HVDT (Eqs. (7.93) and (7.96)) and KVDT (Eqs. (7.99) and (7.100)) are plotted along with the on-axis value of $|E_x/E_o|^2$ (from Eq. (7.93)) for $a/\lambda = 0.5$, 1, 2.5, and 5. As discussed before, for $a/\lambda > 1$ the on-axis intensity values go through a number of minima and maxima, with the total number of minima or maxima roughly equal to the value of a/λ and the positions of the maxima or minima given by

$$z_1 = \frac{1}{2m\pi} \tag{8.1}$$

where the maxima occur for odd integer values of m and the minima occur for even integer values of m. The last maximum (farthest from the aperture) occurs at a normalized axial location of

$$z_1 = \frac{1}{2\pi}. \tag{8.2}$$

For axial locations farther than those of Eq. 8.2 the intensity drops off as the inverse of the square of the distance from the aperture corresponding to the spreading of a spherical wave.

For small z_1, i.e. *in and near the aperture plane*, the HVDT and KVDT results are different from each other and are both different from $|E_x/E_o|^2$ for all the cases of a/λ, as shown in Figure 8.10. As discussed in Chapter 7, HVDT is the more accurate diffraction model that intrinsically produces results which satisfy both the Helmholtz wave equation and Maxwell's equations. However, the HVDT and KVDT results agree with each other for axial locations away from the aperture ($z_1 > 0.2$), showing that even for $a/\lambda \approx 0.5$, KVDT can be used to predict beam shapes at planes sufficiently far away from the aperture.

8.2.2   Diffracted Beam Shapes

To determine if the radial asymmetry observed for $a/\lambda = 0.5$ in Figure 8.9 persists as the beam propagates away from the aperture, the dependence of S_z/S_o on the radial coordinate r_1 (defined in Eq. (7.40)) is plotted along the x and y axes in Figures 8.11(a)–(c) for $a/\lambda = 0.5$ and in Figures 8.11(d)–(f) for $a/\lambda = 5$ for three different values of z_1 (10^{-4}, 0.1 and 1). Figure 8.11(a)–(c) shows that for $a/\lambda = 0.5$, at the aperture plane ($z_1 = 10^{-4}$) the beam is elliptically shaped with the major axis in the x-direction. Results of calculation using the FDTD method (for $a/\lambda = 0.308$, at $z_1 = 0.052$) are shown in Ref. [5]. The plot of transmitted intensity in Figure 8.10 of [5] also shows the radial asymmetry and the elongation of the beam along the x-axis. As the beam propagates, the ellipticity decreases and beyond $z_1 > 0.1$ the beam becomes elongated along the y-axis. The plots in Figures 8.11(d)–(f) show that for larger a/λ values, although the beam is strongly asymmetric near the aperture plane, as it propagates away from the aperture plane it becomes radially symmetric more quickly than for the smaller a/λ case. Figures 8.12 and 8.13 show the detailed two-dimensional distribution of S_z/S_o at the positions of an axial maximum ($z_1 = 1/(6\pi)$) and an axial minimum ($z_1 = 1/(4\pi)$), for $a/\lambda = 5$.

8.3   The Longitudinal Component of the Electric Field, E_z

Due to the fact that the HVDT model is a *vector* diffraction theory, it allows for the calculation of the longitudinal componets of the light fields, E_z and H_z. The incident light field chosen for all calculations in this chapter is that of a plane wave, which has no longitudinal component. However, due to the fact that all of the components of the light fields are coupled together by Maxwell's equations, the diffracted light fields beyond the diffraction plane *can* (and certainly do) have non-zero longitudinal components.

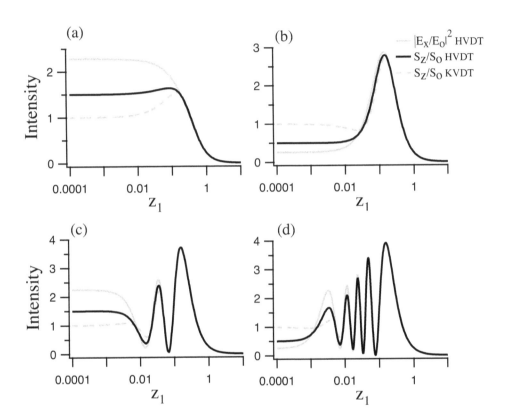

FIGURE 8.10
Calculated on-axis values for the modulus square of the x-component of the electric field (gray line) using HVDT, the z-component of the Poynting vector (black line) using HVDT, and the z-component of the Poynting vector (gray dashed line) using KVDT for: (a) $a/\lambda = 0.5$, (b) $a/\lambda = 1$, (c) $a/\lambda = 2.5$, and (d) $a/\lambda = 5$ [1].

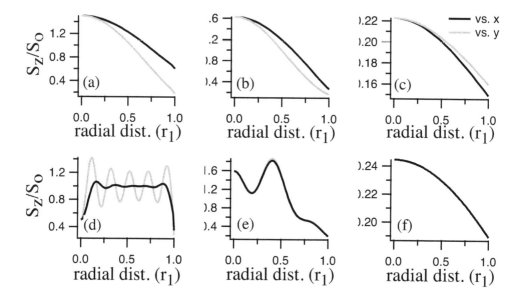

FIGURE 8.11

Calculated S_z/S_o for $a/\lambda = 0.5$ with (a) $z_1 = 10^{-4}$, (b) $z_1 = 0.1$, (c) $z_1 = 1$, and for $a/\lambda = 5$ with (d) $z_1 = 10^{-4}$, (e) $z_1 = 0.1$, and (f) $z_1 = 1$, using the single integral HVDT. The distance r_1 is either x or y normalized to the aperture radius, a [1].

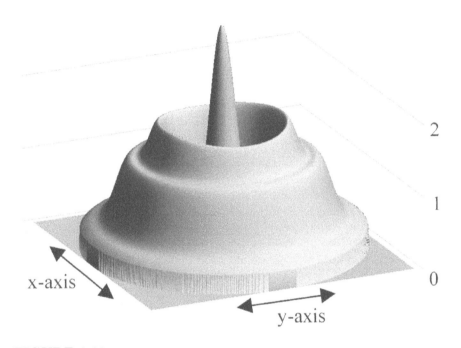

FIGURE 8.12

Calculated z-component of the Poynting vector (S_z/S_o) versus x and y for $z_1 = 1/6\pi$ using the single integral HVDT, $\sqrt{(x/a)^2 + (y/a)^2} < 1$ and $a/\lambda = 5$ [1].

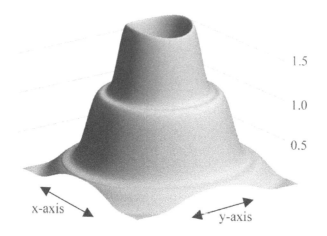

FIGURE 8.13

Calculated z-component of the Poynting vector (S_z/S_o) versus x and y for $z_1 = 1/4\pi$ using the single integral HVDT for $a/\lambda = 5$ [1].

In this section, we present the results of some calculations showing distribution of E_z/E_o at various distances near the aperture. Figures 8.14 and 8.15 show the dependence of E_z/E_o on the radial coordinate r_1 for various values of the angular coordinate θ_1, for $a/\lambda = 0.5$ and 5, respectively, at a distance of $z_1 = 0.05$. Here, $\theta_1 = \tan^{-1} \frac{y_1}{x_1}$ and r_1 has been defined in Eq. 7.40. It is seen that the maximum value of both the real and the imaginary parts of E_z/E_o (for both values of a/λ) are obtained for $\theta_1 = 0$, i.e., along the x-axis. Figures 8.16 and 8.17 show the dependence of E_z/E_o on x_1 (for $y_1 = 0$) at various distances from the aperture ($z_1 = 0.05, 0.1, 0.5,$ and 1) for $a/\lambda = 0.5$ and 5, respectively. It is seen that even for $a/\lambda = 5$, the E_z can be a substan-

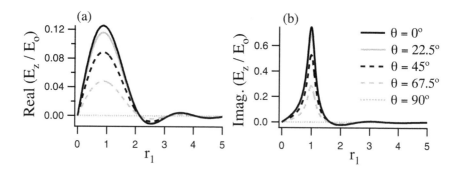

FIGURE 8.14

Calculated real, (a), and imaginary, (b), components of E_z/E_o for $\theta = 0°$ (x-axis), 22.5°, 45°, 67.5°, and 90° (y-axis), for $a/\lambda = 0.5$ and $z_1 = 0.05$ [1].

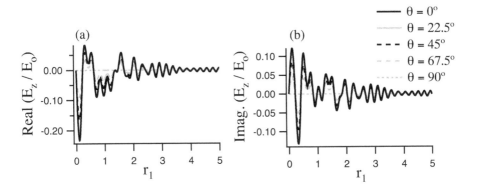

FIGURE 8.15
Calculated real, (a), and imaginary, (b), components of E_z/E_o for $\theta = 0°$
(x-axis), 22.5°, 45°, 67.5°, and 90° (y-axis), for $a/\lambda = 5$ and $z_1 = 0.05$ [1].

tial fraction (20 percent) of the incident field amplitude at certain positions
in front of the aperture.

8.4 Beam Distributions in the Aperture Plane, Elliptical Aperture

In this section we present calculations of some of the rather interesting light
field distributions found within the aperture for plane waves incident upon an
elliptical aperture.

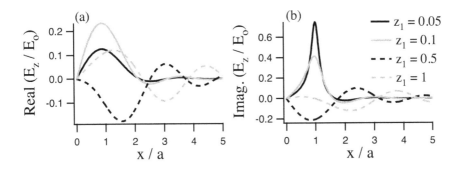

FIGURE 8.16
Calculated real, (a), and imaginary, (b), components of E_z/E_o along the x-axis
for $z_1 = 0.05$, 0.1, 0.5, and 1, for $a/\lambda = 0.5$ [1].

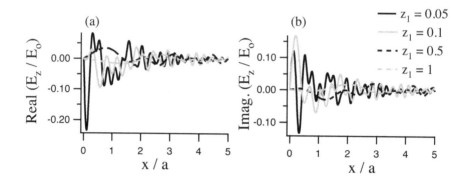

FIGURE 8.17

Calculated real, (a), and imaginary, (b), components of E_z/E_o along the x-axis for $z_1 = 0.05$, 0.1, 0.5, and 1, for $a/\lambda = 5$ [1].

Figure 8.18 is a high-resolution beam intensity profile, in the aperture plane, for the diffraction of a plane wave incident upon an elliptical aperture with a minor axis (of length a) equal to the half of the wavelength of the light and an ellipticity of 2, and the incident polarization parallel to the major axis (of length b) of the ellipse. The surface plot of Figure 8.18 is a calculation of the z-component of the Poynting vector using HVDT. For apertures with dimensions less than the wavelength of light, no interference maxima or minima due to the presence of the aperture are observed in the aperture plane. Only the smooth "leakage" of light through the middle and tapering off towards the edges is observed. Once the aperture dimensions become equal to or larger than the wavelength of light, the perturbation of the light field due to the presence of the aperture becomes the dominant characteristic of the intensity beam profile as illustrated in Figure 8.19. Figure 8.19 is a calculation for an elliptical aperture with the minor axis equal to the wavelength of light and the major axis equal to twice the wavelength of the light.

Figures 8.20 and 8.21 are similar to Figure 8.18 and illustrate two-dimensional intensity profiles for minor axes equal to 2.5 and 5 times the wavelength of the incident light, and the incident polarization parallel to the major axis of the ellipse. Along the minor axis, there are a/λ oscillations in the intensity. Oscillations in the radial direction perpendicular to the incident polarization were also reported in Ref. [1] for a circular aperture. When contrasting diffraction patterns observed for diffraction of light by a circular aperture with the diffraction of light by an elliptical aperture, there are a few noticeable differences between intensity beam profiles. First, for the circular case the radial oscillations were observed along the axis perpendicular to the incident polarization angle, and nearly absent (save for a variation right in the middle) along the axis of polarization (see Figures 8.1, 8.5–8.8). The radial variations in intensity observed in the circular case followed the angular curve

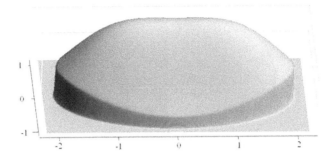

FIGURE 8.18
Intensity distribution of the normalized z-component of the Poynting vector
in the aperture plane for an ellipticity of 2, with $a = \lambda/2$, and $b = \lambda$. The
polarization of the laser is along the major axis [6].

of the aperture edge. Similar to the circular aperture diffraction patterns, the
elliptical aperture patterns in the aperture plane still exhibit radial intensity
fluctuation along the direction of the polarization of the incident plane wave.
In addition, the maxima and minima observed do not strictly follow the angular curve of the rim of the aperture, but rather tend to follow a linear pattern
parallel to the incident polarization angle. This tendency of the intensity maximum and minimum to follow the axis of polarization towards the middle of
the ellipse is clearly evident in the plots of Figures 8.20 and 8.21.

Figure 8.22 shows the effects of an initial polarization (which is not aligned
with the major axis) on the calculated diffraction patterns. Figure 8.22 is a
surface plot of the z-component of the Poynting vector for light incident upon

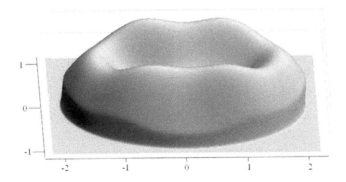

FIGURE 8.19
Intensity distribution of the normalized z-component of the Poynting vector
in the aperture plane for an ellipticity of 2, with $a = \lambda$, and $b = 2\lambda$. The
polarization of the laser is along the major axis [6].

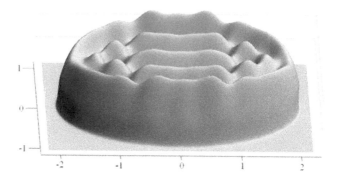

FIGURE 8.20
Intensity distribution of the normalized z-component of the Poynting vector in the aperture plane for an ellipticity of 2, with $a = 2.5\lambda$, and $b = 5\lambda$. The polarization of the laser is along the major axis [6].

the elliptical aperture with a polarization that has an angle of $\pi/4$ with respect to the major axis. The regions within the aperture near the areas where the rim is parallel to the incident polarization show strong intensity modulations, whereas regions near portions of the rim that are perpendicular to the incident polarization show weak intensity modulations, indicating that scattering of the incident field due to the effects of the aperture rim are a significant contributor to the presence of the observed intensity modulations within the aperture. Figure 8.23 is a surface plot for the diffraction of light incident upon the aperture with the initial polarization aligned with the minor axis. Throughout the middle, the intensity modulations tend to align themselves

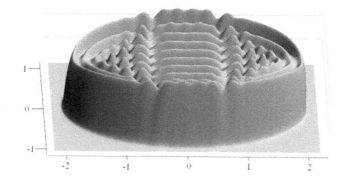

FIGURE 8.21
Intensity distribution of the normalized z-component of the Poynting vector in the aperture plane for an ellipticity of 2, with $a = 5\lambda$, and $b = 10\lambda$. The polarization of the laser is along the major axis [6].

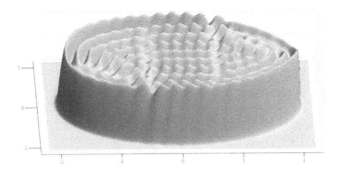

FIGURE 8.22
Intensity distribution of the normalized z-component of the Poynting vector in the aperture plane for an ellipticity of 2, with $a = 5\lambda$, and $b = 10\lambda$. The polarization of the laser is oriented at an angle of $\pi/4$ from the major axis towards the minor axis [6].

parallel to the incident polarization. The persistent alignment of the intensity modulations with the incident polarization indicate that the light scattered off of the aperture rim is interfering with the incident field as well as scattered waves from other points along the rim.

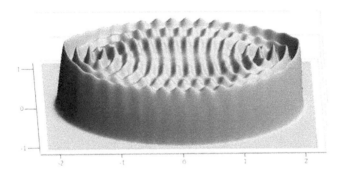

FIGURE 8.23
Intensity distribution of the normalized z-component of the Poynting vector in the aperture plane for an ellipticity of 2, with $a = 5\lambda$, and $b = 10\lambda$. The polarization of the laser is oriented along the minor axis [6].

8.5   Beam Distributions beyond the Aperture Plane for an Elliptical Aperture

The plots in Figures 8.24–8.27 investigate on-axis and radial intensity distributions for planes at locations beyond the aperture plane for an elliptical aperture with an ellipticity of 2 and a minor axis equal to 5 times the wavelength of the incident plane wave. Figure 8.24 is a plot of the on-axis value of the z-component of the Poynting vector. Overall, the on-axis distribution resembles what would be expected for the diffraction of light by a circular aperture with a radius equal to 5 times the wavelength of the incident plane wave where there would be 5 oscillations in the calculated on-axis intensity. The deviation from the expected on-axis intensity distribution for a circular aperture (Figure 8.10) for the elliptical aperture of Figure 8.24 mainly occurs in the range of approximately $0.07 < z_1 < 1$. For a circular aperture, the last (single and smooth) on-axis maximum will occur for $z_1 = 1/(2\pi)$, or 0.16. For the ellipse (with an ellipticity of 2) there are 8 changes in the sign of the second derivative of the on-axis intensity through this axial range, where there would be only 3 for a circular aperture. Figure 8.25 is a high-resolution intensity profile for a plane located at $z_1 = 0.054$, or the location of the fourth maximum from the aperture. Figure 8.26 is an intensity profile for the on-axis minimum located at $z_1 = 0.09$, and Figure 8.27 is for the plane located at $z_1 = 0.45$, or the last on-axis maximum. Strong radial asymmetries and localized field spikes exist throughout the near-field region ($z_1 < 1$) for diffraction by elliptical apertures. Beyond an axial distance of $z_1 = 1$, the intensity dis-

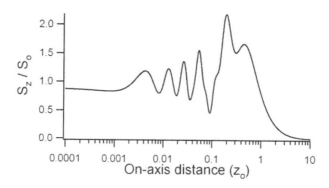

FIGURE 8.24

Calculated normalized intensity of the z-component of the Poynting vector versus on-axis distance from the aperture plane for $a = 5\lambda$ and $b = 10\lambda$, and the variable z_o is defined to be $z_o = 2\pi a^2/\lambda$ [6].

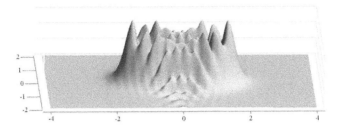

FIGURE 8.25
Calculated normalized intensity distribution of the z-component of the Poynting vector at a plane located at $z = 0.054z_o$, for $a = 5\lambda$ and $b = 10\lambda$ [6].

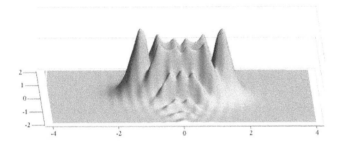

FIGURE 8.26
Calculated normalized intensity distribution of the z-component of the Poynting vector at a plane located at $z = 0.09z_o$, for $a = 5\lambda$ and $b = 10\lambda$ [6].

tributions radially smooth out with a central maximum surrounded by subtle ripples or wings.

8.6 Beam Distributions in the Aperture Plane for a Square Aperture

In Sections 8.6 and 8.7 we present light intensity distributions for a plane wave incident upon a square diffracting aperture. Figure 8.28 is a plot of the z-component of the Poynting vector in the aperture plane for a plane wave incident upon a square aperture with a side width equal to the wavelength of light, and the light polarized parallel to the x-axis. The width of the square apertures in this section are expressed in terms of the variable a, which represents the half-width of one of the sides of the square. The half-width is chosen as the size parameter of the square apertures to allow qualitative comparison

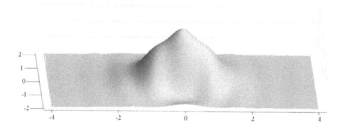

FIGURE 8.27
Calculated normalized intensity distribution of the z-component of the Poynting vector at a plane located at $z = 0.46z_o$, for $a = 5\lambda$ and $b = 10\lambda$ [6].

to the plots illustrated in the previous section where the sizing parameter of circular apertures is expressed in terms of the radius of the circular aperture. Similar to the intensity profiles in the aperture plane for circular apertures, the intensity distributions within an aperture with dimensions equal to, or smaller than, the wavelength of light are composed of a single central maxima that smoothly decreases in amplitude as the edges of the aperture are approached.

Figure 8.29 is a similar plot for a square aperture whose width is equal to two wavelengths of light. Similar to the plots for a circular aperture, if the

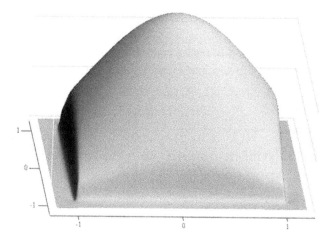

FIGURE 8.28
Intensity distribution of the normalized z-component of the Poynting vector in the aperture plane for a square aperture with $a = \lambda/2$, where a is the half-width of one of the sides.

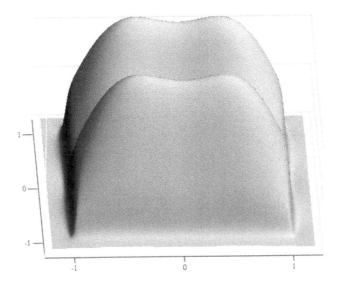

FIGURE 8.29
Intensity distribution of the normalized z-component of the Poynting vector in the aperture plane for a square aperture with $a = \lambda$.

half-width of the square is equal to an integer multiple of the wavelength of light, there exists a minimum at the center of the aperture. Also, in moving away from the center of the aperture in a direction perpendicular to the incident polarization the number of oscillations observed in the intensity profile is equal to the parameter a/λ. The general orientations of the maxima and minima within the aperture plane align themselves with the direction of the incident polarization as they do for the central regions of an elliptical aperture.

Figure 8.30 is an intensity distribution plot for light incident upon a square aperture having a side half-width of one wavelength with the light polarization oriented along the diagonal between the x and y-axes. Four maxima are observed to occur along the diagonal lines of the aperture with a single central minimum. The maxima along the diagonal parallel to the incident polarization are located slightly farther away from the center than the maxima along the diagonal perpendicular to the incident polarization and have a smaller overall amplitude. Figures 8.31 and 8.32 are plots similar to Figure 8.29 where the half-width of the square aperture is equal to 2.5λ.

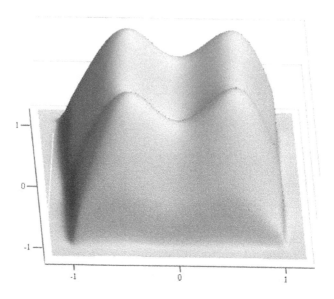

FIGURE 8.30
Intensity distribution of the normalized z-component of the Poynting vector in the aperture plane for a square aperture with $a = \lambda$, and the incident polarization oriented at $45°$.

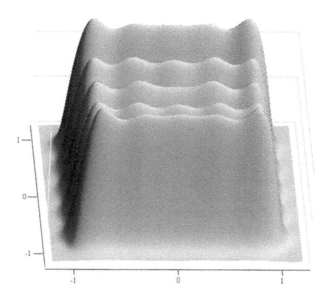

FIGURE 8.31
Intensity distribution of the normalized z-component of the Poynting vector in the aperture plane for a square aperture with $a = 2.5\lambda$.

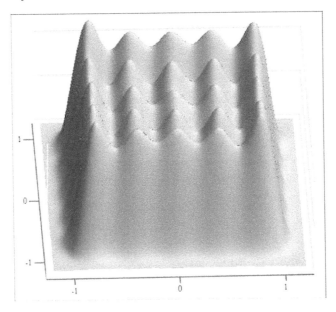

FIGURE 8.32
Intensity distribution of the normalized z-component of the Poynting vector in the aperture plane for a square aperture with $a = 2.5\lambda$, and the incident polarization oriented at $45°$.

8.7 Beam Distributions beyond the Aperture Plane for a Square Aperture

Figure 8.33 is a plot of the z-component of the Poynting vector for on-axis points beyond a square aperture having a side half-width of $a = 5\lambda$. As is the case with a circular aperture, the number of oscillations in the axial direction corresponds to the ratio of a/λ. The amplitude of the oscillations for the circular aperture are larger than those observed for the square aperture due to the differences in the symmetry of the aperture shapes. For the circular case, the equal radial symmetry allows for greater constructive and destructive interference between radially symmetric points in the aperture plane. For the square aperture, the radial symmetry is broken for radial distances from the center of the aperture greater than the half-width of the sides.

Figure 8.34 is an intensity distribution plot for the square aperture $(a/\lambda = 5)$ at the axial location of the second to last on-axis maximum, or $z = 0.053z_o$. Strong radial asymmetries are observed with multiple maxima and minima located throughout the plane. Figures 8.35 and 8.36 are similar

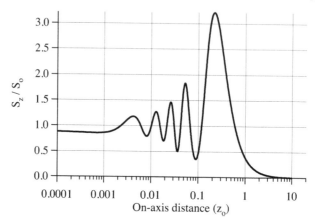

FIGURE 8.33
Calculated normalized z-component of the Poynting vector for a square aperture with $a = 5\lambda$ versus distance from the aperture for on-axis locations.

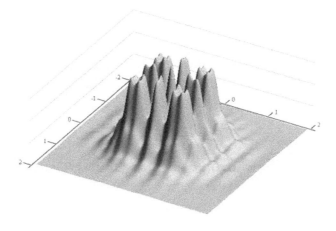

FIGURE 8.34
Intensity distribution of the normalized z-component of the Poynting vector for a square aperture with $a = 5\lambda$ in the plane of $z = 0.053z_o$ (or the second to last on-axis maximum).

intensity distribution plots for the axial location of the last on-axis minimum, or $z = 0.089z_o$, and the last on-axis maximum, $z = 1/(2\pi)$, respectively. Figure 8.36 resembles the traditionally expected diffraction pattern using Fresnel and Fraunhofer diffraction theory.

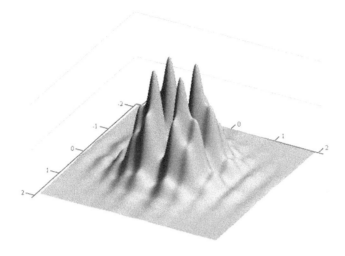

FIGURE 8.35
Intensity distribution of the normalized z-component of the Poynting vector
for a square aperture with $a = 5\lambda$ in the plane of $z = 0.089z_o$ (or the last
on-axis minimum).

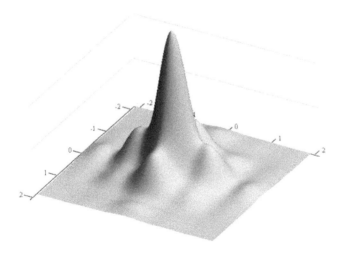

FIGURE 8.36
Intensity distribution of the normalized z-component of the Poynting vector
for a square aperture with $a = 5\lambda$ in the plane of $z = 0.22z_o$ (or the last
on-axis maximum).

Bibliography

[1] S. Guha and G. D. Gillen, "Description of light propagation through a circular aperture using nonparaxial vector diffraction theory," Opt. Express **26**, 1424–1447 (2005).

[2] C. L. Andrews, "Diffraction pattern in a circular aperture measured in the microwave region," J. Appl. Phys. **21**, 761–767 (1950).

[3] M. J. Ehrlich, S. Silver, and G. Held, "Studies of the Diffraction of Electromagnetic waves by circular apertures and complementary obstacles: the near-zone field," J. Appl. Phys. **26**, 336–345 (1955).

[4] G. Bekefi, "Diffraction of electromagnetic waves by an aperture in a large screen," J. App. Phys. **24**, 1123–1130 (1953).

[5] M. Mansuripur, A. R. Zakharian, and J. V. Moloney, "Interaction of light with subwavelength structures," Opt. Photonics News, 56–61 (March 2003).

[6] G. D. Gillen and S. Guha, "Vector diffraction theory of light propagation through nanostructures," Proc. SPIE **5931**, 59310W (2005).

9

Vector Diffraction across a Curved Interface

In this chapter we present mathematical models and calculations for the vector electromagnetic fields for points beyond a spherical boundary between two different linear optical media using vector diffraction theory. The assumptions of the models presented are:

- each of the two media are linear; i.e., the electric permittivity, ϵ, and magnetic permeability, μ, are both constants;

- the spherical boundary between the media is azimuthally symmetric, and spatially limited by a maximum polar angle;

Due to the fact that the interface is spatially limited, diffraction effects play a crucial role in the propagation. Thus, vector diffraction theories are utilized. We split this chapter up into two cases where the light is focused into the second medium:

- Case 1, the light travels through a *convex* boundary into a medium with a *higher* refractive index;

- Case 2, the light travels through a *concave* boundary into a medium with a *lower* refractive index.

9.1 Introduction

Rayleigh-Sommerfeld and scalar Kirchhoff beam propagation formalisms of Chapter 7 are convenient ways to describe the propagation of light from a plane surface (where the light fields are known) to a point of interest beyond that surface. These methods can easily be adapted to include curved optical elements placed along the axis of propagation if the optical elements are "thin" and paraxial conditions apply [1]. However, these beam propagation models cannot be used for high numerical aperture lenses or curved mirrors that are commercially available and used in a wide variety of optical applications. Additionally, the scalar nature of these beam propagation models do not account for the interdependence of the different components of the electromagnetic fields. To accurately model the propagation of light where the usual "thin lens" and paraxial approximations do not apply, a more rigorous approach is needed that incorporates the full vector nature of electromagnetic radiation.

Vector components of the electromagnetic fields of light in the focal region had been determined previously by Richards and Wolf [2] for the case of an "aplanatic," i.e., aberration free, converging beam. Visser and Wiersma [3] and Hsu and Barakat [4] have considered the case of aberrated incident phase fronts, but not the case of refraction between two media. In this chapter we present an extension of the vector diffraction theory of Stratton and Chu [5, 6]. This variation of vector diffraction theory applies the vector analog of Green's theorem to determine the electromagnetic fields at a point inside a closed volume given the fields on the curved surface.

The case of a linearly polarized plane wave of light traveling in one transparent medium and incident on a spherical surface separating the medium from a second one having a different refractive index is considered here. It is assumed that the values of the fields incident on the surface are known. For highly curved surfaces, the angular dependence of the Fresnel transmission coefficients must be taken into consideration as the angle of incidence along the surface can also become large. Using the Fresnel transmission factors and angular-dependent refraction effects, the values of the fields "just inside" the second medium are determined and the vector diffraction integrals are then used to find the values of the fields at any point in the second medium in terms of integrals involving the values of the fields at the surface. The onset of spherical aberration is investigated along with the effects of spherical aberration on the maximum obtainable intensity. There are four main differences of the models presented in this chapter from previous treatments of vector diffraction theory for focused light:

1. the theory and results are expressed in dimensionless units so as to be applicable to a wide range of laser and optical parameters,

2. the *paraxial-type* of approximations used in Eq. (15) of Hsu and Barakat [4] are not invoked,

3. detailed distributions of the longitudinal field components are determined for both the aberrated and unaberrated cases, and

4. the relative contributions of the three different terms of the Stratton-Chu diffraction integral are investigated.

Throughout this chapter it is assumed that the incident light is either a true plane wave, or can be approximated as a plane wave (as discussed in Section 2.4).

9.2 Theoretical Setup, Case 1 vs. Case 2

Within this chapter we consider two different cases for light propagating across a spherical boundary. For both cases, the light is focused as a result of crossing the interface between two different linear optical media.

9.2.1 Case 1, Light Propagation across a Convex Boundary into an Optically Thicker Medium

Case 1 is illustrated in Figure 9.1. The theoretical setup is one where the light is propagating from a medium with refractive index n_1 to a medium with higher refractive index n_2, i.e.,

$$n_2 > n_1. \tag{9.1}$$

Focusing of the electromagnetic waves occurs when the spherical boundary between the two media is ***convex*** from the point of view of the light propagation direction, as shown in Figure 9.1.

9.2.2 Case 2, Light Propagation across a Concave Boundary into an Optically Thinner Medium

Case 2 is illustrated in Figure 9.2. Similar to the theoretical setup for Case 1, Case 2 is of interest because the electromagnetic waves are focused as a result

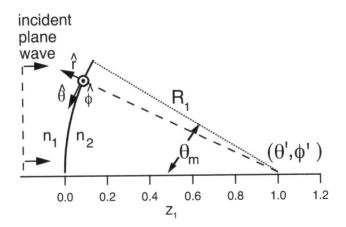

FIGURE 9.1
Illustration of the spherical boundary between media 1 and 2, the radius of curvature, R_1, the maximum angle of the surface, θ_m, and the polar angles to the surface source point, θ' and ϕ', and the unit vectors $\hat{r}$, $\hat{\theta}$, and $\hat{\phi}$ [7].

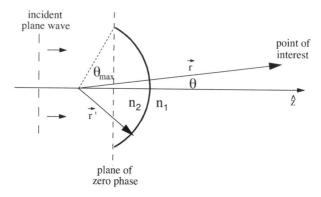

FIGURE 9.2
Theoretical setup for calculating light fields beyond the exit surface of a microlens. The plane wave is incident in the $+\hat{z}$ direction and refracted at the spherical exit surface with maximum half angle θ_{max}, and radius of curvature R [8].

of passing through the boundary between two different linear optical media. However, for Case 2, the light is propagating from a medium with refractive index n_2 into a medium with lower refractive index n_1, i.e.,

$$n_2 > n_1. \tag{9.2}$$

When passing from an optically denser medium to an optically thinner medium, focusing of the electromagnetic waves occurs when the spherical boundary between the two media is **concave** from the point of view of the light propagation direction. Similarly to our treatment for Case 1, we assume in this chapter that the light incident upon the spherical boundary is that of a plane wave.

9.3   Vector Diffraction Theory at a Spherical Surface, Case 1

Suppose n_1 and n_2 denote the refractive indices of the two media, λ the (vacuum) wavelength of a plane wave of light traveling in the first medium, and the light is incident upon a convex spherical surface having a radius of curvature of R_1, as shown in Figure 9.1. We designate the polarization direction of the wave to be the x direction and the direction of propagation to be the z direction of a rectangular Cartesian coordinate system. For convenience, spherical coordinates are also used here, with the origin of the spherical coordinate system on the z axis, located at the center of the spherical surface separating the two media, as shown in Figure 9.1. The coordinates of the source points, i.e., the points on the surface, are denoted by primed variables, and those of the field points, i.e., the points in the second medium, are denoted by unprimed variables. The time independent parts of the electric and magnetic fields incident upon the surface are given by

$$\mathbf{E}\left(\mathbf{r}'\right) = E_o e^{-ik_1 z'}\hat{i}, \tag{9.3}$$

and

$$\mathbf{H}\left(\mathbf{r}'\right) = H_o e^{-ik_1 z'}\hat{j}, \tag{9.4}$$

where $H_o = \dfrac{n_1 E_o}{Z_o}$, with $Z_o = \sqrt{\dfrac{\mu_o}{\epsilon_o}}$, and $k_1 = 2\pi n_1/\lambda$. Expressing the Cartesian unit vectors $\hat{i}, \hat{j}, \hat{k}$ in terms of the spherical coordinates r', θ', ϕ' and unit vectors $\hat{r}, \hat{\theta}, \hat{\phi}$ (also illustrated in Figure 9.1) Eqs. 9.3 and 9.4 can be re-written as

$$\mathbf{E}_-\left(\mathbf{r}'\right) = E_o e^{-ik_1 R_1\left(1+\cos\theta'\right)}$$
$$\times \left(\sin\theta'\cos\phi'\hat{r} + \cos\theta'\cos\phi'\hat{\theta} - \sin\phi'\hat{\phi}\right) \tag{9.5}$$

$$\mathbf{H}_-\left(\mathbf{r}'\right) = H_o e^{-ik_1 R_1\left(1+\cos\theta'\right)}$$
$$\times \left(\sin\theta'\sin\phi'\hat{r} + \cos\theta'\sin\phi'\hat{\theta} + \cos\phi'\hat{\phi}\right) \tag{9.6}$$

where the "$-$" subscripts of $\mathbf{E}$ and $\mathbf{H}$ represent the fields just outside the spherical surface, i.e., in the incident medium.

At every source point (say, denoted by R_1, θ', and ϕ'), the spherical surface may be considered locally to be a plane surface and the Fresnel transmission factors obtained for a plane wave incident upon a plane surface [9] can be applied. Since the propagation direction of the incident plane wave is in the z direction and the normal to the surface is along the radial ($\hat{r}$) direction, the plane of incidence is defined by the $\hat{r}$ and $\hat{k}$ vectors. By definition, the $\hat{\theta}$ vector lies in this plane and the $\hat{\phi}$ vector is perpendicular to this plane. Therefore, in Eq. 9.5, the $\hat{r}$ and $\hat{\theta}$ components are parallel to the plane of incidence (p-polarized) and the $\hat{\phi}$ component is perpendicular to the plane of incidence (s-polarized).

Applying the appropriate Fresnel transmission coefficients from Ref. [9], and using the appropriate refraction effects on the plane wave's s-polarization and p-polarization components, the transmitted fields can be written as

$$\mathbf{E}_+\left(\mathbf{r}'\right) = E_o e^{-ik_1 R_1\left(1+\cos\theta'\right)}$$
$$\times \left(T_p \sin\theta_t'\cos\phi'\hat{r} + T_p \cos\theta_t'\cos\phi'\hat{\theta} - T_s \sin\phi'\hat{\phi}\right), \tag{9.7}$$

and

$$\mathbf{H}_+ \left(\mathbf{r}\, ' \right) = \frac{n_2}{Z_o} E_o e^{-ik_1 R_1 \left(1 + \cos \theta' \right)}$$
$$\times \left(T_s \sin \theta_t' \sin \phi' \hat{r} + T_s \cos \theta_t' \sin \phi' \hat{\theta} + T_p \cos \phi' \hat{\phi} \right), \tag{9.8}$$

where the "+" subscripts denote the fields just inside the second medium (i.e., just after the light crosses the spherical boundary.) The Fresnel transmission coefficients in Eqs. 9.7 and 9.8 are

$$T_p = \frac{2n_1 \cos \theta'}{n_1 \cos \theta' + n_2 \cos \theta_t'}, \tag{9.9}$$

$$T_s = \frac{2n_1 \cos \theta'}{n_2 \cos \theta' + n_1 \cos \theta_t'}, \tag{9.10}$$

for the p and s-polarizations, respectively, and θ_t' is the local angle of refraction into the transmitted medium, given by Snell's Law; i.e.,

$$\theta_t' = \sin^{-1} \left(\frac{n_1}{n_2} \sin \theta' \right). \tag{9.11}$$

Care needs to be taken in evaluating the inverse sine in Eq. 9.11 to ensure that θ_t' is obtuse if the incident angle θ' is obtuse.

According to Huygen's principle, the fields in a bound, source-free region of space can be thought of as arising from spherical waves emanated by every point on the surface bounding that space. A generalization of this principle to the vector case is given in Eq. 22 in Stratton [6]. For the choice of the time-independent component of the electromagnetic fields assumed in the treatment here, the "vector Huygen's principle" formulae in Rothwell and Cloud [10] are appropriate:

$$\mathbf{E} \left(\mathbf{r} \right) = - \int_S da' \left[-i\omega\mu \left(\hat{n}' \times \mathbf{H}_+ \right) G \right.$$
$$\left. + \left(\hat{n}' \times \mathbf{E}_+ \right) \times \boldsymbol{\nabla}' G + \left(\hat{n}' \cdot \mathbf{E}_+ \right) \boldsymbol{\nabla}' G \right], \tag{9.12}$$

and

$$\mathbf{H} \left(\mathbf{r} \right) = - \int_S da' \left[i\omega\epsilon \left(\hat{n}' \times \mathbf{E}_+ \right) G \right.$$
$$\left. + \left(\hat{n}' \times \mathbf{H}_+ \right) \times \boldsymbol{\nabla}' G + \left(\hat{n}' \cdot \mathbf{H}_+ \right) \boldsymbol{\nabla}' G \right], \tag{9.13}$$

where the integrations extend over the surface S enclosing the field point $\mathbf{r}$ and the values of $\mathbf{E}_+$ and $\mathbf{H}_+$ in the integrands are those evaluated just inside the second medium and given by Eqs. 9.7 and 9.8. The Green's function, G, used in Eqs. 9.12 and 9.13 is given by

$$G = \frac{e^{-ik_2 \rho}}{4\pi \rho}, \tag{9.14}$$

where $k_2 = 2\pi n_2/\lambda$, and

$$
\begin{aligned}
\rho^2 &= \left| \mathbf{r}\,(x, y, z) - \mathbf{r}\,'\,(x', y', z')\,\right|^2 \\
&= \left(x - x'\right)^2 + \left(y - y'\right)^2 + \left(z - z'\right)^2,
\end{aligned} \tag{9.15}
$$

with $\mathbf{r}\,'$ denoting the position vector of the source point, and $\mathbf{r}$ denoting the position vector of the field point. The components of $\mathbf{r}$ and $\mathbf{r}\,'$ can be expressed in terms of the spherical coordinates as:

$$
\begin{aligned}
x &= r \sin\theta \cos\phi \\
y &= r \sin\theta \sin\phi \\
z &= r \cos\theta
\end{aligned} \tag{9.16}
$$

and

$$
\begin{aligned}
x' &= R_1 \sin\theta' \cos\phi' \\
y' &= R_1 \sin\theta' \sin\phi' \\
z' &= R_1 \cos\theta',
\end{aligned} \tag{9.17}
$$

where r is the distance of the field point from the origin.

9.4 Normalization and Simplification, Case 1

Dimensionless parameters, dimensionless functions, and normalized distances are used in this treatment for mathematical convenience, and to allow the treatment to be applicable to a variety of chosen wavelengths and radii of curvature. The normalized distances are defined as

$$
x_1 \equiv \frac{x}{R_1}, \qquad y_1 \equiv \frac{y}{R_1}, \qquad z_1 \equiv \frac{z}{R_1}, \tag{9.18}
$$

$$
x_1' \equiv \frac{x'}{R_1}, \qquad y_1' \equiv \frac{y'}{R_1}, \qquad z_1' \equiv \frac{z'}{R_1}, \tag{9.19}
$$

and

$$
\rho_1 \equiv \frac{\rho}{R_1}. \tag{9.20}
$$

Using a dimensionless parameter and a dimensionless function

$$
p_2 \equiv \frac{2\pi n_2 R_1}{\lambda}, \qquad G_1 \equiv 4\pi R_1 G = \frac{e^{-ip_2\rho_1}}{\rho_1}, \tag{9.21}
$$

we obtain

$$
\nabla' G = \frac{ip_2 G_2}{4\pi R_1^2} \mathbf{r}_{11}, \tag{9.22}
$$

where

$$\mathbf{r}_{11} \equiv (x_1 - x_1')\,\hat{i} + (y_1 - y_1')\,\hat{j} + (z_1 - z_1')\,\hat{k} \tag{9.23}$$

and

$$G_2 \equiv -\frac{1}{ip_2\rho_1}\frac{\partial G_1}{\partial \rho_1} = \frac{G_1}{\rho_1}\left(1 + \frac{1}{ip_2\rho_1}\right). \tag{9.24}$$

We also further define

$$n \equiv \frac{n_2}{n_1} \quad , \quad t_1 \equiv \frac{T_s}{T_p} = \frac{\cos\theta' + n\cos\theta_t'}{\cos\theta_t' + n\cos\theta'} \tag{9.25}$$

as the "normalized" refractive index and transmission coefficient and

$$f_2 \equiv \exp\left[-ik_1R_1\left(1 + \cos\theta'\right)\right] = \exp\left[-ip_2\left(1 + \cos\theta'\right)/n\right] \tag{9.26}$$

as the phase of the fields just outside the spherical interface between the media. The fields in Eqs. 9.7 and 9.8 may now be rewritten as:

$$\mathbf{E}_+\left(\mathbf{r}\right) = E_o f_2 T_p \mathbf{e}_1\left(\mathbf{r}\right), \tag{9.27}$$

$$\mathbf{H}_+\left(\mathbf{r}\right) = \frac{n_2}{Z_o} E_o f_2 T_p \mathbf{h}_1\left(\mathbf{r}\right), \tag{9.28}$$

where

$$\mathbf{e}_1\left(\mathbf{r}\,'\right) \equiv e_r\hat{r} + e_\theta\hat{\theta} + e_\phi\hat{\phi}, \tag{9.29}$$

$$\mathbf{h}_1\left(\mathbf{r}\,'\right) \equiv h_r\hat{r} + h_\theta\hat{\theta} + h_\phi\hat{\phi}, \tag{9.30}$$

with

$$\begin{aligned} e_r &= \sin\theta_t{}'\cos\phi' \\ e_\theta &= \cos\theta_t{}'\cos\phi' \\ e_\phi &= -t_1\sin\phi' \end{aligned} \tag{9.31}$$

and

$$\begin{aligned} h_r &= t_1\sin\theta_t{}'\sin\phi' \\ h_\theta &= t_1\cos\theta_t{}'\sin\phi' \\ h_\phi &= \cos\phi'. \end{aligned} \tag{9.32}$$

Substituting Eqs. 9.22, 9.27, and 9.28 into Eqs. 9.12 and 9.13 and using the relationships

$$\frac{i\omega\mu G n_2}{Z_o}E_o = \frac{ip_2 G_1}{4\pi R_1^2}E_o \tag{9.33}$$

$$i\omega\epsilon G E_o = \frac{ip_2 G_1}{4\pi R_1^2 Z_o}n_2 E_o \tag{9.34}$$

we obtain

$$\mathbf{E}\left(\mathbf{r}\right) = \frac{ip_2 E_o}{4\pi R_1^2} \int da' T_p \left[G_1 f_2 \left(\hat{r} \times \mathbf{h}_1\right)\right.$$

$$\left. -G_2 f_2 \left(\hat{r} \times \mathbf{e}_1\right) \times \mathbf{r}_{11} - G_2 f_2 \left(\hat{r} \cdot \mathbf{e}_1\right) \mathbf{r}_{11}\right] \tag{9.35}$$

and

$$\mathbf{H}\left(\mathbf{r}\right) = \frac{ip_2 n_2 Z_o E_o}{4\pi R_1^2} \int da' T_p \left[G_1 f_2 \left(\hat{r} \times \mathbf{e}_1\right)\right.$$

$$\left. +G_2 f_2 \left(\hat{r} \times \mathbf{h}_1\right) \times \mathbf{r}_{11} + G_2 f_2 \left(\hat{r} \cdot \mathbf{h}_1\right) \mathbf{r}_{11}\right]. \tag{9.36}$$

Since the surface element is $da' = R_1^2 \sin\theta' d\theta' d\phi'$, the expressions in Eqs. 9.35 and 9.36 can be simplified to

$$E\left(\mathbf{r}\right) \equiv A\mathbf{U}\left(\mathbf{r}\right), \tag{9.37}$$

where

$$\mathbf{U}\left(\mathbf{r}\right) = \int_0^{2\pi} \int_\pi^{\pi-\theta_m} \sin\theta' d\theta' d\phi' \left[\mathbf{u}_1 + \mathbf{u}_2 + \mathbf{u}_3\right] \tag{9.38}$$

and

$$H\left(\mathbf{r}\right) \equiv A\frac{n_2}{Z_o}\mathbf{V}\left(\mathbf{r}\right), \tag{9.39}$$

where

$$\mathbf{V}\left(\mathbf{r}\right) = \int_0^{2\pi} \int_\pi^{\pi-\theta_m} \sin\theta' d\theta' d\phi' \left[\mathbf{v}_1 + \mathbf{v}_2 + \mathbf{v}_3\right] \tag{9.40}$$

with

$$A = \frac{-ip_2}{4\pi} E_o. \tag{9.41}$$

θ_m is the half-angle aperture, and the vectors $\mathbf{u}_1$, $\mathbf{u}_2$, $\mathbf{u}_3$, $\mathbf{v}_1$, $\mathbf{v}_2$, and $\mathbf{v}_3$ represent the three vectors of the three individual terms of $\mathbf{E}$ and $\mathbf{H}$ in Eqs. 9.12 and 9.13, respectively. Each of these vectors have the following components:

$$\begin{aligned} u_{1x} &= T_p f_2 G_1 \left(\cos\theta' \cos^2\phi' + t_1 \cos\theta_t' \sin^2\phi'\right) \\ u_{1y} &= T_p f_2 G_1 \sin\phi' \cos\phi' \left(\cos\theta' - t_1 \cos\theta_t'\right) \\ u_{1z} &= -T_p f_2 G_1 \sin\theta' \cos\phi', \end{aligned} \tag{9.42}$$

$$\begin{aligned} u_{2x} &= T_p f_2 G_2 \left[\left(z_1 - z_1'\right)\alpha_{1y} - \left(y_1 - y_1'\right)\alpha_{1z}\right] \\ u_{2y} &= T_p f_2 G_2 \left[\left(x_1 - x_1'\right)\alpha_{1z} - \left(z_1 - z_1'\right)\alpha_{1x}\right] \\ u_{2z} &= -T_p f_2 G_2 \left[\left(y_1 - y_1'\right)\alpha_{1x} + \left(x_1 - x_1'\right)\alpha_{1y}\right], \end{aligned} \tag{9.43}$$

$$u_{3x} = T_p f_2 G_2 \sin\theta_t' \cos\phi' (x_1 - x_1')$$
$$u_{3y} = T_p f_2 G_2 \sin\theta_t' \cos\phi' (y_1 - y_1') \qquad (9.44)$$
$$u_{3z} = T_p f_2 G_2 \sin\theta_t' \cos\phi' (z_1 - z_1'),$$

$$v_{1x} = T_p f_2 G_1 \sin\phi' \cos\phi' (t_1 \cos\theta' - \cos\theta_t')$$
$$v_{1y} = T_p f_2 G_1 \left(t_1 \cos\theta' \sin^2\phi' + \cos\theta_t' \cos^2\phi'\right) \qquad (9.45)$$
$$v_{1z} = T_p f_2 G_1 \left(t_1 \sin\theta' \sin\phi'\right),$$

$$v_{2x} = T_p f_2 G_2 \left[(z_1 - z_1')\beta_{1y} - (y_1 - y_1')\beta_{1z}\right]$$
$$v_{2y} = T_p f_2 G_2 \left[(x_1 - x_1')\beta_{1z} - (z_1 - z_1')\beta_{1x}\right] \qquad (9.46)$$
$$v_{2z} = T_p f_2 G_2 \left[(y_1 - y_1')\beta_{1x} - (x_1 - x_1')\beta_{1y}\right],$$

and

$$v_{3x} = T_p f_2 G_2 t_1 \sin\theta_t' \sin\phi' (x_1 - x_1')$$
$$v_{3y} = T_p f_2 G_2 t_1 \sin\theta_t' \sin\phi' (y_1 - y_1') \qquad (9.47)$$
$$v_{3z} = T_p f_2 G_2 t_1 \sin\theta_t' \sin\phi' (z_1 - z_1').$$

$\boldsymbol{\alpha}_1$ and $\boldsymbol{\beta}_1$, respectively, denote the $\hat{r} \times \mathbf{e}_1$ and $\hat{r} \times \mathbf{h}_1$ terms occurring in Eqs. 9.35 and 9.36. The components of $\boldsymbol{\alpha}_1$ and $\boldsymbol{\beta}_1$ are given by:

$$\alpha_{1x} = \sin\phi' \cos\phi' (t_1 \cos\theta' - \cos\theta_t')$$
$$\alpha_{1y} = \cos\theta_t' \cos^2\phi' + t_1 \cos\theta' \sin^2\phi' \qquad (9.48)$$
$$\alpha_{1z} = -t_1 \sin\theta' \sin\phi',$$

and

$$\beta_{1x} = -\cos\theta' \cos^2\phi' - t_1 \cos\theta_t' \sin^2\phi'$$
$$\beta_{1y} = \sin\phi' \cos\phi' (t_1 \cos\theta_t' - \cos\theta') \qquad (9.49)$$
$$\beta_{1z} = \sin\theta' \cos\phi'.$$

The coordinates x_1, y_1, z_1, x_1', y_1', and z_1' can be expressed in terms of the spherical coordinates as

$$x_1 = r_1 \sin\theta \cos\phi$$
$$y_1 = r_1 \sin\theta \sin\phi \qquad (9.50)$$
$$z_1 = r_1 \cos\theta,$$

and

$$x_1' = \sin\theta' \cos\phi'$$
$$y_1' = \sin\theta' \sin\phi' \qquad (9.51)$$
$$z_1' = \cos\theta',$$

where $r_1 \equiv r/R_1$.

The Poynting vector of the wave in the second medium can be written as

$$
\begin{aligned}
\mathbf{S}_2 &= Re\,(\mathbf{E} \times \mathbf{H}^\star) \\
&= \frac{n_2}{Z_o}|A|^2 Re\,(\mathbf{U} \times \mathbf{V}^\star),
\end{aligned}
\tag{9.52}
$$

while the Poynting vector associated with the plane wave traveling in the first medium is given by

$$
\begin{aligned}
\mathbf{S}_1 &= \frac{n_1}{Z_o}|E_o|^2 \hat{z} \\
&= S_1 \hat{z}.
\end{aligned}
\tag{9.53}
$$

9.5 Calculation of Electromagnetic Fields and Poynting Vectors, Case 1

In Figures 9.3 and 9.4, the maximum values of the z-component of $\mathbf{S}_2$, normalized to $\mathbf{S}_1$, are plotted against the half-angular aperture, i.e., θ_m. The maximization of S_{2z} is done with respect to the x_1, y_1, and z_1 coordinates for each value of θ_m. S_{2z} maximizes on-axis, i.e., at $x_1 = y_1 = 0$, but not necessarily at the same values of z_1 for different θ_m. The dashed lines in Figures 9.3 and 9.4 show the values of the on-axis maximum irradiance obtained using the paraxial approximation, which is given by

$$
I_p = \left(\frac{2}{1+n}\right)^2 \left(\frac{p_2 \sin\theta_m}{2 f_1}\right)^2
\tag{9.54}
$$

where

$$
f_1 = \frac{n}{n-1},
\tag{9.55}
$$

and the on-axis maximum is obtained at $z_1 = f_1$. Eq. 9.54 can be derived from the treatment in Ref. [1].

In Figure 9.3, the value of p_2 chosen is 10^6 for two choices of n. In Figure 9.4 the value of n is chosen to be 1.5 for two choices of p_2. It is seen that the on-axis irradiance maximizes at a particular value of θ_m, beyond which the effects of abberation become pronounced and the irradiance is limited to a lower value than that at the maximum. Thus, when a spherical surface is used for focusing light, for given values of the refractive indices, wavelength of light, and radius of curvature of the surface, there exists an optimum aperture angle at which the highest irradiance can be obtained by focusing. For $p_2 = 10^6$ and $n = 1.5$, the optimum value of θ_m is about 8.6°. If the radius of curvature is reduced by a factor of 2, i.e. for $p_2 = 5 \times 10^5$ and $n = 1.5$, the optimum value of θ_m is about 10°. In Figure 9.5, the dependence of the (normalized) on-axis

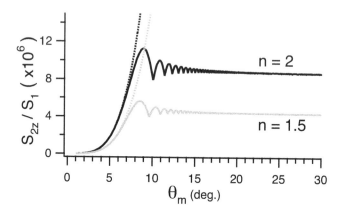

FIGURE 9.3
The maximum values of the normalized on-axis irradiance versus the half angular aperture, θ_m, for p_2 equal to 1×10^6. The upper and lower plots are for $n = 1.5$ and 2, respectively [7]. The dashed lines are calculated using the paraxial approximation, Eq. 9.54.

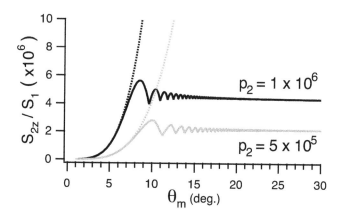

FIGURE 9.4
The maximum values of the normalized on-axis irradiance for $n = 1.5$. The upper plots are for a p_2 value of 1×10^6, and the lower plots are for a p_2 value of 5×10^5 [7]. The dashed lines are the paraxial values calculated using Eq. 9.54.

irradiance is plotted against the distance from the vertex for two values of θ_m. At $\theta_m = 5°$, the wave focuses at a normalized distance of 3, in accordance with the paraxial result in Eq. 9.55 as well as that obtained from the ray tracing [11]. At $\theta_m = 20°$ however, the on-axis irradiance is not as sharply

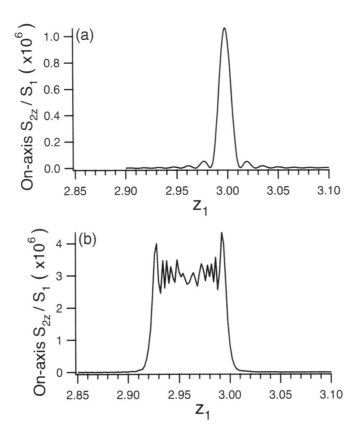

FIGURE 9.5
The calculated on-axis irradiance versus axial position for (a) $\theta_m = 5°$ and (b) $\theta_m = 20°$, with p_2 value of 1×10^6, and $n = 1.5$ [7].

focused, and the region of high irradiance extends over a larger axial region ($z_1 = 2.94$ to 3).

In Figure 9.6, the maximum values of the normalized energy density associated with the *longitudinal* field component E_z are plotted as a function of z_1. For each z_1 shown in Figure 9.6, the value of $|E_z|^2$ is maximized with respect to x_1 and y_1 variables. For $\theta_m = 5°$, it is found that $|E_z|^2$ maximizes at the axial position of the geometrical focus (i.e., at $z_1 \approx 3$). As the beam comes into focus, the longitudinal component of the Poynting vector increases. The value of the longitudinal component maximizes slightly off-axis, and is zero for all axial points by symmetry. For $\theta_m = 20°$, spherical aberration has become significant and $|E_z|^2$ maximizes over a much larger axial range, reaching a global maximum at a closer location of $z_1 \simeq 2.93$. Figure 9.6 also shows that at $\theta_m = 20°$, the normalized value of $|E_z|^2$ reaches a maximum value of ~ 4350, whereas the maximum value of S_z is about 4×10^6 for $p_2 = 10^6$, $n = 1.5$, and $z_1 = 2.927$.

The radial extents of the transverse and the longitudinal fields (and irradiances) are shown in Figures 9.7 and 9.8. These figures show that both $|E_x|^2$ and $|E_z|^2$ are mostly confined to within $x_1 \approx 10^{-4}$, although $|E_x|^2$ maximizes on-axis, whereas $|E_z|^2$ is zero for on-axis points and has its maximum value on the x-axis, i.e., $x_1 \approx 5 \times 10^{-5}$ and $y_1 = 0$.

To determine the relative contributions of the three terms in the integrands of Eqs. 9.38 and 9.40, which correspond to the three terms in the integrands of Eqs. 9.12 and 9.13, we show in Tables 9.1–9.4 the squares of the absolute values of all the Cartesian components of u_1, u_2, u_3, and v_1, v_2, v_3 maximized with respect to x_1, y_1, and z_1, for $p_2 = 10^6$, $n = 1.5$, and $\theta_m = 5°$ and $20°$. From the tables we find that the dominant contributions to the fields near the focus come from the field components parallel to the incident polarization directions (i.e., E_x and H_y). However, the components perpendicular to the incident fields (i.e., E_y, E_z, H_x, and H_z) also achieve non-negligible values in the region near the focus. We also find that for E_x and H_y, both of the first two terms (i.e., u_1 and u_2, v_1 and v_2) are important, while the contribution of the third term is small. However, for the longitudinal components, the third terms (u_3 and v_3) contribute dominantly.

Tables 9.1 and 9.2 provide the values of the modulus square of the three Cartesian components of each of the three terms for **E** and **H** in Eqs. 9.38 and 9.40, for $\theta_m = 5°$ and $20°$. The parameters chosen are: $p_2 = 10^6$ and $n = 1.5$. The values of the modulus squares of the individual components of the terms given in Tables 9.1 and 9.2 are much different from the values of the modulus square of the fields in Figures 9.3– 9.8 because the individual components interfere constructively or destructively. Tables 9.1 and 9.2 also illustrate the increases of each of the components of the three vector terms for **E** and **H** for larger collection angles. For the electric field, the increase of the x-component is disproportionately larger for the third integral term, whereas, the increase in the y component is disproportionately larger for the first and third integral term. For the magnetic field, the increase of the y-component

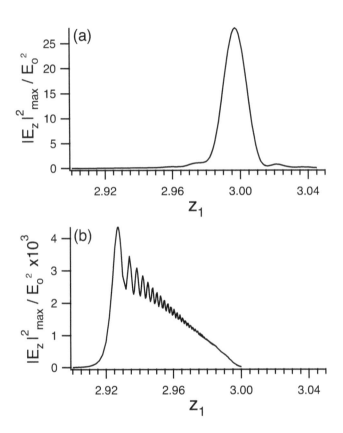

FIGURE 9.6
The maximum calculated value of the modulus square of the longitudinal
component of the electric field (E_z) as a function of axial position for (a)
$\theta_m = 5°$ and (b) $\theta_m = 20°$, with $n = 1.5$ and $p_2 = 10^6$. The value of $|E_z|^2/E_o^2$
displayed for each axial position is maximized with respect to x_1 and y_1 [7].

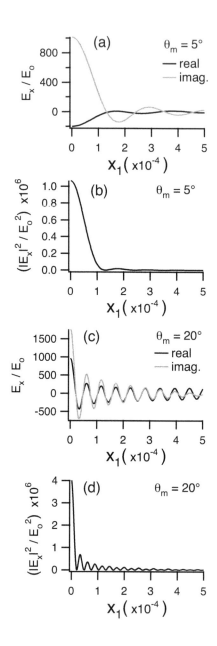

FIGURE 9.7

Plots of the real (black) and imaginary (grey) parts, and the modulus squared (black) of the x-component of the electric field versus x_1 for θ_m equal to ((a) and (b)) 5 degrees and ((c) and (d)) 20 degrees, respectively [7]. All plots are for $p_2 = 10^6$ $n = 1.5$, and $y_1 = 0$. z_1 for each plot is chosen such that the on-axis value of $|E_x|^2$ is maximum at that z_1; i.e., for $\theta_m = 5°$, $z_1 = 2.997$ and for $\theta_m = 20°$, $z_1 = 2.927$.

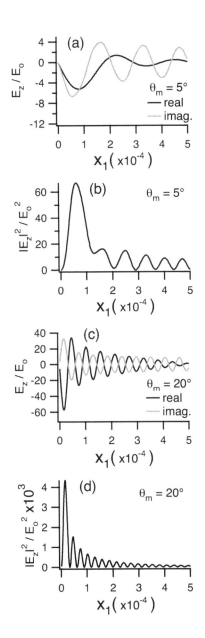

FIGURE 9.8
Plots of the real part (black) and imaginary part (grey) and modulus squared
(black) of the normalized longitudinal component, E_z/E_o, for ((a) and (b))
$\theta_m = 5°$ and ((c) and(d)) 20°, respectively [7]. For each plot $p_2 = 10^6$, $n = 1.5$
and $y_1 = 0$. In each graph, z_1 is chosen such that $|E_z|^2$ is maximum; i.e., for
$\theta_m = 5°$, $z_1 = 2.997$, and for $\theta_m = 20°$, $z_1 = 2.927$.

is disproportionately larger for the third integrand term, and the increase in the x component is disproportionately larger for the first and third integrand term. The increase in the longitudinal component of both the electric and the magnetic field is fairly even from all three integral terms. Thus, the effects of the increasing spherical aberration on the beam distributions in the focal region cannot be attributed to a single term in Eqs. 9.12 or 9.13, but rather manifest themselves from all three terms.

Tables 9.3 and 9.4 illustrate the locations of the global maxima of the Cartesian components of all three of the integral terms for $\mathbf{E}$ and $\mathbf{H}$ given in Tables 9.1 and 9.2.

9.6   Summary, Case 1

We have derived and evaluated the integrals describing the electromagnetic fields when a plane wave of light is refracted by a spherical surface. The integrals are expressed in terms of dimensionless variables and dimensionless physical parameters, so that the results are applicable over a wide range of wavelengths. The result allows the user to calculate the maximum values of irradiances achievable in the presence of spherical aberrations and also shows the distribution and the relative contributions of the longitudinal component of the fields.

9.7   Introduction, Case 2

As mentioned in Section 9.2, "Case 2" denotes the situation where light is propagating from a medium with a higher refractive index into a medium with a lower refractive index through a spherically concave boundary between the media. Ray-tracing and geometrical optics are traditionally used to locate the focal regions of lenses. Although these techniques are useful for quick calculations of focal distances and the magnitudes and regions of spherical aberration and coma, they are limited if the vector components of the light fields and/or detailed beam distributions are desired. Additionally, traditional geometrical optics fails when either the spherical collection angle is large, or the dimensions of the lens are not many orders of magnitude larger than the wavelength of the light. Examples of such conditions would be for low-numerical aperture lenses, and microlenses (which have both low numerical apertures and sizes on the millimeter or sub-millimeter scale). In this latter half of Chapter 9 we apply vector diffraction theory to the special case of a microlens in order to calculate light distributions and beam profiles in the

Component	Global Max. $\theta_m = 5°$	Global Max. $\theta_m = 20°$	Ratio
$\|u_{1x}\|^2$	1.14×10^5	3.76×10^5	3.31×10^0
$\|u_{2x}\|^2$	3.24×10^5	1.09×10^6	3.37×10^0
$\|u_{3x}\|^2$	7.19×10^{-2}	2.62×10^2	3.64×10^3
$\|u_{1y}\|^2$	4.01×10^{-3}	1.71×10^1	4.27×10^3
$\|u_{2y}\|^2$	5.89×10^2	4.12×10^3	7.00×10^0
$\|u_{3y}\|^2$	1.41×10^{-2}	6.19×10^1	4.38×10^3
$\|u_{1z}\|^2$	1.12×10^2	1.70×10^4	1.51×10^2
$\|u_{2z}\|^2$	4.40×10^1	6.99×10^3	1.59×10^2
$\|u_{3z}\|^2$	1.76×10^2	2.69×10^4	1.53×10^2

TABLE 9.1
The global maxima of the x, y, and z components of the three vector terms of the calculated electric fields for $p_2 = 10^6$ and $n = 1.5$, and for $\theta_m = 5°$ and $20°$ [7]. Ratio denotes the ratio of the global maximum of that component for $\theta = 20°$ to the global maximum value for $\theta_m = 5°$.

focal region. Using the radiation and vector diffraction theory of Stratton [6], the diffraction integrals are numerically evaluated using average desktop computers. The onset and effects of spherical aberration and coma on one and two-dimensional beam profiles is investigated for various microlens and light parameters.

9.8 Theoretical Setup, Case 2

In the following sections we explore the application of vector diffraction theory across a spherical surface to the propagation of light beyond a microlens. We

Component	Global Max. $\theta_m = 5°$	Global Max. $\theta_m = 20°$	Ratio
$|v_{1x}|^2$	4.57×10^{-2}	1.95×10^2	4.27×10^3
$|v_{2x}|^2$	9.36×10^3	5.59×10^4	5.97×10^0
$|v_{3x}|^2$	1.41×10^{-2}	6.19×10^1	4.38×10^3
$|v_{1y}|^2$	5.73×10^5	2.22×10^6	3.88×10^0
$|v_{2y}|^2$	3.23×10^5	1.31×10^6	4.05×10^0
$|v_{3y}|^2$	7.19×10^{-2}	2.62×10^2	3.64×10^3
$|v_{1z}|^2$	5.69×10^2	8.50×10^4	1.49×10^2
$|v_{2z}|^2$	2.21×10^1	3.55×10^3	1.61×10^2
$|v_{3z}|^2$	1.76×10^2	2.68×10^4	1.53×10^2

TABLE 9.2
The global maxima of the x, y, and z components of the three vector terms of
the calculated magnetic fields for $p_2 = 10^6$ and $n = 1.5$, and for $\theta_m = 5°$ and
$20°$ [7]. Ratio denotes the ratio of the global maximum of that component for
$\theta = 20°$ to the global maximum value for $\theta_m = 5°$.

will first begin with the theoretical setup of *normally incident* light as depicted
in Figure 9.9. Then we adapt the theoretical model for the case when the light
is *not* incident normal to the optical axis of the microlens, as illustrated in
Figure 9.10. Finally, we treat the situation where two beams are incident upon
the lens at different angles. Optical phenomena investigated are: on-axis light
distributions, spherical aberrations, off-axis focussing, and coma.

Component	Global Max. Location (x_1, y_1, z_1) $\theta_m = 5°$	Global Max. Location (x_1, y_1, z_1) $\theta_m = 20°$
$\|u_{1x}\|^2$	$(0, 0, 2.997)$	$(0, 0, 2.9717)$
$\|u_{2x}\|^2$	$(0, 0, 2.997)$	$(0, 0, 2.9717)$
$\|u_{3x}\|^2$	$(0, 0, 2.997)$	$(0, 0, 2.9272)$
$\|u_{1y}\|^2$	$(8.8 \times 10^{-5}, 8.6 \times 10^{-5}, 2.997)$	$(1.9 \times 10^{-5}, 1.9 \times 10^{-5}, 2.9273)$
$\|u_{2y}\|^2$	$(9.6 \times 10^{-5}, 9.5 \times 10^{-5}, 2.997)$	$(1.9 \times 10^{-5}, 1.9 \times 10^{-5}, 2.9274)$
$\|u_{3y}\|^2$	$(8.7 \times 10^{-5}, 8.8 \times 10^{-5}, 2.997)$	$(1.9 \times 10^{-5}, 1.9 \times 10^{-5}, 2.9273)$
$\|u_{1z}\|^2$	$(7.9 \times 10^{-5}, 0, 2.997)$	$(1.62 \times 10^{-5}, 0, 2.9273)$
$\|u_{2z}\|^2$	$(7.9 \times 10^{-5}, 0, 2.997)$	$(1.62 \times 10^{-5}, 0, 2.9272)$
$\|u_{3z}\|^2$	$(7.9 \times 10^{-5}, 0, 2.997)$	$(1.61 \times 10^{-5}, 0, 2.9272)$

TABLE 9.3
The normalized locations, (x_1, y_1, z_1), of the global maxima of the three vector terms of the calculated electric fields for $p_2 = 10^6$ and $n = 1.5$, and for $\theta_m = 5°$ and $20°$ [7].

9.9 Theory, Case 2

The model presented in this section is fundamentally similar to that of Section 9.3. However, there are subtle differences between the values of the refractive indices, location of the origin, the Fresnel coeficients, etc. Thus, for completeness of the model presented for Case 2, we present here an entire set of equations and parameters for light focusing via Case 2.

The vector electromagnetic fields within a given source-free volume of space can be determined from the vector analog of Green's second identity using Stratton and Chu's method [5]. The fields at point **r** can be expressed in terms of the field values on a surface, S, surrounding the point **r** using the

Component	Global Max. Location (x, y, z) $\theta_m = 5°$	Global Max. Location (x, y, z) $\theta_m = 20°$
$\lvert v_{1x} \rvert^2$	$\left(8.8 \times 10^{-5}, 8.8 \times 10^{-5}, 2.997\right)$	$\left(1.9 \times 10^{-5}, 1.9 \times 10^{-5}, 2.9272\right)$
$\lvert v_{2x} \rvert^2$	$\left(9.6 \times 10^{-5}, 9.6 \times 10^{-5}, 2.997\right)$	$\left(1.9 \times 10^{-5}, 1.9 \times 10^{-5}, 2.9272\right)$
$\lvert v_{3x} \rvert^2$	$\left(8.7 \times 10^{-5}, 8.8 \times 10^{-5}, 2.997\right)$	$\left(1.9 \times 10^{-5}, 1.9 \times 10^{-5}, 2.9272\right)$
$\lvert v_{1y} \rvert^2$	$(0, 0, 2.997)$	$(0, 0, 2.9272)$
$\lvert v_{2y} \rvert^2$	$(0, 0, 2.997)$	$(0, 0, 2.9272)$
$\lvert v_{3y} \rvert^2$	$(0, 0, 2.997)$	$(0, 0, 2.9272)$
$\lvert v_{1z} \rvert^2$	$\left(0, 8.0 \times 10^{-5}, 2.997\right)$	$\left(0, 1.62 \times 10^{-5}, 2.9275\right)$
$\lvert v_{2z} \rvert^2$	$\left(0, 7.5 \times 10^{-5}, 2.997\right)$	$\left(0, 1.52 \times 10^{-5}, 2.9275\right)$
$\lvert v_{3z} \rvert^2$	$\left(0, 8.0 \times 10^{-5}, 2.997\right)$	$\left(0, 1.62 \times 10^{-5}, 2.9275\right)$

TABLE 9.4
The normalized locations, (x_1, y_1, z_1), of the global maxima of the three vector terms of the calculated magnetic fields for $p_2 = 10^6$ and $n = 1.5$, and for $\theta_m = 5°$ and $20°$ [7].

formulae of Rothwell and Cloud [10] as

$$
\begin{aligned}
\mathbf{E}(\mathbf{r}) = - \int_S da' \, [&-i\omega\mu \, (\hat{n}' \times \mathbf{H}(\mathbf{r}'))\, G \\
&+ (\hat{n}' \times \mathbf{E}(\mathbf{r}')) \times \nabla'G + (\hat{n}' \cdot \mathbf{E}(\mathbf{r}'))\, \nabla'G] ,
\end{aligned}
\tag{9.56}
$$

$$
\begin{aligned}
\mathbf{H}(\mathbf{r}) = - \int_S da' \, [&i\omega\epsilon \, (\hat{n}' \times \mathbf{E}(\mathbf{r}'))\, G \\
&+ (\hat{n}' \times \mathbf{H}(\mathbf{r}')) \times \nabla'G + (\hat{n}' \cdot \mathbf{H}(\mathbf{r}'))\, \nabla'G] ,
\end{aligned}
\tag{9.57}
$$

where $\hat{n}'$ is the normal to the input surface pointing into the volume, $\mathbf{r}'$ denotes a point on the surface S, and da' is the surface area element pointing in the direction of $\hat{n}'$. The integrations extend over the input surface S and the values of $\mathbf{E}(\mathbf{r}')$ and $\mathbf{H}(\mathbf{r}')$ in the integrands are those evaluated *just inside* the volume of space after the input surface. The Green's function, G, used in Eqs. 9.56 and 9.57 is given by

$$G = \frac{e^{-ik_1\rho}}{4\pi\rho},$$
(9.58)

where $k_1 = 2\pi n_1/\lambda$, n_1 being the refractive index of the medium between $\mathbf{r}$ and $\mathbf{r}'$, and

$$\rho^2 = |\mathbf{r}(x,y,z) - \mathbf{r}'(x',y',z')|^2.$$
(9.59)

When the surface separating the two media is spherical in shape, it is convenient to use a spherical coordinate system to evaluate the field integrals. For an incident plane wave traveling parallel to the z-axis, the values of the electromagnetic field components $\mathbf{E}(\mathbf{r}')$ and $\mathbf{H}(\mathbf{r}')$ *just outside* of the spherical surface can be obtained from the surface curvature, refraction effects, and the appropriate Fresnel transmission coefficients. If E_o denotes the amplitude of the electric field of the incident plane wave, assumed to be polarized along the x-direction,

$$\begin{aligned}
\mathbf{E}(\mathbf{r}') &= E_o e^{-ip_2\cos\theta'} \left(T_p \sin\theta_t' \cos\phi'\hat{r} + T_p \cos\theta_t' \cos\phi'\hat{\theta} \right. \\
&\quad \left. - T_s \sin\phi'\hat{\phi} \right),
\end{aligned}$$
(9.60)

and

$$\begin{aligned}
\mathbf{H}(\mathbf{r}') &= \frac{n_1}{Z_o} E_o e^{-ip_2\cos\theta'} \left(T_s \sin\theta_t' \sin\phi'\hat{r} + T_s \cos\theta_t' \sin\phi'\hat{\theta} \right. \\
&\quad \left. + T_p \cos\phi'\hat{\phi} \right),
\end{aligned}$$
(9.61)

where $Z_o = \sqrt{\dfrac{\mu_o}{\epsilon_o}}$, $p_2 = 2\pi n_2 R/\lambda$, and R is the radius of curvature. θ' and ϕ' denote the angular coordinates of the point $\mathbf{r}'$. T_p is the Fresnel transmission function for p-polarized light,

$$T_p = \frac{2n_1 \cos\theta'}{n_1 \cos\theta' + n_2 \cos\theta_t'},$$
(9.62)

and T_s is the Fresnel transmission function for s-polarized light,

$$T_s = \frac{2n_1 \cos\theta'}{n_2 \cos\theta' + n_1 \cos\theta_t'}.$$
(9.63)

The angle of transmission, θ_t', is given through Snell's Law, or

$$\theta_t' = \sin^{-1}\left(\frac{n_2}{n_1}\sin\theta'\right).$$
(9.64)

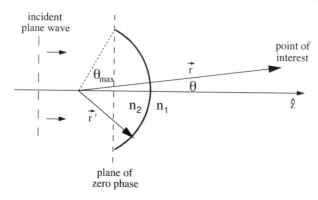

FIGURE 9.9
Theoretical setup for calculating light fields beyond the exit surface of a microlens [8]. The plane wave is incident in the $+\hat{z}$ direction and refracted at the spherical exit surface with maximum half angle θ_{max}, and radius of curvature R.

9.9.1 Normally Incident Light, Case 2

The theoretical setup for this investigation is illustrated in Figure 9.9. A plane wave is traveling within the microlens medium having a refractive index of n_2 and incident upon the spherical exit surface of the microlens. The curvature of the spherical exit surface is assumed to have a maximum half-angle of θ_{max}. Beyond the exit surface, the refractive index is assumed to be n_1, where $n_2 > n_1$.

For incident plane waves traveling parallel to the z-axis, the electric and magnetic fields can be expressed as

$$E\left(\mathbf{r}\right) \equiv A\mathbf{U}\left(\mathbf{r}\right), \tag{9.65}$$

and

$$H\left(\mathbf{r}\right) \equiv A\frac{n_1}{Z_o}\mathbf{V}\left(\mathbf{r}\right), \tag{9.66}$$

where

$$\mathbf{U}\left(\mathbf{r}\right) = \int_0^{2\pi} \int_0^{\theta_m} T_p \sin\theta' d\theta' d\phi' \left[\mathbf{u}_1 + \mathbf{u}_2 + \mathbf{u}_3\right] \tag{9.67}$$

and

$$\mathbf{V}\left(\mathbf{r}\right) = \int_0^{2\pi} \int_0^{\theta_m} T_p \sin\theta' d\theta' d\phi' \left[\mathbf{v}_1 + \mathbf{v}_2 + \mathbf{v}_3\right] \tag{9.68}$$

with

$$A = \frac{-ip_1}{4\pi}E_o, \tag{9.69}$$

where $p_1 = 2\pi n_1 R/\lambda$.

The angle θ_m is the maximum half-angle of the aperture, and $\mathbf{u}_1$, $\mathbf{u}_2$, $\mathbf{u}_3$, $\mathbf{v}_1$, $\mathbf{v}_2$, and $\mathbf{v}_3$ represent the three vectors of the three individual terms of $\mathbf{E}$ and $\mathbf{H}$ in Eqs. 9.56 and 9.57, respectively. Each of these vectors have the following components:

$$
\begin{aligned}
u_{1x} &= T_p f_2 G_1 \left(\cos\theta' \cos^2\phi' + t_1 \cos\theta'_t \sin^2\phi'\right) \\
u_{1y} &= T_p f_2 G_1 \sin\phi' \cos\phi' \left(\cos\theta' - t_1 \cos\theta'_t\right) \\
u_{1z} &= -T_p f_2 G_1 \sin^2\theta' \cos\phi',
\end{aligned}
\tag{9.70}
$$

$$
\begin{aligned}
u_{2x} &= T_p f_2 G_2 \left[(z_1 - z'_1)\,\alpha_{1y} - (y_1 - y'_1)\,\alpha_{1z}\right] \\
u_{2y} &= T_p f_2 G_2 \left[(x_1 - x'_1)\,\alpha_{1z} - (z_1 - z'_1)\,\alpha_{1x}\right] \\
u_{2z} &= T_p f_2 G_2 \left[(y_1 - y'_1)\,\alpha_{1x} - (x_1 - x'_1)\,\alpha_{1y}\right],
\end{aligned}
\tag{9.71}
$$

$$
\begin{aligned}
u_{3x} &= T_p f_2 G_2 \sin\theta'_t \cos\phi' \,(x_1 - x'_1) \\
u_{3y} &= T_p f_2 G_2 \sin\theta'_t \cos\phi' \,(y_1 - y'_1) \\
u_{3z} &= T_p f_2 G_2 \sin\theta'_t \cos\phi' \,(z_1 - z'_1),
\end{aligned}
\tag{9.72}
$$

$$
\begin{aligned}
v_{1x} &= T_p f_2 G_1 \sin\phi' \cos\phi' \,(t_1 \cos\theta' - \cos\theta'_t) \\
v_{1y} &= T_p f_2 G_1 \left(t_1 \cos\theta' \sin^2\phi' + \cos\theta'_t \cos^2\phi'\right) \\
v_{1z} &= T_p f_2 G_1 \left(t_1 \sin\theta' \sin\phi'\right),
\end{aligned}
\tag{9.73}
$$

$$
\begin{aligned}
v_{2x} &= T_p f_2 G_2 \left[(z_1 - z'_1)\,\beta_{1y} - (y_1 - y'_1)\,\beta_{1z}\right] \\
v_{2y} &= T_p f_2 G_2 \left[(x_1 - x'_1)\,\beta_{1z} - (z_1 - z'_1)\,\beta_{1x}\right] \\
v_{2z} &= T_p f_2 G_2 \left[(y_1 - y'_1)\,\beta_{1x} - (x_1 - x'_1)\,\beta_{1y}\right],
\end{aligned}
\tag{9.74}
$$

and

$$
\begin{aligned}
v_{3x} &= T_p f_2 G_2 t_1 \sin\theta'_t \sin\phi' \,(x_1 - x'_1) \\
v_{3y} &= T_p f_2 G_2 t_1 \sin\theta'_t \sin\phi' \,(y_1 - y'_1) \\
v_{3z} &= T_p f_2 G_2 t_1 \sin\theta'_t \sin\phi' \,(z_1 - z'_1).
\end{aligned}
\tag{9.75}
$$

The function f_2 is the phase of the electromagnetic field at the point $\mathbf{r}'$, or

$$
f_2 = e^{-ip_2 \cos\theta'},
\tag{9.76}
$$

and G_1 and G_2 are the functions:

$$
G_1 \equiv 4\pi R G = \frac{e^{-ip_2\rho_1}}{\rho_1}, \quad \text{and}
\tag{9.77}
$$

$$
G_2 \equiv -\frac{1}{ip_2\rho_1}\frac{\partial G_1}{\partial \rho_1} = \frac{G_1}{\rho_1}\left(1 + \frac{1}{ip_2\rho_1}\right).
\tag{9.78}
$$

The parameter t_1 is the ratio of the Fresnel transmission function for s-polarized light to the transmission function for p-polarized light, or

$$t_1 \equiv \frac{T_s}{T_p} = \frac{n_1 \cos \theta' + n_2 \cos \theta'_t}{n_1 \cos \theta'_t + n_2 \cos \theta'}. \tag{9.79}$$

The components of α_1 and β_1 introduced in Eqs. 9.72 and 9.75 are given by:

$$\begin{aligned}
\alpha_{1x} &= \sin \phi' \cos \phi' (t_1 \cos \theta' - \cos \theta'_t) \\
\alpha_{1y} &= \cos \theta'_t \cos^2 \phi' + t_1 \cos \theta' \sin^2 \phi' \\
\alpha_{1z} &= -t_1 \sin \theta' \sin \phi',
\end{aligned} \tag{9.80}$$

and

$$\begin{aligned}
\beta_{1x} &= -\cos \theta' \cos^2 \phi' - t_1 \cos \theta'_t \sin^2 \phi' \\
\beta_{1y} &= \sin \phi' \cos \phi' (t_1 \cos \theta'_t - \cos \theta') \\
\beta_{1z} &= \sin \theta' \cos \phi'.
\end{aligned} \tag{9.81}$$

The coordinates x_1, y_1, z_1, x'_1, y'_1, and z'_1 can be expressed in terms of the spherical coordinates as

$$\begin{aligned}
x_1 &= r_1 \sin \theta \cos \phi \\
y_1 &= r_1 \sin \theta \sin \phi \\
z_1 &= r_1 \cos \theta,
\end{aligned} \tag{9.82}$$

and

$$\begin{aligned}
x'_1 &= \sin \theta' \cos \phi' \\
y'_1 &= \sin \theta' \sin \phi' \\
z'_1 &= \cos \theta',
\end{aligned} \tag{9.83}$$

where $r_1 \equiv r/R$.

9.9.2 Light Incident at an Angle, Case 2

Figure 9.10 illustrates the theoretical setup for light incident upon the spherical surface at an angle of γ with respect to the z-axis. For ease of calculation, a second (double-primed) coordinate system is introduced such that $\hat{x}''$ and $\hat{z}''$ are rotated about $\hat{y}'$ by the angle of incidence, γ, with respect to $\hat{x}'$ and $\hat{z}'$, and $\hat{y}'' = \hat{y}'$.

All of the equations described previously for normally incident plane waves are valid for light incident at an angle, except for the fact that they apply to the double-primed coordinate system; i.e., all of the "primes" in Eqs. 9.56–9.84 are replaced with "double-primes." The difficulty in using the double-primed coordinate system arises in the limits of the integrations. The limits

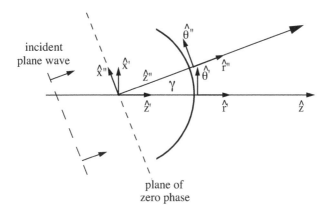

FIGURE 9.10
Theoretical setup for calculating light fields beyond the exit surface of a microlens [8]. The plane wave is incident at an angle of γ with respect to the $+\hat{z}$ direction.

of the integrations are determined by the shape of the exit surface in that coordinate system. In the primed coordinate system of Figure 9.10, the limits of integration are $\theta' = 0$ to θ_m, and $\phi' = 0$ to 2π. In the double-primed coordinate system, the limits of integration are a complicated function of θ'' and ϕ'' due to the fact that z'' does not pass through the center of the exit surface and the exit surface is not angularly symmetric about r'' for all values of θ''. Mathematically and computationally it would be desirable to perform the integrations over the primed coordinates. If an expression is found that transforms the double-primed coordinates to the primed coordinates, then the integration can be performed in the primed coordinate system where all of the variables and functions of $\mathbf{U}$ and $\mathbf{V}$ (Eqs. 9.67 and 9.68) are expressed in the primed coordinate system.

The relationships between the primed and double-primed spherical coordinates can be found by first expressing the spherical primed coordinates in terms of the Cartesian primed coordinates, or

$$
\begin{aligned}
x' &= \sin\theta' \cos\phi' \\
y' &= \sin\theta' \sin\phi' \\
z' &= \cos\theta'.
\end{aligned}
\tag{9.84}
$$

Then the double-primed coordinate system is rotated about $\hat{y}''$ by an angle $-\gamma$ to become the primed Cartesian coordinates:

$$
\begin{aligned}
x'' &= x' \cos\gamma - z' \sin\gamma \\
y'' &= y' \\
z'' &= x' \sin\gamma + z' \cos\gamma.
\end{aligned}
\tag{9.85}
$$

Substituting Eq. 9.84 into 9.85, we get

$$
\begin{aligned}
x'' &= (\sin\theta'\cos\phi')\cos\gamma - (\cos\theta')\sin\gamma \\
y'' &= \sin\theta'\sin\phi' \\
z'' &= (\sin\theta'\cos\phi')\sin\gamma + (\cos\theta')\cos\gamma.
\end{aligned}
\tag{9.86}
$$

Expressing the double-primed spherical coordinates in terms of the double-primed Cartesian coordinates results in

$$
\theta'' = \cos^{-1}\left(\cos\theta'\cos\gamma + \sin\theta'\cos\phi'\sin\gamma\right)
\tag{9.87}
$$

and

$$
\phi'' = \tan^{-1}\left(\frac{\sin\theta'\sin\phi'}{\sin\theta'\cos\phi'\cos\gamma - \cos\theta'\sin\gamma}\right).
\tag{9.88}
$$

Equations 9.56–9.84 can now be used for light incident upon a concave interface at an angle of γ if the right-hand sides of Eqs. 9.87 and 9.88 are substituted for all of the "θ'" and "ϕ'" variables, respectively, in Eqs. 9.56–9.84.

9.10   Normal Incidence Calculations, Case 2

Figure 9.11 is a calculation of the modulus square of the electric field versus the on-axis distance from the center of curvature. The parameters chosen for this calculation represent the focusing of light by a small tightly focusing lens, a microlens. They are the same as those for the microlenses by Dumke et al. [12] to optically trap cold ^{85}Rb atoms. The microlens array is an array of spherically curved surfaces on one side of a fused silica substrate ($n = 1.454$ at $\lambda = 780$ nm). The light enters the other (flat) side of the substrate and exits each spherical surface. The spherical surface has a radius of curvature of $R = 312.5$ μm and a maximum half-angle of $\theta_m = 11.5°$. Using geometrical optics, the focal spot should be observed at an axial distance of $3.2\,R$ from the center of curvature. The peak intensity calculated in Figure 9.11 is found at an axial distance of $3.11\,R$. Figure 9.12 is a plot of the modulus square of the x-component of the electric field versus the radial position, at the focal plane ($z = 3.11\,R$). One interesting feature of the on-axis intensity is the asymmetry before and after the focal region. The on-axis intensity modulations before the primary focal plane resemble the on-axis intensity characteristics for the diffraction of light by a circular aperture without a focusing medium [13, 14]. The on-axis intensity features before the focal spot indicate that light distributions in this region are influenced by the diffractive nature of the circular edge of the exit surface as well as the focal nature of the spherical shape of the exit surface. To further investigate the features before the focal plane, Figures 9.13 and 9.14 are plots of the radial intensity at the locations of the on-axis minimum and maximum occurring just before the focal plane, respectively. The beam profiles calculated in Figures 9.13 and 9.14 very closely resemble those of beam profiles calculated for diffraction of light by a simple circular aperture using scalar diffraction theory [13], and vector diffraction theory [14, 15].

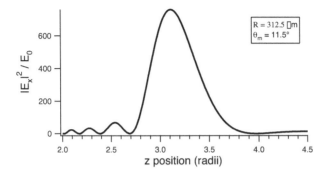

FIGURE 9.11
Modulus square of the x-component of the electric field versus axial position for a microlens with $2\pi nR/\lambda = 3660$ and a maximum half-angle of $11.5°$ [8].

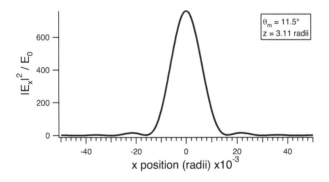

FIGURE 9.12
Modulus square of the x-component of the electric field versus radial position for a microlens with $2\pi nR/\lambda = 3660$ and a maximum half-angle of $11.5°$ [8].

9.11 Spherical Aberration, Case 2

In this section we investigate the effects of spherical aberration for the case of curvature dimensions much, much larger than the wavelength of light, param-

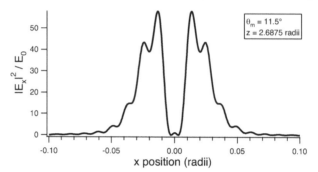

FIGURE 9.13
Modulus square of the x-component of the electric field versus radial position
for a microlens with $2\pi n R/\lambda = 3660$, a maximum half-angle of 11.5°, and
an axial position of the on-axis minimum closest to the focal plane, or $z = 2.69R$ [8].

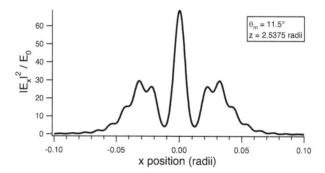

FIGURE 9.14
Modulus square of the x-component of the electric field versus radial position
for a microlens with $2\pi n R/\lambda = 3660$, a maximum half-angle of 11.5°, and an
axial position of the first localized on-axis maximum closest to the focal plane,
or $z = 2.54R$ [8].

eters commonly found in optics labs with lenses, and optical elements with
dimensions in the millimeter to centimeter range.

Figure 9.15 is a plot of the on-axis intensity for a lens with a focal parameter of $p_2 = 2\pi n_2 R/\lambda = 5 \times 10^5$ for a variety of maximum half-angles. A lens
parameter of $p_2 = 5 \times 10^5$ would correspond to a radius of curvature of 2.8
cm for 532–nm light. For $p_2 = 5 \times 10^5$ the onset of the effects of spherical
aberration are evident for even a maximum half-angle of $\theta_m = 7.5°$. As θ_m is
increased from 5 to 7.5° the maximum value of the on-axis intensity increases.
Beyond a half-angle of 7.5°, the maximum on-axis intensity no longer increases
in magnitude with increasing θ_m. For $\theta_m = 10°$, significant deviations exist in

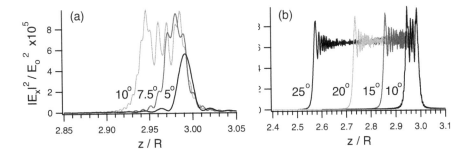

FIGURE 9.15

Calculated values of the intensity (normalized to the incident intensity) along the axis of the lens (normalized to the radius of curvature of the lens) for various values of the maximum half-angle θ_m (5°, 7.5°, 10°, 15°, 20°, and 25°) of the lens surface for a normally incident plane wave with $2\pi nR/\lambda = 5 \times 10^5$ [8].

the on-axis intensity profile from that of an aberration-free lens such as that of the $\theta_m = 5°$ on-axis intensity plot. Note that this was not the case for a microlens with $p_2 = 3660$ and $\theta_m = 11.5°$, discussed in the previous section, illustrating that the onset of spherical aberration is dependent upon the ratio of the curvature of radius to the wavelength of light. The higher the ratio, the earlier the onset of spherical aberration for increasing maximum half-angles. In Figure 9.15(b) the maximum half-angle is increased even further to investigate the effects of severe spherical aberration. Once the maximum on-axis intensity has saturated (around 7.5° for $p_2 = 5 \times 10^5$), the region of higher intensities elongates along the axis of propagation towards the exit surface of the lens. The growth of the "focal spot" towards the exit surface of the lens can be explained using geometrical optics. According to geometrical optics, if a series of parallel rays are refracted by a spherical surface then the paraxial rays focus at the expected focal spot. Non-paraxial rays will be refracted by the spherical surface such that they will cross the optical axis at a point closer to the exit surface. The farther the initial ray is located from the optical axis, the closer to the surface of the lens the ray will cross the optical axis.

Figures 9.16(a) and (b) are plots of the radial intensity profile at the focal plane of a plane wave refracted by a spherical exit surface as a function of the maximum half-angle on a linear and a logarithmic scale, respectively. The intensity axis of Figure 9.16 is normalized to the incident intensity before the spherical surface. Once the on-axis maximum intensity saturation point is reached for increasing θ_m, no significant changes in the radial beam profile is observed on a linear scale for increasing θ_m. Figure 9.16(b) is the same plot as Figure 9.16(a), except that the intensity scale is logarithmic. On a logarithmic scale it is apparent that the light collected by the outer rings

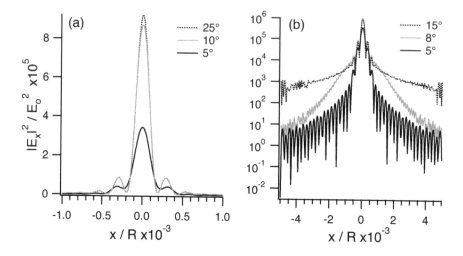

FIGURE 9.16

Calculated values of the focal plane intensity distributions (normalized to the incident intensity) along the x-axis (normalized to the radius of curvature of the lens) on a linear and long scale, respectively, for various values of the maximum half-angle θ_m (5°, 8°, 10°, 15°, and 25°) of the lens surface [8]. The light is a normally incident plane wave with $2\pi nR/\lambda = 5 \times 10^5$, and the *focal plane* occurs for $z/R = 2.984$.

of the spherical surface crosses the focal plane at off-axis locations; i.e., the energy collected by the outer regions of the spherical surface is propagated to off-axis locations in the focal plane. Away from the central focal region, the intensity of a spherically aberrated beam can be orders of magnitude higher than that of a non-aberrated beam.

9.12   Off-Axis Focusing and Coma, Case 2

9.12.1   Single Beam

Figure 9.17 is a calculation of the intensity of the light field versus x and y in the focal plane ($z/R = 3.15$ for $2\pi n_2 R/\lambda = 10^4$) and incident angles, γ, of 0, 7, and 15°. With an incident angle of 7° the effects of coma are observed as subtle partial rings on the far side of the focal spot, and a slight elongation of the focal spot in the direction of the incident angle. For an incident angle of 15°, severe coma is observed as a broadening of the focal spot in both the x and y-directions in addition to conical beam fanning as

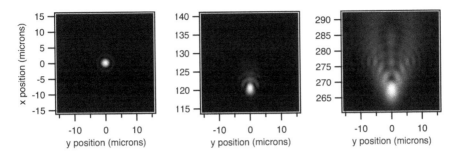

FIGURE 9.17

Calculated intensity distributions in the focal plane for $2\pi n R/\lambda = 10^4$ with $\theta_m = 11.54°$, $n = 1.454$, and incident angles of $\gamma = 0$, $7°$ and $15°$, respectively [8].

expected according to geometrical optics. However, contrary to geometrical optics (which would predict a smooth intensity decay throughout the beam fanning) intensity calculations presented here predict the interference patterns arising from the wave nature of light. The degree of coma observed is also a function of the radius of curvature to wavelength ratio. Similar to Figure 9.17, Figure 9.18 is a series of intensity versus x and y images, where the only difference is that the value of p_2 has been decreased from 10^4 to 3660. For smaller p_2, the effects of coma for $\gamma = 7°$ have essentially been eliminated, and the coma effects for $\gamma = 15°$ have been significantly reduced, as compared to Figure 9.17.

Figure 9.19 is a calculation of the beam intensity profile (along the x-axis) for the same parameters as that of Figure 9.17, but a series of angles of incident leading up to $7°$. The relative ranges for each graph are kept constant to aid

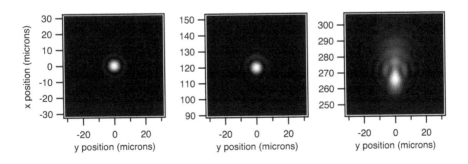

FIGURE 9.18

Calculated intensity distributions in the focal plane for $2\pi n R/\lambda = 3660$ with $\theta_m = 11.54°$, $n = 1.454$, and incident angles of $\gamma = 0$, $7°$ and $15°$, respectively [8].

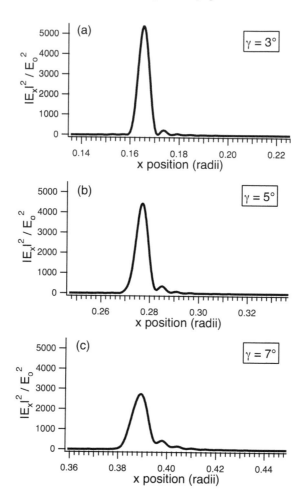

FIGURE 9.19
Effects of coma on the radial beam profile of a plane wave incident on a microlens spherical exit surface with $2\pi nR/\lambda = 10^4$ and $z/R = 3.15$ at an angle of (a) 3°, (b) 5°, and (c) 7° [8]. All horizontal and vertical scales are over the same ranges.

in the visual comparison of the absolute intensities and radial beam fanning. Even for an angle of incidence of 3°, asymmetries in the beam profile are observed for $2\pi nR/\lambda = 10^4$. As the angle of incidence is increased, the overall maximum intensity decreases and the width of the focal spot increases as well as the amount of energy observed outside of the primary focal spot.

9.12.2 Two Beams

In this section we investigate the use of two orthogonally polarized beams incident upon the spherical exit surface of a microlens with one beam having a variable angle of incidence. Because the polarizations of the beams are orthogonal, no interference occurs and a double focal spot is possible. This particular setup is of interest to the cold atom trapping and quantum computation community [12]. Using the dipole force [16], cold neutral atoms can be attracted to and trapped in localized light intensity maxima. One of the goals of quantum computation is to trap two atoms in separate dipole traps, bring them together to create an entangled state, and then separate the traps yielding a pair of distinguishable and locally independent entangled atoms. Dumke et al. [12] experimentally demonstrated that a single dipole trap can be split into two traps using a pair of laser beams with a variable angle of incidence between them. Here we numerically calculate the radial light fields in the focal region using vector diffraction theory and investigate the splitting of a single focal spot into two distinguishable focal spots as a function of the variable angle of incidence of one of the beams.

We begin with two coherent beams of equal intensity. One is polarized along the x-direction and traveling along the central axis of the microlens (z-direction). The other is polarized along the y-direction and traveling in the x-z plane with an adjustable angle of incidence, γ, with respect to the z-axis. The electromagnetic fields of each beam are calculated individually at the point of interest. The two fields are then superimposed, yielding the net electromagnetic fields and intensity at the point of interest. Figure 9.20 is a collection of intensity profiles of the net intensity in the focal plane versus position along the x-axis for four different values of γ. All of the plots are over the same horizontal and vertical range to aid in the reader's visual interpretation of the data. For angles as small as 0.3°, splitting of the single focal spot into two resolvable peaks is observed. By an angle of 1.2° the peaks are separated by 19μm (for $R = 312.5\mu$m) and the intensity between the peaks is 1% of of its maximum value.

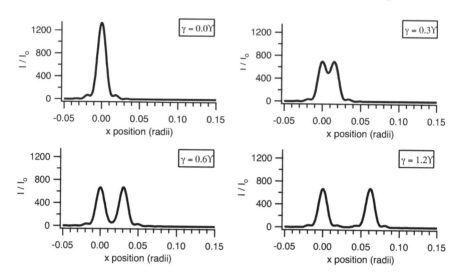

FIGURE 9.20

Calculated total intensity versus the radial x position for two orthogonally polarized beams. One beam is incident along the optical axis of the microlens, and the other is incident at a variable angle, γ. The parameters used for the spherical surface are $2\pi n_2 R/\lambda = 3660$, where $R = 312.5\mu\text{m}$.

Bibliography

[1] J. W. Goodman, *Fourier Optics* (Roberts & Company Publishers, Englewood, Colorado, 2005), chapter 5.

[2] B. Richards and W. Wolf, "Electromagnetic diffraction in optical systems II," Proc. R. Soc. London Ser. A **253**, 353–379 (1959).

[3] T. D. Visser and S. H. Wiersma, "Diffraction of converging electromagnetic waves," J. Opt. Soc. Am. A **9**, 2034–2047 (1992).

[4] W. Hsu and R. Barakat, "Stratton-Chu vectorial diffraction of electromagnetic fields by apertures with application to small-Fresnel-number systems," J. Opt. Soc. Am. A **11**, 623 (1994).

[5] J. A. Stratton and L. J. Chu, "Diffraction theory of electromagnetic waves," Phys. Rev. **56**, 99–107 (1939).

[6] J. A. Stratton, *Electromagnetic Theory* (McGraw Hill Book Company, New York, 1941).

[7] S. Guha and G. D. Gillen, "Vector diffraction theory of refraction of light by a spherical surface," J. Opt. Soc. Am. B **24**, 1–8 (2007).

[8] G. D. Gillen and S. Guha, "Description of spherical aberration and coma of a microlens using vector diffraction theory," Proc. SPIE **6342**, 63420B (2006).

[9] M. Born and E. Wolf, *Principles of Optics* (Cambridge University Press, Cambridge, UK, 2003.)

[10] E. J. Rothwell and E. J. Cloud, *Electromagnetics* (CRC Press, Boca Raton, Florida, 2001.)

[11] F. A. Jenkins and H. E. White, *Fundamentals of Optics*, (McGraw Hill, New York, 1957.)

[12] R. Dumke, M. Volk, T. Müther, F. B. J. Buchkremer, G. Birkl, and W. Ertmer, "Micro-optical realization of arrays of selectively addressable dipole traps: a scalable configuration for quantum computation with atomic qubits," Phys. Rev. Lett. **89**, 097903 (2002).

[13] G. D. Gillen and S. Guha, "Modeling and propagation of near-field diffraction patterns: a more complete approach," Am. J. Phys. **72**, 1195–1201 (2004).

[14] S. Guha, and G. D. Gillen, "Description of light propagation through a circular aperture using nonparaxial vector diffraction theory," Opt. Express **26**, 1424–1447 (2005).

[15] G. D. Gillen, S. Guha, and K. Christandl, "Optical dipole traps for cold atoms using diffracted laser light," Phys. Rev. A **73**, 013409 (2006).

[16] J. E. Bjorkholm, R. R. Freeman, A. Ashkin, and D. B. Pearson, "Observation of focusing of neutral atoms by the dipole forces of resonance-radiation pressure," Phys. Rev. Lett. **41**, 1361–1364 (1978).

10

Diffraction of Gaussian Beams

In this chapter we present mathematical models for calculating the vector fields for Gaussian beams passing through a spatially limiting aperture. Two vector diffraction theories are presented:

- Gaussian Hertz vector diffraction theory (GHVDT), and

- Luneberg's vector diffraction theory for Gaussian beams

GHVDT is valid for all points within and beyond the diffracting aperture. Luneberg's model is mathematically simpler, but invalid for points within the aperture plane, or very close to it. Additionally, we present an analytical model that enables the user to quickly calculate a diffracted Gaussian laser beam's electric field and intensity along the optical axis. In the last section we present some experimental measurements of diffracted Gaussian laser beams, and compare the results to those predicted by the models presented in this chapter.

10.1 Gaussian Hertz Vector Diffraction Theory, GHVDT

The vector electromagnetic fields, $\mathbf{E}$ and $\mathbf{H}$, of a propagating light field can be determined from a vector polarization potential, $\mathbf{\Pi}$, using [1, 2]

$$\mathbf{E} = k^2 \mathbf{\Pi} + \mathbf{\nabla}\left(\mathbf{\nabla} \cdot \mathbf{\Pi}\right), \qquad (10.1)$$

and

$$\mathbf{H} = ik\sqrt{\frac{\epsilon_o}{\mu_o}}\mathbf{\nabla} \times \mathbf{\Pi}, \qquad (10.2)$$

where $k = 2\pi/\lambda$ is the wave number, and ϵ_o and μ_o are the permittivity and permeability of free space, respectively. As discussed in Sec. 1.6, the vector potential $\mathbf{\Pi}$, also known as the Hertz vector [2], must satisfy the wave equation at all points in space.

In HVDT,[1] a Hertz vector for the incident field, $\mathbf{\Pi}_i$, is chosen at a plane

---

[1] Hertz Vector Diffraction Theory (HVDT) for the diffraction of incident plane waves is presented and discussed in detail in Section 7.5.

and then the Hertz vector beyond the diffraction surface, $\mathbf{\Pi}$, is determined from the propagation of the incident vector field using Neumann boundary conditions [1]. The complete six-component electromagnetic field can be determined from the x-component of the Hertz vector at the point of interest beyond the aperture plane, given by [1]

$$\Pi_x(\mathbf{r}) = -\frac{1}{2\pi} \iint \left(\frac{\partial \Pi_{ix}(\mathbf{r_o})}{\partial z_o} \right)_{z_o=0} \frac{e^{-ik\rho}}{\rho} dx_o dy_o. \qquad (10.3)$$

where $\mathbf{r}$ is a function of the coordinates of the point of interest beyond the aperture plane, (x, y, z), $\mathbf{r_o}$ is a source point in the diffraction plane, (x_o, y_o, z_o), and

$$\rho = |\mathbf{r} - \mathbf{r_o}|. \qquad (10.4)$$

The limits of integration of Eq. (10.3) are the limits of the open area in the aperture plane. The complete $\mathbf{E}$ and $\mathbf{H}$ vector fields at the point of interest are then determined by substitution of Eq. (10.3) into Eqs. (10.1) and (10.2).

The x-component of an incident Gaussian electric field polarized in the x-direction can be written as [3]

$$E_{ix}(\mathbf{r_o}) = E_0 \exp\left[-i\left(-i\ln\left(1 + \frac{z_0}{q_0}\right) + \frac{kr_0^2}{2(q_0 + z_0)} + kz_0 \right) \right], \qquad (10.5)$$

or equivalently,

$$E_{ix}(\mathbf{r_o}) = \frac{E_o}{\left(1 + \dfrac{z_0}{q_0}\right)} \exp\left[-\frac{ikr_o^2}{2(q_o + z_o)} - ikz_o \right], \qquad (10.6)$$

where

$$q_0 = i\frac{k}{2}w_0^2, \qquad (10.7)$$

$$r_o^2 = x_o^2 + y_o^2, \qquad (10.8)$$

and ω_o is the e^{-1} minimum half-width of the incident electric field.

The Hertz vector of the incident field of a Gaussian laser beam with the minimum beam waist located at $z = z_G$ can be chosen as

$$\Pi_{ix}(\mathbf{r_o}) = \frac{E_o}{k^2} \frac{1}{\left(1 + \dfrac{z_o - z_G}{q_o}\right)} \exp\left[-\frac{ikr_o^2}{2(q_o + z_o - z_G)} - ik(z_o - z_G) \right]. \qquad (10.9)$$

The normal partial derivative of the incident Hertz vector evaluated in the aperture plane becomes

$$\left(\frac{\partial \Pi_{ix}}{\partial z_o} \right)_{z_o=0} = \frac{E_o}{k^2} A(x_o, y_o) \qquad (10.10)$$

where

$$A(x_o, y_o) = \frac{1}{1 - \frac{z_G}{q_o}} \left(\frac{ikr_o^2}{2(q_o - z_G)^2} - \frac{1}{q_o - z_G} - ik \right)$$

$$\times \exp \left(\frac{-ikr_o^2}{2(q_o - z_G)} \right). \tag{10.11}$$

Substitution of Eqs. (10.11) and (10.10) into Eq. (10.3) yields the Hertz vector at the point of interest. The components of the electromagnetic field at the point of interest are then determined by substitution of the Hertz vector at the point of interest into Eqs. (10.1) and (10.2), yielding

$$E_x = k^2 \Pi_x + \frac{\partial^2 \Pi_x}{\partial x^2},$$

$$E_y = \frac{\partial^2 \Pi_x}{\partial y \partial x},$$

$$E_z = \frac{\partial^2 \Pi_x}{\partial z \partial x}, \tag{10.12}$$

and

$$H_x = 0,$$

$$H_y = ik \sqrt{\frac{\epsilon_o}{\mu_o}} \frac{\partial \Pi_x}{\partial z},$$

$$H_z = -ik \sqrt{\frac{\epsilon_o}{\mu_o}} \frac{\partial \Pi_x}{\partial y}. \tag{10.13}$$

It is computationally convenient to express all of the parameters in dimensionless form. To do so, all tangential distances and parameters are normalized to the minimum beam waist, ω_o, and all axial distances and parameters are normalized to a quantity $z_n \equiv k\omega_o^2 = p_1\omega_o$, where the parameter p_1 is defined as $p_1 = k\omega_o$. All normalized variables, parameters, and functions are denoted with a subscript "1." The normalized variables and parameters are

$$x_1 \equiv \frac{x}{\omega_o}, \quad y_1 \equiv \frac{y}{\omega_o}, \quad z_1 \equiv \frac{z}{z_n}, \tag{10.14}$$

$$x_{01} \equiv \frac{x_0}{\omega_o}, \quad y_{01} \equiv \frac{y_0}{\omega_o}, \quad z_{G1} \equiv \frac{z_G}{z_n}, \quad \text{and} \quad q_1 \equiv \frac{q_o}{z_n}. \tag{10.15}$$

Substitution of Eqs. (10.11) & (10.10) into (10.3) and (10.3) into Eqs. (10.12) & (10.13), carrying out the differentiations, and using normalized variables and functions (quite the analytical challenge), the electric and magnetic fields at the point of interest can be expressed as

$$E_x(\mathbf{r_1}) = -\frac{E_o}{2\pi p_1} \iint A_1 f_1$$
$$\times \left[(1 + s_1) - (1 + 3s_1) \frac{(x_1 - x_{01})^2}{\rho_1^2} \right] dx_{01} dy_{01}, \tag{10.16}$$

$$E_y(\mathbf{r_1}) = \frac{E_o}{2\pi p_1} \iint A_1 f_1 (1 + 3s_1) \frac{(x_1 - x_{01})(y_1 - y_{01})}{\rho_1^2} dx_{01} dy_{01}, \tag{10.17}$$

$$E_z(\mathbf{r_1}) = \frac{E_o z_1}{2\pi} \iint A_1 f_1 (1 + 3s_1) \frac{(x_1 - x_{01})}{\rho_1^2} dx_{01} dy_{01}, \tag{10.18}$$

and

$$H_x(\mathbf{r_1}) = 0, \tag{10.19}$$

$$H_y(\mathbf{r_1}) = -\frac{ip_1 z_1 H_0}{2\pi} \iint A_1 f_1 s_1 dx_{01} dy_{01}, \tag{10.20}$$

$$H_z(\mathbf{r_1}) = \frac{iH_0}{2\pi} \iint A_1 f_1 s_1 (y_1 - y_{01}) dx_{01} dy_{01}, \tag{10.21}$$

where

$$A_1 = \frac{1}{1 + 2iz_{G1}} \left(\frac{ir_{01}^2}{2(q_1 - z_{G1})^2} - \frac{1}{(q_1 - z_{G1})} - ip_1^2 \right)$$
$$\times \exp\left(\frac{-ir_{01}^2}{2(q_1 - z_{G1})} \right), \tag{10.22}$$

$$f_1 = \frac{e^{-ip_1\rho_1}}{\rho_1}, \tag{10.23}$$

$$s_1 = \frac{1}{ip_1\rho_1} \left(1 + \frac{1}{ip_1\rho_1} \right), \tag{10.24}$$

$$\rho_1^2 = (x_1 - x_{01})^2 + (y_1 - y_{01})^2 + p_1^2 z_1^2, \tag{10.25}$$

and

$$r_{01}^2 = x_{01}^2 + y_{01}^2. \tag{10.26}$$

At this point, to keep one's sanity, it is best to enter Eqs. 10.16–10.26 into a numerical integration computational program in order to calculate the electromagnetic fields at the point of interest.

10.2 Validation of GHVDT

In this section, we apply Eqs. 10.16–10.26 to a practical case of apertured Gaussian laser beam propagation to test the validity of the model. A Gaussian

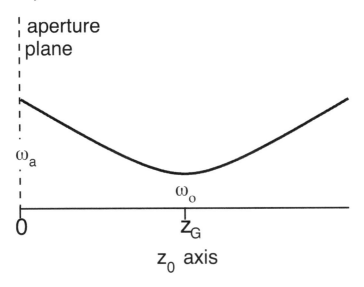

FIGURE 10.1
Theoretical setup for most calculations, where ω_a is the e^{-1} width of the electric field of the incident Gaussian field in the aperture plane, ω_o is the minimum beam waist, and z_G is the on-axis location of the Gaussian focal plane [4].

laser beam (TEM$_{00}$ mode) having a wavelength of 780 nm and focused to a spot size ω_o (e^{-1} Gaussian half width of the electric field) of 5 μm is chosen for this example. The Rayleigh range, $z_R = \pi \omega_o^2 / \lambda$, for this beam is 100.7 μm.

Figure 10.1 illustrates the theoretical setup for the calculations conducted using GHVDT. The focused beam is incident upon the aperture and propagating in the $+z$ direction. The origin of the coordinate system is at the center of a circular aperture placed normal to the beam such that the center of the aperture coincides with the center of the beam. For the GHVDT presented in this chapter, the parameter z_G is left as an independent variable such that the axial location of the unperturbed focal plane is independent with respect to the location of the aperture plane. The beam waist in the aperture plane is ω_a, and determined by [3]

$$\omega_a^2 = \omega_o^2 \left(1 + \frac{z_G^2}{z_R^2}\right). \tag{10.27}$$

By allowing z_G to remain an independent variable, calculations can be performed for the focal plane before, coplanar, or after the aperture plane depending upon the value used for z_G.

Three regimes are studied in this investigation:

1. pure Gaussian behavior where the incident beam is unperturbed by the aperture size, or $a \gg \omega_a$,

2. pure diffraction behavior where the incident beam is highly clipped, or $a \ll \omega_a$, and

3. the diffracted-Gaussian regime where $0.01 < \dfrac{a}{\omega_a} < 3$.

The regime where the radius of the aperture is significantly larger than the beam width in the aperture plane corresponds to a regime where a Gaussian beam is expected to continue behaving as an unperturbed Gaussian beam. Under these conditions, the physical aperture or optic is significantly larger than the beam, and an incident focused Gaussian TEM_{00} will continue through the aperture as an unperturbed TEM_{00} beam. To model this regime, an a/ω_a ratio of 4 is chosen, where for a centered beam, the relative intensity at the edge of the the the aperture is $\sim 10^{-14}$. Calculations using the GHVDT model presented here are directly compared to calculations using a purely Gaussian beam propagation model presented and discussed in Chapter 6 (henceforth referred to as the "PG" model.)

Figure 10.2 is a collection of calculations of the relative intensity of the light field for $a = 4\omega_a$ using the complete GHVDT outlined previously, and using the PG model. The intensity calculated in the plots, S_z/S_o, is the normalized z-component of the Poynting vector of the electromagnetic fields. All intensities of Figure 10.2 are normalized to the peak intensity in the focal plane. Figures 10.2(a) and (b) are calculations where $z_G = 0$, or the focal plane is coplanar with the aperture plane. When the focal plane is coplanar with the aperture plane, calculations for the intensity versus radial position, Figure 10.2(a), reproduce the unperturbed incident Gaussian beam using the PG model in the aperture plane. Figure 10.2(b) shows that the calculated on-axis intensity for axial positions beyond the aperture plane using GHVDT also reproduce the expected axial behavior for diverging TEM_{00} purely Gaussian beams.

Figures 10.2(c) and (d) are calculations for $z_G = 0.01$ m, where the focal plane is located 1 cm beyond the aperture plane. The incident Hertz vector (Eq. 10.9) in the aperture plane is one that yields a converging focused Gaussian beam with a width in the aperture plane which follows that of Eq. (10.27), or $\omega_a = 99\omega_o$, and the normalized peak intensity in the aperture plane is 1×10^{-4}. Using the full GHVDT and a z_G value of 1 cm, Figure 10.2(c) is a calculation of the normalized intensity versus radial position for a distance of 1 cm beyond the aperture plane. Note that calculations using GHVDT for the regime of $a \ll \omega_a$ reproduce calculated behavior using the PG model. For an unperturbed converging Gaussian beam, the on-axis intensity is expected to follow a Lorentzian function with an offset maximum occurring at the axial location of the focal plane, and having a normalized intensity of 1. Using GHVDT and input parameters of $z_G = 0.01$ m and $a = 4\omega_a$, the expected on-axis Gaussian behavior is reproduced, as illustrated in Figure 10.2(d).

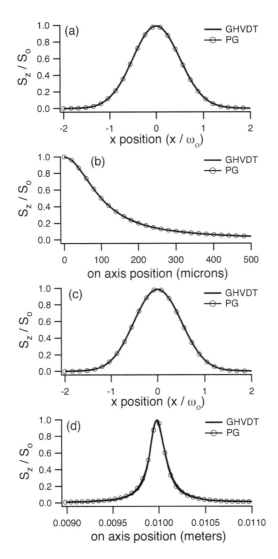

FIGURE 10.2

Calculated Gaussian behavior for $a/\omega_a = 4$ using GHVDT [4]. Figures (a) and (b) are the normalized intensity versus x in the aperture plane and versus z, respectively, for $z_G = 0$. Figures (c) and (d) are the normalized intensity versus x in the focal plane and versus z, respectively, for $z_G = 0.01$ m. Calculated intensity distributions using a purely Gaussian (PG) beam propagation model are included for comparison purposes.

For the pure diffraction regime, the parameters chosen are $z_G = 0.01$ m, and values of $a/\omega_a < 0.016$. For these values of a/ω_a, the variation between the intensity at the center of the aperture and the edge of the aperture is less than 0.05%. Thus the incident Hertz vector in the aperture region is virtually uniform, and calculations using GHVDT should reproduce those for a plane wave incident upon a circular aperture. Thus, the model chosen for comparison is that of Hertz vector diffraction theory, HVDT, applied to the diffraction of a plane wave incident upon a circular aperture.

As discussed in Section 7.5, an important parameter in the pure diffraction regime is the ratio of the aperture radius to the wavelength of the light, a/λ. According to HVDT applied to plane wave diffraction, the on-axis intensity will oscillate as a function of increasing axial distance. The number of oscillations of the on-axis intensity will be equal to the a/λ ratio [5]. The maximum on-axis intensity will have a value of 4 times the normalized incident intensity and will be located at a position of $z = a^2/\lambda$. In the aperture plane, the central intensity will modulate about the normalized incident intensity value as a function of the aperture to wavelength ratio due to scattering effects of the aperture edge [1]. Modulations in the fields within the aperture plane were both experimentally measured and calculated using HVDT by Bekefi in 1953 [1]. The intensity value at the center of the aperture plane will be a maximum for integer values of the a/λ ratio, and will be a minimum for half integer values. Using HVDT, the z-component of the Poynting vector at the center of the aperture will have a normalized maximum value in the aperture plane of 1.5 and a minimum value of 0.5 [5].

Figure 10.3 is a calculation of the normalized on-axis z-component of the Poynting vector using GHVDT for a/λ ratios of (a) 5, (b) 5.5, and (c) 10, and an unperturbed focal plane location of $z_G = 0.01$ m. Note that the intensity is normalized to the expected peak Gaussian intensity in the focal plane of an unperturbed beam, which makes the normalized intensity incident upon the aperture 10^{-4}. Figure 10.3 also includes plots for the calculated on-axis intensity distributions according to HVDT of a plane wave incident upon a circular aperture. Note that all aspects of the calculations in Figure 10.3 for a highly clipped, converging, focused Gaussian beam reproduce those expected for the diffraction of a plane wave even though the GHVDT still contains all of the phase information for a converging focused Gaussian TEM_{00} within input light field.

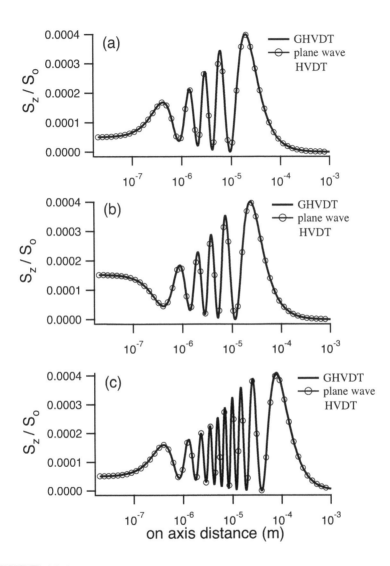

FIGURE 10.3
Calculated plane-wave diffraction behavior using GHVDT [4]. Calculations are of the normalized intensity versus on-axis distance from the aperture for aperture to wavelength ratios, a/λ, of (a) 5, (b) 5.5, and (c) 10. The incident peak intensity in the aperture plane is 10^{-4}, normalized to the theoretical peak intensity in the focal plane at $z = 0.01$ m. Calculated intensity distributions using a purely plane wave vector diffraction theory, HVDT, are included for comparison purposes.

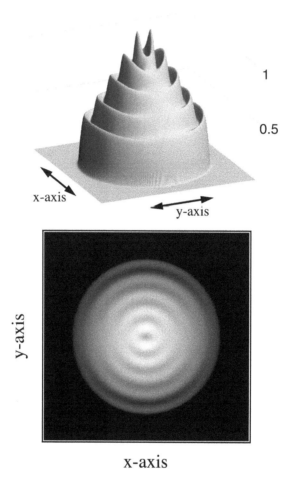

FIGURE 10.4
Calculation of the z-component of the Poynting vector versus both x and y [4]. Calculations are in the aperture plane for an incident Gaussian beam with the beam waist, ω_o, in the aperture plane and an aperture radius of $a = 5\lambda = 0.78\omega_o$.

10.3 Calculations of Clipped Gaussian Beams Using GHVDT

10.3.1 The Region between Pure-Diffraction and Unperturbed Behavior

In the diffracted-Gaussian regime, the a/ω_a ratio enters the region of ~ 0.1 to ~ 2, and the behavior of the light field within and beyond the aperture plane is neither pure diffractive behavior nor pure Gaussian behavior. This is a regime where both traditional diffraction models *and* Gaussian beam propagation models are invalid. Thus, comparison of light field distributions for the model presented here to other vector diffraction models is currently not possible. An example of a calculation only possible with the GHVDT presented here is the normalized intensity illustrated in Figure 10.4. Figure 10.4 is the z-component of the Poynting vector as a function of both x and y in the aperture plane for the focal plane coplanar with the aperture plane ($z_G = 0$), an aperture radius of $a = 5\lambda = 0.78\omega_o$, and the polarization of the incident light along the x-axis. For an a/ω_a ratio of 0.78, the intensity of the incident light on the edge of the aperture is 29.6% of the value at the center of the aperture. The resulting light distribution is one of a partial Gaussian beam profile that is perturbed by the scattering effects of the incident light field on the aperture itself. According to experimental measurements [1] and calculations using HVDT [5], the scattering modulations are strongest along the axis perpendicular to the incident light polarization, and weakest along the axis parallel to the incident light polarization, and the number of oscillations from the center of the aperture to the aperture edge is equal to the a/λ ratio.

Figure 10.5 shows the calculated values of the on-axis intensity around the focal plane (where $z_G = 0.01$ m) for various aperture sizes, normalized to the values of the peak intensity obtained in the absence of the aperture. The ratios of the aperture radius to the beam waist ω_a chosen are 0.16, 0.31, 0.5, and 1. These values were chosen as they illustrate calculations in the regime between the fully diffractive and the fully Gaussian regimes. The aperture radius is also expressed in multiples of the wavelength. In the diffraction regime, the location of the on-axis peak shifts with an increasing aperture radius and is located at an on-axis position of a^2/λ. In the Gaussian beam propagation regime, the on-axis location of the peak intensity is constant and located in the focal plane. For the chosen laser parameters, these two values are equal when $a = 113\lambda$. As observed in Figure 10.5, for values of $a < 113\lambda$ the peak on-axis intensity is shifted closer to the aperture plane, and for values of $a > 113\lambda$ the peak on-axis intensity remains at a distance of 1 cm from the aperture even as the aperture radius is increased. Also note that as the aperture radius is increased the peak intensity significantly grows in magnitude and axially narrows due to the increasing contribution of the focal properties of the incident converging beam.

10.3.2 Effect of Diffraction on the M² Parameter

Using the GHVDT model of full wave equation vector diffraction of clipped focused-Gaussian beams, presented in Section 10.1 and Ref. [4], radial intensity distributions are calculated for a variety of axial distances from the aperture plane, and for a variety of clipping ratio values. Calculated radial intensity distributions are then analyzed using a Gaussian fitting routine to obtain representative beam widths. For each clipping ratio value, the beam

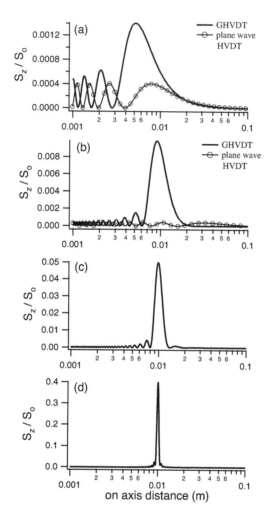

FIGURE 10.5

Calculations of the normalized on-axis intensity for the regime between pure diffractive and pure Gaussian behavior for $z_G = 0.01$ m and (a) $a = 100\lambda = 0.16\omega_a$, (b) $a = 200\lambda = 0.31\omega_a$, (c) $a = 318\lambda = 0.5\omega_a$, and (d) $a = 636\lambda = \omega_a$ [4]. All intensities are normalized to the peak unperturbed focal spot intensity. Calculated intensity distributions using a purely plane wave vector diffraction theory, HVDT, are included to illustrate the transition of the GHVDT model from the diffraction regime to the focused diffracted-Gaussian regime.

width versus axial distance data is then compared to the M^2 beam propagation model. The M^2 propagation model used is that described in Section 6.6,

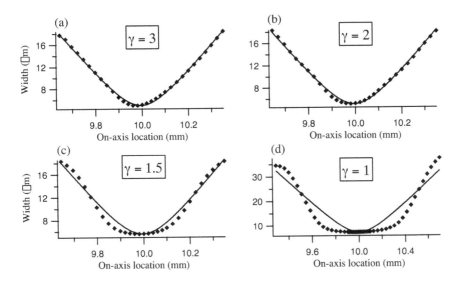

FIGURE 10.6

Calculated e^{-2} beam half-widths of the intensity of a clipped focused-Gaussian beam using GHVDT (diamonds) and fits to the calculations using a traditional M^2 model (solid lines) for various clipping ratios, γ, between values of 3 and 1 [7]. All figures are for $w_o = 5\mu m$, $z_G = 10$ mm, and $\lambda = 780$ nm. The values of M^2 fit for each of the four cases are (a) 1.00, (b) 1.02,(c) 1.12, and (d) 1.32.

and in section 11.3 of Ref. [6]. Figure 10.6 illustrates a collection of e^{-2} beam half-widths versus on-axis location graphs and their corresponding M^2 fits for clipping ratios of 3, 2, 1.5, and 1. The parameters chosen for Figure 10.6 calculations are a wavelength of $\lambda = 780$ nm, a minimum beam waist of $w_o = 5$ μm, and an aperture to focal plane distance of $z_G = 99$ $z_R = 1$ cm. Identical results are obtained if Eq. 10.49 is used instead of GHVDT. As illustrated in Figure 10.6, for clipping ratios less than $\sim$2, the M^2 model begins to significantly deviate from calculated beam propagation behavior, and is thus not a valid model when describing the propagation of clipped focused-Gaussian beams. For reference, clipping ratios of 2, 1.5, and 1 correspond to only 0.5%, 3.4%, and 15.7% of the energy being blocked by the aperture, respectively. As the clipping ratio decreases for values less than 2, the effects of diffraction on the beam manifest themselves as an elongation of the focal region with larger and larger minimum beam waists. Having a longer collimated region can be useful in nonlinear optical frequency conversion or in measurement of third order optical nonlinearities. For clipping ratios of $\lesssim$1 the on-axis beam width behavior demonstrates asymmetric behavior away from the focal region as diffraction effects become a dominant influence upon beam propagation.

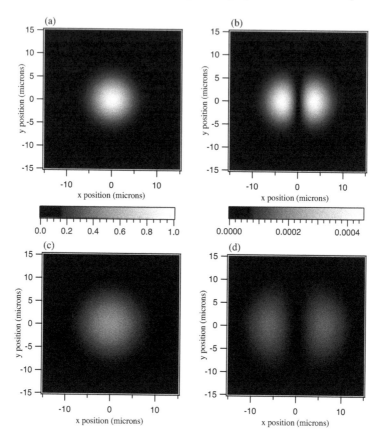

FIGURE 10.7

Beam intensity profiles in the focal plane, $z = z_G$, for (a) $|E_x|^2$ with $\gamma = 3$, (b) $|E_z|^2$ with $\gamma = 3$, (c) $|E_x|^2$ with $\gamma = 1$, and (d) $|E_z|^2$ with $\gamma = 1$ [7]. The gray-scale shading of (a) and (c) are both normalized to the peak intensity of (a), while the shading scale of (b) and (d) are both normalized to the peak intensity of (b). All figures are for $\omega_o = 5\mu m$, $z_G = 10$ mm and $\lambda = 780$ nm.

10.3.3   Radial and Longitudinal Intensities

Figure 10.7 investigates the effects of clipping on the beam intensity distributions in the focal plane for an unperturbed focused Gaussian beam, $\gamma = 3$, and a clipped focused Gaussian beam, $\gamma = 1$. Figures 10.7(a) and (c) are two-dimensional images of the modulus square of the x-component of the electric field calculated using Eq. 10.58, and the modulus square of the z-component of the electric field calculated using Eq. 10.59, respectively. The parameters used for the calculations of Figure 10.7 are the same as those used for Figure 10.6(a) and (d), or $\omega_o = 5\mu m$, $z_G = 10$ mm, and $\lambda = 780$ nm. The gray-scale shading

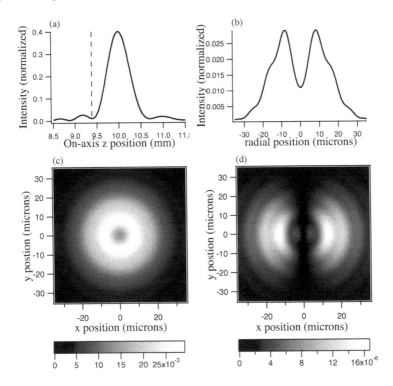

FIGURE 10.8
(a) On-axis intensity, (b) radial intensity for an on-axis position of the on-axis minimum at $z = 9.385$ mm, (c) beam profile of $|E_x|^2$ for $z = 9.385$ mm, and (d) the beam profile of $|E_z|^2$ for $z = 9.385$ mm [7]. All figures are for $\gamma = 1$, $\omega_o = 5\mu$m, $z_G = 10$ mm, and $\lambda = 780$ nm.

of both $|E_x|^2$ distributions are normalized to the peak value of $|E_x|^2$ for the unperturbed beam. Similarly, the shanding scale of both $|E_z|^2$ distributions are normalized to the peak value of $|E_z|^2$ for the unperturbed beam. As the clipping ratio decreases, the peak values of the intensity profiles decrease, and the overall width of both components of the electric field increases. Notice that the peak values for the clipped beam do not decrease equally for $|E_x|^2$ and $|E_z|^2$. The peak value of part (c) is 40% of that of part (a), whereas the peak value of part (d) is only 21.7% of that of part (b). Similar intensity fluctuations for clipped Gaussian beams were investigated and observed by Campillo et al. [8].

Figure 10.8 shows the effects of clipping on the beam intensity distributions for an off-focal plane axial location. The parameters used for the calculations of Figure 10.8 are the same as those used for Figures 10.6(d) and 10.7(c) & (d), or $\gamma = 1$, $\omega_o = 5\mu$m, $z_G = 10$ mm, and $\lambda = 780$ nm. Part (a) is a

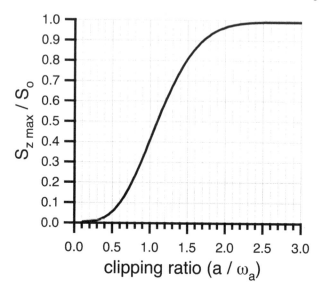

FIGURE 10.9

Calculation of the maximum normalized intensity beyond the aperture as a function of the clipping ratio, a/ω_a, where $z_G = 0.01$ m [7].

calculation of the on-axis intensity of the beam. Due to diffraction of the beam, the on-axis intensity distribution is no longer pure Lorentzian as it is for an unperturbed focused Gaussian beam. Away from the focal plane, oscillations are observed in the on-axis intensity distribution of part (a). Calculations performed for parts (b)–(d) are for the axial location of the on-axis minimum closest to the focal plane, or $z = 9.385$ mm, as indicated by the dashed line in part (a). Figure 10.8(b) is a radial beam distribution illustrating that the radial behavior of the beam is no longer Gaussian, and actually has a localized minimum in the middle of the beam, which is better observed in the two-dimensional beam profile of $|E_x|^2$ calculated in part (c). Part (d) is a calculation of the longitudinal component of the intensity, $|E_z|^2$. The effects of diffraction on the propagation behavior differences between focal plane and off-focal plane locations is observed via the comparison of Figures 10.7(c) and (d), and Figures 10.8(c) and (d).

10.3.4   Maximum Obtainable Irradiance for Clipped Gaussian Beams

One of the important parameters of focused Gaussian laser beams is the maximum achievable intensity value. In the diffracted-Gaussian regime the maximum achievable intensity value of a converging clipped beam increases with an increasing aperture radius. This behavior is observed in Figures 10.7

and 10.12 as the maximum normalized intensity increases by a factor of 1,000 from 4×10^{-4} for $a \ll \omega_a$ to 0.4 for a clipping ratio $a/\omega_a = 1$. Figure 10.9 is a calculation of the peak on-axis intensity as a function of the clipping ratio, for $z_G = 0.01$ m, and for a wide range of clipping ratios. The mathematical fit to the data calculated in Figure 10.9 empirically follows that of an error function, where

$$\frac{S_z}{S_o} = 0.49 + 0.50 \operatorname{erf} \left[-0.266 \left(\frac{a}{\omega_a} \right)^2 + 2.25 \frac{a}{\omega_a} - 2.14 \right]. \qquad (10.28)$$

Another important parameter of focused Gaussian laser beams is the beam width in the focal plane. For unperturbed Gaussian beam propagation the minimum beam waist is commonly known as the diffraction-limited spot size. Figure 10.10 is a plot of the Gaussian width of the calculated intensity distribution in the focal plane as a function of the clipping ratio ranging from 0.01 to 3. For clipping ratios larger than 2, the minimum beam width in the focal plane asymptotically approaches that of the unperturbed minimum beam width. In the diffraction-Gaussian regime, the beam width in the focal plane is empirically found to follow a log-log relationship for clipping ratios less than ~ 0.6, where for the parameters investigated here,

$$\log \left(\frac{\omega}{\omega_o} \right) = -\log \left(\frac{a}{\omega_a} \right) + \log \left(1.33 \right). \qquad (10.29)$$

Figures 10.9 and 10.10, and Eqs. (10.28) and (10.29), can be used to predict the theoretical maximum intensity and theoretical minimum beam waist for a clipped focused Gaussian beam. For example, if an unperturbed Gaussian beam having a wavelength of 780 nm and a minimum spot size of 5μm passes through an aperture located 1 cm before the focal spot with a diameter of 1 mm ($a/\omega_a = 1$), then the new theoretical maximum intensity would be 40% of that of a diffraction-limited Gaussian beam, and the new theoretical beam waist would be 46% wider than that of an unperturbed beam. If the aperture diameter is increased to 1.5 mm, then the theoretical maximum intensity would double to 80%, and the minimum beam waist would shrink to only 12.4% wider than that of a diffraction-limited beam.

10.4 Longitudinal Field Component in the Unperturbed Paraxial Approximation

The integral expressions for all the components of the electric and magnetic field provided in Eqs. 10.16 to 10.21 constitute the key results of this chapter. The integrals can be evaluated in analytic form in special cases, such as for on-axis ($x_1 = y_1 = 0$) or in the paraxial approximation (PA). Since an expression

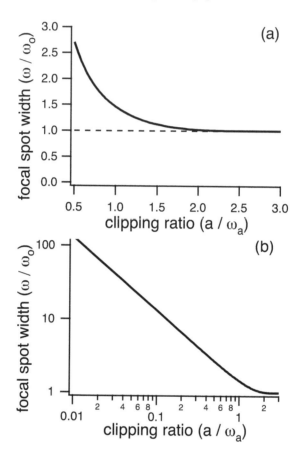

FIGURE 10.10
Calculation of the e^{-2} width of the beam intensity profile in the focal plane as a function of the clipping ratio, a/ω_a, where $z_G = 0.01$ m [7].

for the longitudinal component (E_z) of the electric field for focused Gaussian beams is not easily available in the literature, we provide here the expression in the paraxial approximation for the unperturbed Gaussian propagation regime $(a \gg \omega_a)$. The on-axis value of the longitudinal component is zero.

In the paraxial approximation, when ρ_1 occurs in the exponent, it can be simplified to

$$\rho_1 \approx p_1 z_1 + \frac{(x_1 - x_{01})^2 + (y_1 - y_{01})^2}{2 p_1 z_1}. \tag{10.30}$$

For all occurrences of ρ_1 in the denominators, it is replaced by $p_1 z_1$. Under the paraxial approximation, the integral in Eq. 10.18 can be analytically evaluated, leading to

$$E_z(\vec{r_1}) = \frac{E_0}{4} \frac{(1 + 3s_{11})}{p_1^3 z_1^2 d_1^2} (\kappa_1 + \kappa_2 + \kappa_3) e^{-ip_1^2 z_1} e^{-c_1(x_1^2 + y_1^2)}, \qquad (10.31)$$

where

$$\kappa_1 = \frac{ax_1}{d_1^4} (b_1 - 2a_1) \qquad (10.32)$$

$$\kappa_2 = \frac{ab_1 x_1}{d_1^4} \qquad (10.33)$$

$$\kappa_3 = \frac{bx_1}{d_1^2} \left[\frac{aa_1^2}{d_1^4} (x_1^2 + y_1^2) + b \right] \qquad (10.34)$$

$$s_{11} = \frac{1}{ip_1^2 z_1} \left(1 + \frac{1}{ip_1^2 z_1} \right) \qquad (10.35)$$

with

$$a = \frac{i}{2(1 + 2iz_{G1})(q_1 - z_{G1})^2} \qquad (10.36)$$

$$b = -\frac{1}{1 + 2iz_{G1}} \left(\frac{1}{q_1 - z_{G1}} + ip_1^2 \right) \qquad (10.37)$$

$$a_1 = \frac{i}{2z_1} \qquad (10.38)$$

$$b_1 = \frac{i}{2(q_1 - z_{G1})} \qquad (10.39)$$

$$c_1 = \frac{a_1 b_1}{a_1 + b_1} \qquad (10.40)$$

and

$$d_1 = \sqrt{(a_1 + b_1)}. \qquad (10.41)$$

10.5 Gaussian Beam Propagation Using Luneberg's Vector Diffraction Theory

Beginning with Green's scalar theorem for any two scalar functions, U and V,

$$\iint_S (U\boldsymbol{\nabla}V - V\boldsymbol{\nabla}U) \cdot \hat{n}\, ds = \iiint_\nu (U\nabla^2 V - V\nabla^2 U)\, dv, \qquad (10.42)$$

the diffraction integrals for scalar diffraction theory can be derived [9]. If the region of space of interest is restricted to be free of charges, the diffracting aperture plane is that of surface S_o, V is chosen to be the scalar electric field of the incident light, and U is chosen to be a Green's function, then Eq. 10.42 can be expressed as [10]

$$E\left(\mathbf{r}\right) = -\frac{1}{2\pi} \iint_{S_o} E\left(\mathbf{r_o}\right) \left(\frac{\partial G}{\partial z}\right)_{S_o} ds_o \qquad (10.43)$$

where G is the Green's function,

$$G = \frac{e^{-ik\rho}}{\rho} \qquad (10.44)$$

and ρ is the distance from a point in the aperture plane, (x_o, y_o, z_o), to the point of interest, (x, y, z), or

$$\rho^2 = |\mathbf{r} - \mathbf{r_o}|^2 = (x - x_o)^2 + (y - y_o)^2 + (z - z_o)^2 . \qquad (10.45)$$

The partial derivative with respect to the axial location of the point of interest, z, of Eq. 10.43 is

$$\frac{\partial G}{\partial z} = -ikz \left(1 + \frac{1}{ik\rho}\right) \frac{e^{-ik\rho}}{\rho^2} . \qquad (10.46)$$

Scalar diffraction theories derived from Green's theorem generally choose the light field within the aperture plane, $E\left(\mathbf{r_o}\right)$, to be the same as that of the incident light field. Here we will choose that the incident light field is that of a focused TEM$_{00}$ Gaussian beam with an electric field represented as [3]

$$E\left(x_o, y_o, z_o\right) = \frac{E_o}{\left(1 + \frac{z_o - z_G}{q_o}\right)} e^{-\frac{ik\left(x_o^2 + y_o^2\right)}{2\left(q_o + z_o - z_G\right)} - ik(z_o - z_G)} \qquad (10.47)$$

where E_o is the incident field amplitude, z_G is the axial location of the focal plane, and q_o is the complex focal parameter

$$q_o = \frac{i\pi\omega_o^2 n}{\lambda} \qquad (10.48)$$

with ω_o representing the e^{-1} half-width of the electric field, or the e^{-2} half-width of the intensity in the focal plane, n is the refractive index of the medium the beam is traveling through, and λ is the wavelength of the light field. The parameters z_G and ω_o are illustrated in Figure 10.11, the theoretical setup used in this investigation, along with the additional parameters, ω_a and r_c representing the unperturbed beam width in the aperture plane and the clipping aperture radius, respectively.

If $z_G = z_o$, the focal plane is coplanar with the aperture plane, an assumption frequently used for other scalar and vector diffraction models for Gaussian beams. Although this assumption greatly simplifies the mathematics for calculating fields for points beyond the aperture plane, it limits the use of these

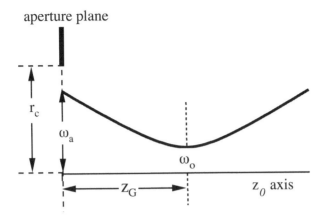

FIGURE 10.11
Theoretical setup used in this investigation where w_o is the unperturbed minimum beam half-width, w_a is the unperturbed beam half-width in the aperture plane, r_c is the radius of the clipping aperture, and z_G is axial the distance from the aperture to the unperturbed beam waist [7].

models to the diffraction of Gaussian beams where the diffraction plane can only occur at the location of the beam's minimum waist. Throughout this chapter, z_G is retained as an independent parameter. In addition to the value of $z_o - z_G$ describing the axial location of the unperturbed focal plane, the sign of $z_o - z_G$ determines whether or not the incident Gaussian light field is converging or diverging. The rest of this chapter will assume that $z_G - z_o$ has a positive value representing the diffraction of a converging focused-Gaussian light field. This particular choice for z_G is such that the diffraction effects of a clipped focused-Gaussian beam can be studied and the effects of diffraction on the focal behavior of the light fields can be quantified in order to present an analytical beam propagation model for clipped-Gaussian beams.

Substitution of Eqs. 10.46 and 10.47 into Eq. 10.43, and defining the location of the $z_o = 0$ plane to be that of the diffracting surface, we get

$$E_x\left(x, y, z\right) = \frac{ikzE_o e^{ikz_G}}{2\pi\left(1 - \frac{z_G}{q_o}\right)} \iint e^{-\frac{ik\left(x_o^2 + y_o^2\right)}{2\left(q_o - z_G\right)}} \frac{e^{-ik\rho}}{\rho^2}\left(1 + \frac{1}{ik\rho}\right) dx_o dy_o \quad (10.49)$$

where it has been assumed that the incident light field is polarized in the x-direction. The limits of the integral in Eq. 10.49 are the edges of the open area of the aperture.

Equation 10.49 has been obtained through scalar diffraction theory based upon Green's scalar theorem and is frequently referred to as the Rayleigh-Sommerfeld diffraction integral. Similarly, starting with Green's scalar theorem Luneberg developed a vector diffraction theory using the electromagnetic wave property that $\nabla \cdot \mathbf{E} = 0$ [10]. Using Luneberg's variation of vector diffraction theory, the x-component of the field at the point of interest is Eq. 10.43, and the y and z-components can be expressed as

$$E_y(\mathbf{r}) = -\frac{1}{2\pi} \iint_{S_o} E_y(\mathbf{r_o}) \left(\frac{\partial G}{\partial z}\right)_{S_o} ds_o, \qquad (10.50)$$

and

$$E_z(\mathbf{r}) = \frac{1}{2\pi} \iint_{S_o} \left[E_x(\mathbf{r_o}) \left(\frac{\partial G}{\partial x}\right)_{S_o} + E_y(\mathbf{r_o}) \left(\frac{\partial G}{\partial y}\right)_{S_o} \right] ds_o, \qquad (10.51)$$

respectively (see Eq. 45.76 of ref. [10]). Following the same procedure as Eq. 10.49, and assuming that the incident light is polarized in the x-direction, the field components E_y and E_z become

$$E_y(x, y, z) = 0 \qquad (10.52)$$

and

$$E_z(x, y, z) = \frac{ikE_o e^{ikz_G}}{2\pi \left(1 - \dfrac{z_G}{q_o}\right)} \qquad (10.53)$$

$$\times \iint e^{-\frac{ik\left(x_o^2 + y_o^2\right)}{2(q_o - z_G)}} \frac{e^{-ik\rho}}{\rho^2} \left(1 + \frac{1}{ik\rho}\right) (x - x_o) \, dx_o dy_o.$$

Equations 10.49, 10.52, and 10.54 are the general integral results of this section for the diffraction of focused Gaussian beams using Luneberg vector diffraction theory.

10.6   Analytical Model for Clipped Gaussian Beams

One of the objectives of this section is to develop an analytical beam propagation model for focused TEM$_{00}$ Gaussian beams clipped by a circular aperture (or other spatial limitation on the radial size of the beam.) To simplify the mathematics two mathematical assumptions will be employed. It will first be assumed that $\rho \gg \lambda$. This assumption forces the parenthesis at the end of Eq. 10.49 to be 1. Using cylindrical coordinates, Eqs. 10.49 and 10.54 become

$$E_x(r, \theta, z) = \frac{ikzE_o e^{ikz_G}}{2\pi \left(1 - \dfrac{z_G}{q_o}\right)} \iint e^{-\frac{ikr_o^2}{2(q_o - z_G)}} \frac{e^{-ik\rho}}{\rho^2} r_o dr_o d\theta_o, \qquad (10.54)$$

and

$$E_z(r, \theta, z) = \frac{ikE_o e^{ikz_G}}{2\pi \left(1 - \dfrac{z_G}{q_o}\right)} \tag{10.55}$$

$$\times \iint e^{-\frac{ikr_o^2}{2(q_o - z_G)}} \frac{e^{-ik\rho}}{\rho^2} (r\cos\theta - r_o\cos\theta_o) r_o dr_o d\theta_o.$$

The second assumption in this section will use the Fresnel approximation to simplify the integrals. Here it is assumed that $z > r_c$; i.e., the distance from the aperture plane to the axial distance of the point of interest is greater than the radius of the diffracting aperture. This assumption corresponds to optical systems with f numbers of 0.5 or higher, which is true for most practical systems. Thus, ρ can be expanded as a Taylor series. For terms in Eq. 10.54 and 10.56 where ρ appears in the phase, the first two terms of the Taylor series expansion are retained. For terms where ρ appears in the amplitude, only the first term of the Taylor series expansion is used. Application of these assumptions and approximations to Eqs. 10.54 and 10.56 yield integral expressions for the x and z-components of the electric field as

$$E_x(r, \theta, z) = \frac{ikE_o e^{ik(z_G - z)} e^{-\frac{ikr^2}{2z}}}{z\left(1 - \dfrac{z_G}{q_o}\right)}$$

$$\times \int e^{-ik\left(\frac{1}{2(q_o - z_G)} + \frac{1}{2z}\right) r_o^2} J_0\left(\frac{krr_o}{z}\right) r_o dr_o. \tag{10.56}$$

and

$$E_z(r, \theta, z) = \frac{ikE_o \cos\theta\, e^{ik(z_G - z)} e^{-\frac{ikr^2}{2z}}}{z^2\left(1 - \dfrac{z_G}{q_o}\right)} \int e^{-ik\left(\frac{1}{2(q_o - z_G)} + \frac{1}{2z}\right) r_o^2}$$

$$\times \left[r J_0\left(\frac{krr_o}{z}\right) - ir_o J_1\left(\frac{krr_o}{z}\right) \right] r_o dr_o. \tag{10.57}$$

Equations 10.56 and 10.57 constitute the Fresnel approximations of the general integral solutions, Eqs. 10.49 and 10.54, of the model presented in this chapter.

For integration purposes, Eqs. 10.56 and 10.57 can be more conveniently expressed as

$$E_x(r, \theta, z) = Ae^{-\frac{ikr^2}{2z}} \int e^{-ar_o^2} J_0(\beta r_o) r_o dr_o \tag{10.58}$$

and

$$E_z\left(r,\theta,z\right) = Ae^{-\frac{ikr^2}{2z}}\frac{\cos\theta}{z}\int e^{-ar_o^2}\left[rJ_0\left(\beta r_o\right) - ir_oJ_1\left(\beta r_o\right)\right]r_odr_o, \quad (10.59)$$

where

$$A = \frac{ikE_oe^{ik(z_G-z)}}{z\left(1-\dfrac{z_G}{q_o}\right)}, \quad (10.60)$$

$$a = ik\left(\frac{1}{2\left(q_o - z_G\right)} + \frac{1}{2z}\right), \quad \text{and} \quad \beta = \frac{kr}{z}. \quad (10.61)$$

For unperturbed focused Gaussian beam propagation, the upper limit of the radial integral goes to infinity, and analytical solutions to Eqs. 10.58 and 10.59 can be found to be

$$E_x\left(r,\theta,z\right) = \frac{E_o}{\left(1+\dfrac{z-z_G}{q_o}\right)}e^{-\frac{ikr^2}{2\left(q_o+z-z_G\right)}-ik(z-z_G)} \quad (10.62)$$

and

$$E_z\left(r,\theta,z\right) = \frac{E_or\cos\theta}{q_o\left(1+\dfrac{z-z_G}{q_o}\right)^2}e^{-\frac{ikr^2}{2\left(q_o+z-z_G\right)}-ik(z-z_G)}$$

or

$$E_z\left(r,\theta,z\right) = E_x\left(r,\theta,z\right)\frac{r\cos\theta}{\left(q_o+z-z_G\right)}. \quad (10.63)$$

Note that after using paraxial approximations and integration, the x-component of the model presented here yields the exact expression as Eq. 10.47 (the scalar beam propagation model of Refs. [3, 11]) for an unperturbed focused Gaussian beam.

Equation 10.63 allows a rough estimation of the magnitude of the longitudinal field component with respect to the dominant transverse component E_x by normalizing E_z to E_x, or

$$\frac{E_z\left(x,y,z\right)}{E_x\left(x,y,z\right)} = \frac{x}{\left(q_o+z-z_G\right)}. \quad (10.64)$$

In the focal plane, $z = z_G$, the ratio of the longitudinal to the transverse field components becomes

$$\frac{E_z}{E_x} = -\frac{ix}{z_R}. \quad (10.65)$$

where the term z_R is the Rayleigh range [3], and is related to the complex focal parameter q_o by $q_o = iz_R$ or

$$z_R = \frac{\pi\omega_o^2 n}{\lambda}. \quad (10.66)$$

The maximum value of E_z in the focal plane occurs when $x = \frac{w_o}{\sqrt{2}}$, and the ratio of the longitudinal component to the transverse component becomes

$$\left(\frac{E_z}{E_x}\right)_{max} = \frac{-i\lambda}{\pi\sqrt{2}w_o}. \tag{10.67}$$

For a very tightly focused beam such that $w_o \approx \lambda$, this ratio becomes $\sim \dfrac{-i}{\pi\sqrt{2}}$; i.e., the peak intensity of the longitudinal component can be upwards of 5% of the transverse component. The effects of the longitudinal component in nonlinear optical experiments with tight focusing may be non-negligible.

The intensity, or modulus square, of each field component can be found to be

$$|E_x|^2 = \frac{E_o^2}{1 + \dfrac{(z - z_G)^2}{Z_R^2}} e^{-2\frac{r^2}{w^2(z)}}, \tag{10.68}$$

and

$$|E_z|^2 = |E_x|^2 \frac{r^2 \cos^2\theta}{z_R^2\left(1 + \dfrac{(z - z_G)^2}{z_R^2}\right)}. \tag{10.69}$$

For clipped focused Gaussian beams the x and z-components of the field are determined by Eqs. 10.58 and 10.59 where the upper limit of the integral is the radius of the clipping aperture, r_c. In order to obtain a simplified analytical model, it will now be assumed that the points of interest beyond the aperture plane reside along the axis of propagation, or $r = x = y = 0$. For on-axis locations, the x-component of the field becomes

$$E_x(0, \theta, z) = A \int_0^{r_c} e^{-ar_o^2} r_o \, dr_o. \tag{10.70}$$

By definition of Eq. 10.52, the y-component of the electric field, E_y, is equal to 0. Substitution of $r = 0$ into the integral for the z-component of the electric field, Eq. 10.59, is also zero for all on-axis positions, or

$$E_z(0, \theta, z) = 0. \tag{10.71}$$

With a straight-forward integration by parts, the complex electric field of Eq. 10.70 can be analytically represented as

$$E_x(z) = E_G(z)\left[1 - e^{-ar_c^2}\right]. \tag{10.72}$$

where

$$E_G(z) = \frac{A}{2a} = \frac{E_o}{1 + \dfrac{z - z_G}{q_o}} e^{-ik(z - z_G)} \tag{10.73}$$

is the function for the on-axis field of an unperturbed focused Gaussian beam.

Computing the modulus square of Eq. 10.72 and simplifying, the analytical on-axis intensity for a clipped focused Gaussian beam can be expressed as

$$I(z) = I_G(z) \left[1 + e^{-2\gamma^2} - 2e^{-\gamma^2} \cos\left(\frac{kr_c^2}{2} \left[\frac{1}{R_a} - \frac{1}{z} \right] \right) \right] \qquad (10.74)$$

where

$$I_G(z) = \frac{E_o^2}{1 + \frac{(z-z_G)^2}{z_R^2}} \qquad (10.75)$$

is the exact expression for the on-axis intensity function for an unperturbed TEM$_{00}$ Gaussian beam. The variables γ and R_a in Eq. 10.74 are the clipping ratio and the radius of curvature of the wavefront at the focal plane, respectively, given by

$$R_a = z_G \left(1 + \frac{z_R^2}{z_G^2} \right) \qquad (10.76)$$

and

$$\gamma = \frac{r_c}{\omega_a}, \qquad (10.77)$$

where ω_a is defined by Eq. 10.27.

Equations 10.72 and 10.74 are expressed in the form where the on-axis electric field and intensity functions of a clipped Gaussian beam are equal to the unperturbed field and intensity functions multiplied by diffraction-dependent functions. Equations 10.72 and 10.74 constitute the general analytical solutions for this section. Equation 10.74 provides the analytical expression for the on-axis behavior of clipped focused-Gaussian beams as a function of the axial distance from the aperture plane.

A simpler version of Eq. 10.74 can be obtained if we restrict ourselves to on-axis locations near the focal plane, $z \approx z_G$, and restrict the aperture plane to be far from the focal plane with respect to the Rayleigh range, i.e., $z \gg z_R$ such that $R_a \approx z_G$. Thus, the maximum obtainable field intensity is just a function of the clipping ratio, or

$$I_{max}(\gamma)_{z \approx z_G} = I_{G\,max} \left[1 + e^{-2\gamma^2} - 2e^{-\gamma^2} \right]. \qquad (10.78)$$

Equation 10.78 represents a simple analytical model that can be used to quickly determine the maximum theoretical intensity in the focal plane for a clipped focused-Gaussian beam. The limitations of using Eq. 10.78 are discussed further in the next section.

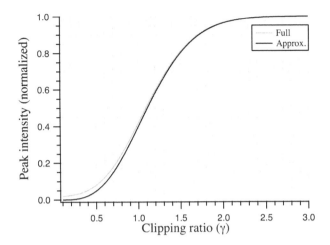

FIGURE 10.12

Calculated peak intensities versus the clipping ratio using the full analytical expression, Eq. 10.74, and the simpler approximated expression, Eq. 10.78 [7].

10.7 Calculations and Measurements for Clipped Gaussian Beams

The main theoretical result of Section 10.6 is the analytical expression for the on-axis intensity, Eq. 10.74, and its simpler form, Eq. 10.78. Using Eqs. 10.74 and 10.78, the peak obtainable on-axis intensity can be calculated. As the clipping ratio decreases and more and more of the wings of the incident Gaussian beam are blocked by the aperture, mirror, lens edges, etc., diffraction effects on the beam's propagation and focusability become non-negligible and limit the theoretical peak obtainable intensity of the beam. Figure 10.12 is a plot of peak obtainable intensities using either Eq. 10.74 or Eq. 10.78, normalized to the peak obtainable intensity for an unperturbed focused TEM_{00} Gaussian beam. The parameters chosen for calculation of Figure 10.12 are a wavelength of $\lambda = 780$ nm, a minimum beam waist of $\omega_o = 10.1$ μm, and an aperture to focal plane distance of $z_G = 20$ $z_R = 8.2$ mm. These particular parameters were chosen as they match experimental parameters measured in the laboratory [7]. Significant deviations occur between the full analytical expression, Eq. 10.74, and the simpler approximated expression, Eq. 10.78, for clipping ratios $\lesssim 1$.

The primary reason for the deviation is the fact that the values plotted for the *full* analytical solution in Figure 10.12 are the peak intensities obtained for *any* on-axis position. The values calculated using the *approximated* expression are the peak intensities in the focal plane (fixed $z = z_G$ value). For purely

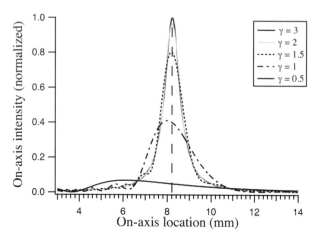

FIGURE 10.13
Calculated on-axis intensities using the analytical model for clipping ratios of
3, 2, 1.5, 1, and 0.5 [7]. The dashed vertical line corresponds to the unperturbed
focal plane.

Gaussian beams (large γ values), the peak on-axis intensity always occurs at
the focal plane. For purely diffracted beams, $(0 \lesssim \gamma \ll 1)$, the peak on-axis
intensity value occurs for $z = r_c^2/\lambda$. As the aperture gets smaller, the diffraction peak intensity moves closer and closer to the aperture. This behavior is
observed in Figure 10.13, which is a collection of on-axis calculations using
Eq. 10.74 for various clipping ratios. For example, the difference between the
full and approximated values for $\gamma = 1$ in Figure 10.12 is due to the shift of
the peak intensity from the unperturbed focal plane towards the aperture as
observed in the $\gamma = 0.5$ line of Figure 10.13. In addition to the shift of the
peak intensity towards the aperture for smaller clipping ratios, another important effect of the diffraction on the focal behavior of clipped beams is the
observed asymmetry of the on-axis intensity. As diffraction effects play a more
significant role for smaller values of γ, the on-axis behavior of the intensity
deviates from the symmetrical Lorentzian distributions for pure TEM$_{00}$ Gaussian beams, and begins to resemble the oscillatory and asymmetrical behavior
of diffracted plane waves [5].

The theoretical results were experimentally verified using a diode laser as
the light source with the output tuned to a wavelength of 780.24 nm. As is
common with most diode-based laser cavities the output of the system is not
that of a pure TEM$_{00}$ Gaussian beam. The output of the laser head contains
some higher-mode Hermite-Gaussian contributions and is slightly elliptical. To
remove the higher-order Hermite-Gaussian modes, a spatial filter was built and
optimized. The output of the spatial filter is near TEM$_{00}$ but still contains
some ellipticity. In order to remove the ellipticity of the beam profile and

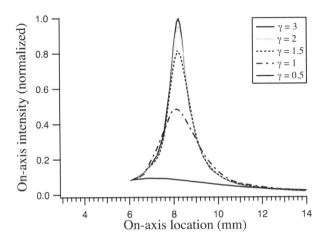

FIGURE 10.14
Experimentally measured on-axis intensities for clipping ratios of 3, 2, 1, 5, 1, and 0.5 [7]. The abrupt cutoff $\sim$ 6 mm is due to physical limitations between the aperture mount and the face of the beam profiler. All intensities are normalized to the peak of an unperturbed beam.

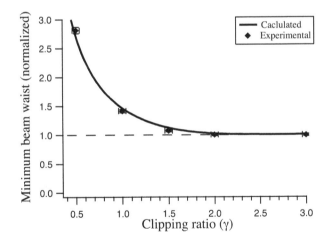

FIGURE 10.15
Calculated and experimentally measured minimum beam e^{-2} half-widths of the intensity [7]. Calculations were performed using the general integral solution Eq. 10.49.

to improve the beam overall to establish a baseline circular TEM_{00} mode, the beam is coupled into a coiled single-mode fiber optic cable. The output of the single-mode fiber is collimated using an aspheric lens, which yields a

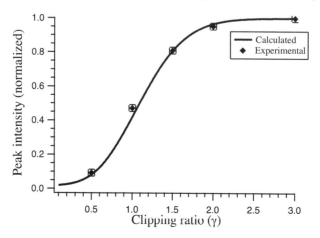

FIGURE 10.16

Calculated and experimentally measured peak intensities [7]. Calculations were performed using the full analytical expression, Eq. 10.74, and experimentally measured peak intensities are normalized to the peak intensity of an unperturbed beam.

near-perfect TEM_{00} Gaussian beam with an average ellipticity of $\cong 0.99$, and an unperturbed beam propagation factor of $M^2 = 1.0$. All beam width and intensity measurements were obtained using an optical beam profiler mounted on a long translation stage in order to obtain beam profile data as a function of position in the beam propagation direction to calculate M^2 beam quality values. Beam widths and profiles obtained using the beam profiler were also verified using scanning slit and pinhole techniques. Using a 75-mm focal length lens, a minimum beam waist of $\omega_o = 10.1$ μm was obtained with a Rayleigh range of $z_R = 411$ μm [7]. The location of the aperture plane was chosen to be $z_G = 20z_R = 8.2$ mm.

Figure 10.14 is a collection of experimentally measured on-axis intensity distributions for the same clipping ratios used in the calculations for Figure 10.13. The measured unperturbed beam width in the location of the aperture is $\omega_a = 201$ μm. Commercially available precision pinholes from Melles Griot were used to provide various clipping ratios. The abrupt cutoff observed in Figure 10.14 around 6 mm is a result of the physical limitations of the minimum distance between the pinhole foil and the scanning slits of the beam profiler due to the pinhole mount and holder and the front housing of the beam profiler. All recorded intensity values are normalized to the peak intensity of the unperturbed focused beam. For clipping ratios of $\gamma \lesssim 1$, the shifting of the peak intensity towards the aperture plane as well as the asymmetrical on-axis predicted intensity distributions are observed in accordance with the theoretical calculations shown in Figure 10.13.

Figure 10.15 is a comparison of theoretically predicted and experimentally measured minimum beam widths of clipped focused-Gaussian beams as a function of the clipping ratio [7]. The width is calculated/measured for the minimum beam width at different axial locations beyond the aperture. For clipping ratios > 1 the minimum occurs at the unperturbed focal plane. For clipping ratios $\lesssim 1$ the minimum beam waist moves closer to the aperture. Both the calculated and experimentally measured width values are normalized to the width of an unperturbed focused-Gaussian beam, or ω_o. The theoretically predicted normalized e^{-2} widths of the intensity for clipping ratios of 2, 1.5, 1, and 0.5 are 1.02, 1.12, 1.46, and 2.7, respectively. Similarly, Figure 10.16 is a comparison of theoretically predicted peak intensity values compared to experimentally measured peak intensities. Theoretical calculations are performed using the full analytical expression, Eq. 10.74. The peak intensities graphed are the maximum intensities beyond the aperture plane, and normalized to the peak obtainable intensity for an unperturbed beam. For lower clipping ratio values where the full, Eq. 10.74, and approximated, Eq. 10.78, expressions deviate from one another, the experimentally measured peak intensities best match that of the full analytical on-axis model. The axial locations of both the calculated and measured peak intensities coincide with the axial location of the minimum beam waist. A closer inspection of the beam width and peak intensity values for a clipping ratio of $\gamma = 1$ demonstrates the strong effects of diffraction on the focusability of clipped beams. For a clipping ratio of 1 only 15.7% of the energy of the beam is being blocked by the aperture, yet the beam width has increased by 46% and the peak obtainable intensity has decreased by 58%. As Figures 10.15 and 10.16 illustrate, good agreement is observed between theoretically predicted values using the model presented here, and experimentally measured values in a laboratory setting.

Bibliography

[1] G. Bekefi, "Diffraction of electromagnetic waves by an aperture in a large screen," J. Appl. Phys. **24**, 1123–1130 (1953).

[2] M. Born and E. Wolf, *Principles of Optics*, Seventh Edition (Cambridge University Press, Cambridge, UK, 1999.)

[3] A. Yariv, *Quantum Electronics*, Third Edition (John Wiley & Sons, New York, 1989.)

[4] G. D. Gillen, K. Baughman, and S. Guha, "Application of Hertz vector diffraction theory to the diffraction of focused Gaussian beams and calculations of focal parameters," Opt. Express **17**, 1478–1492 (2009).

[5] S. Guha and G. D. Gillen, "Description of light propagation through a circular aperture using nonparaxial vector diffraction theory," Opt. Express **13**, 1424–1447 (2005).

[6] W. T. Sifvast, *Laser Fundamentals* (Cambridge University Press, New York, NY, 1996.)

[7] G. D. Gillen, C. M. Seck, and S. Guha, "Analytical beam propagation model for clipped focused-Gaussian beams using vector diffraction theory," Opt. Express **18**, 4023–4040 (2010).

[8] A. J. Campillo, J. E. Pearson, S. L. Shapiro, and N. J. Terrell, Jr., "Fresnel diffraction effects in the design of high-power laser systems," Appl. Phys. Lett. **23**, 85–87 (1973).

[9] G. D. Gillen and S. Guha, "Modeling and propagation of near-field diffraction patterns: a more complete approach," Am. J. Phys. **72**, 1195–1201 (2004).

[10] R. K. Luneberg, *Mathematical Theory of Optics* (University of California Press, Berkeley, California, 1964.)

[11] H. Kogelnik and T. Li, "Laser beam resonators," Proc. IEEE **54**, 1312–1329 (1966).

11

Trapping Cold Atoms with Laser Light

> In this chapter we present an interesting application of the propagation of laser light through linear optical media: the ability to spatially confine (trap) cold atoms within localized light intensity patterns. We first present some of the fundamentals for understanding how light can trap an atom. Depending upon the relationship between the properties of the atom and the properties of the laser light, atoms can be trapped in either localized dark spots, or localized bright spots. We will describe several light patterns for trapping atoms:
>
> - The tightly localized bright and dark regions in the near-field diffraction patterns presented in Chapters 7 and 8,
>
> - The near-field patterns projected to a location far away from the diffracting aperture, and
>
> - Polarization-dependent movable light field traps based on the near-field diffraction patterns.

11.1 Introduction to Trapping Atoms Using Light Fields

When an atom is exposed to the oscillating electric field of laser light, the light field can exert an optical dipole force on the atom. The first experimental observation of the optical dipole force was achieved in 1978 by Bjorkholm et al. [1]. Since that time, several fields of study have developed that exploit the optical dipole force on cold atoms and molecules. One application of the optical dipole force is to confine and trap cold neutral atoms within spatially varying light field distributions [2]. If the frequency of the laser light is less than the resonant frequency of the atom, the dipole potential energy is a minimum for localized high-intensity fields and a red-detuned atomic trap (RDT) is created. Conversely, if the frequency of the laser light is greater than the resonant frequency of the atom, the dipole potential energy is a minimum for localized low-intensity fields and a blue-detuned atomic trap (BDT) is created [3]. The difference between the various pursuits within the field of trapping cold neutral atoms is determined by the optical methods employed to create laser light fields with localized maxima or minima. Various methods to create localized high or low intensity fields have included: the focusing of a single laser beam to create a single trap at the focal spot [4], using the periodic interference pattern of counter propagating beams to create one-dimensional [5], two-dimensional [6], and three-dimensional [7] arrays of optical traps, and using axicons [8] or Laguerre-Gaussian beams [9] to create localized dark regions. All of these methods are some of the more common ones employed for neutral atom traps. Detailed descriptions of each are easily found in the published literature. In this chapter we present a few of the lesser-known proposed patterns for neutral atom trapping. One less common method of recent interest for creating light fields with localized maxima and minima for trapping atoms is to use the diffracted light in and around spatially limiting apertures [10, 11, 12, 13]. Another diffraction-based approach uses the polarization dependence of light traps to confine atoms in a particular magnetic substate [14].

11.2   Optical Dipole Trapping Potential Energy

When an atom is exposed to an electromagnetic field the interaction of the light and the induced dipole moment of the atom results in a dipole potential energy and its associated dipole force. In this section, we will follow the steps from [2] to derive an equation relating the dipole potential energy to the intensity pattern of the light field.

The potential energy of a dipole in an electric field $\mathbf{E}$ is given by

$$U = -\mathbf{p} \cdot \mathbf{E}. \tag{11.1}$$

Here, the electric dipole moment $\mathbf{p}$ is defined as

$$\mathbf{p} = q\mathbf{d}, \tag{11.2}$$

where q is the amount of charge at either end of the dipole, and $\mathbf{d}$ is the displacement vector from the negative charge to the positive charge. In this section, we will use a classical model in which the light is treated as a harmonic driving field of angular frequency ω,

$$\mathbf{E} = E_o \hat{x} e^{-i\omega t}, \tag{11.3}$$

and the atom is treated as an electron undergoing a driven, damped oscillation around the positive core. Here, we have assumed linear light polarization. For the corresponding semi-classical equation that applies to any light polarization, see Section 11.5. For this model we have $q = e$, the elementary charge, and $\mathbf{d} = -\mathbf{x}$, where $\mathbf{x}$ is the position vector of the electron relative to the core.

We can find the position x of the electron as a function of time by solving the differential equation

$$m_e \frac{d^2 x}{dt^2} + m_e \Gamma_\omega \frac{dx}{dt} + m_e \omega_o^2 x = -eE_o e^{-i\omega t}, \tag{11.4}$$

where m_e is the mass of the electron, Γ_ω is the radiative decay rate of the energy of an electron oscillating at an angular frequency of ω in an atom (see [2]), and ω_o is the resonance angular frequency of the atom. The solution of this differential equation is

$$x = -\frac{e}{m_e} \frac{1}{(\omega_o^2 - \omega^2 - i\omega \Gamma_\omega)} E_o e^{-i\omega t}. \tag{11.5}$$

With this, the induced electric dipole moment is

$$\mathbf{p} = -e\mathbf{x} = \frac{e^2}{m_e} \frac{1}{(\omega_o^2 - \omega^2 - i\omega \Gamma_\omega)} E_o \hat{x} e^{-i\omega t}. \tag{11.6}$$

This can be written as

$$\mathbf{p} = \alpha \mathbf{E}, \tag{11.7}$$

with

$$\alpha = \frac{e^2}{m_e} \frac{1}{(\omega_o^2 - \omega^2 - i\omega \Gamma_\omega)}, \tag{11.8}$$

the complex polarizability of the atom. Because both the electric field and the induced electric dipole moment of the atom are oscillating rapidly, the relevant expression for the electric dipole potential energy is the time average of Eq. 11.1, so

$$U = -\frac{1}{2} \langle \mathbf{p} \cdot \mathbf{E} \rangle. \tag{11.9}$$

Note that the factor of $1/2$ is introduced to account for the fact that this is an induced dipole, not a permanent one [2]. To calculate this product, we need to revert back from the complex notation. The real part of the electric field is

$$Re\,(\mathbf{E}) = E_o \hat{x} \cos \omega t. \tag{11.10}$$

The real part of the electric dipole moment is

$$Re\,(\mathbf{p}) = Re\,(\alpha)\,Re\,(\mathbf{E}) + Im\,(\alpha)\,Im\,(\mathbf{E})\,. \tag{11.11}$$

Using

$$Im\,(\mathbf{E}) = -E_o \hat{x} \sin \omega t \tag{11.12}$$

this becomes

$$Re\,(\mathbf{p}) = E_o \hat{x}\,(Re\,(\alpha) \cos \omega t - Im\,(\alpha) \sin \omega t)\,. \tag{11.13}$$

Plugging these expressions into Eq. 11.9 and carrying out the time average yields

$$U = -\frac{1}{4} Re\,(\alpha)\,E_o^2 \tag{11.14}$$

or

$$U = -\frac{1}{4}\frac{e^2}{m_e}\frac{\omega_o^2 - \omega^2}{(\omega_o^2 - \omega^2)^2 + \omega^2\Gamma_\omega^2}E_o^2. \tag{11.15}$$

Equation 11.15 can be expressed in terms of the linewidth, Γ, and the saturation field, E_s, both properties of the atomic resonance. The linewidth Γ_ω of an electron oscillating about an atomic core with angular frequency ω is [2]

$$\Gamma_\omega = \frac{e^2\omega^2}{6\pi\epsilon_o m_e c^3}. \tag{11.16}$$

For the atomic resonance, $\omega = \omega_o$, so

$$\Gamma = \frac{e^2\omega_o^2}{6\pi\epsilon_o m_e c^3} \tag{11.17}$$

and

$$\Gamma_\omega = \frac{\omega^2}{\omega_o^2}\Gamma. \tag{11.18}$$

The saturation intensity is given by [15]

$$I_s = \frac{\pi h c}{3\lambda^3 \tau} \tag{11.19}$$

where h is Planck's constant, c is the speed of light in vacuum, λ is the wavelength corresponding to the atomic resonance, and τ is the lifetime of the atomic resonance. Using the relationship between light intensity, I, and electric field amplitude, E_o,

$$I = \frac{1}{2}\epsilon_o c E_o^2, \tag{11.20}$$

we find for the square of the saturation field

$$E_s^2 = \frac{2}{3}\frac{\pi h}{\epsilon_o \lambda^3 \tau}. \tag{11.21}$$

Plugging in

$$\tau = \frac{1}{\Gamma} \tag{11.22}$$

and

$$\lambda = \frac{2\pi c}{\omega_o}, \tag{11.23}$$

we find

$$E_s^2 = \frac{h\Gamma\omega_o^3}{12\pi^2\epsilon_o c^3} \tag{11.24}$$

or

$$E_s^2 = \frac{\hbar\Gamma^2 m_e \omega_o}{e^2}. \tag{11.25}$$

Recasting Eq. 11.15 in terms of the linewidth and saturation field of the atomic transition results in

$$U = -\frac{\hbar\Gamma^2\omega_o}{4} \frac{\omega_o^2 - \omega^2}{(\omega_o^2 - \omega^2)^2 + \left(\frac{\omega^3}{\omega_o^2}\right)^2 \Gamma^2} \frac{E_o^2}{E_s^2}. \tag{11.26}$$

If the driving angular frequency meets the condition

$$(\omega_o^2 - \omega^2)^2 \gg \left(\frac{\omega^3}{\omega_o^2}\right)^2 \Gamma^2, \tag{11.27}$$

we can approximate Eq. 11.26 as

$$U = -\frac{\hbar\Gamma^2\omega_o}{4} \frac{1}{\omega_o^2 - \omega^2} \frac{E_o^2}{E_s^2}. \tag{11.28}$$

This can further be rewritten as

$$U = -\frac{\hbar\Gamma^2}{8} \left(\frac{1}{(\omega_o - \omega)} + \frac{1}{(\omega_o + \omega)} \right) \frac{E_o^2}{E_s^2}. \tag{11.29}$$

If the driving angular frequency and the resonance angular frequency are on the same order of magnitude,

$$\omega \approx \omega_o, \tag{11.30}$$

the first term well exceeds the second term, so we further approximate this as

$$U = -\frac{\hbar\Gamma^2}{8} \frac{1}{(\omega_o - \omega)} \frac{E_o^2}{E_s^2}. \tag{11.31}$$

This approximation is known as the rotating wave approximation. Defining the laser detuning, Δ, as

$$\Delta = \omega - \omega_o \tag{11.32}$$

we find

$$U = \frac{\hbar\Gamma}{8} \frac{\Gamma}{\Delta} \frac{E_o^2}{E_s^2} \tag{11.33}$$

for the electric dipole potential energy for an atom driven at an angular frequency that meets conditions Eq. 11.27 and Eq. 11.30. Therefore, a spatially varying light field can be used to trap atoms in the potential energy minima formed by localized minima ($\Delta > 0$) or maxima ($\Delta < 0$) of electric field amplitude, $E_o(\mathbf{r})$, or intensity, $I(\mathbf{r})$.

Thus far, both the light and the atom were treated classically. However, because the atom has multiple resonances with resonant frequencies determined by its quantum mechanical energy levels, this expression for the dipole

potential energy has to be modified. To continue using the simplified single
resonance expression, the laser detuning must be either very small compared
to the splittings between the multiple atomic levels, so that the other levels
can be ignored for being much farther detuned, or very large, so that multiple
levels can be approximated as one. Thus, Eq. 11.33 holds for detunings that
are large compared to the excited state hyperfine splitting ($\approx 2\pi \times 100$ MHz or
less[1]), but small compared to the fine structure splitting of the excited state
($\approx 2\pi \times 10 - 10000$ GHz)[2], except for one modification. To account for the
existence of the other fine structure excited state, the electric dipole potential energy must be multiplied by the line-strength factor, β, of the atomic
transition. For this range of detunings, the condition in Eq. 11.27 is fulfilled.
For commonly trapped atomic species, the range of linewidths of the atomic
resonances is[3]

$$\frac{\Gamma}{2\pi} \approx 10^6 - 10^8 Hz. \tag{11.34}$$

The range of resonance frequencies is

$$\frac{\omega_o}{2\pi} \approx 10^{14} - 10^{15} Hz. \tag{11.35}$$

Even for very large detunings encountered in atom trapping experiments, the
driving angular frequencies are on the same order of magnitude as the resonance angular frequency, so $\omega \approx \omega_o$. With this, we can estimate the minimum
detuning needed for the approximated expression, Eq. 11.33 to be valid. Taking the positive square root of Eq. 11.27 leads to

$$\left| \omega_o^2 - \omega^2 \right| \gg \left(\frac{\omega^2}{\omega_o^2} \right) \omega\Gamma \tag{11.36}$$

or

$$(\omega_o + \omega) \left| \omega_o - \omega \right| \gg \left(\frac{\omega^2}{\omega_o^2} \right) \omega\Gamma. \tag{11.37}$$

Using $\omega \approx \omega_o$, this simplifies to

$$2\omega_o \left| \omega_o - \omega \right| \gg \omega\Gamma \tag{11.38}$$

and

$$\left| \omega_o - \omega \right| \gg \frac{\Gamma}{2}. \tag{11.39}$$

Thus, we find for the order of magnitude of the detuning

$$\left| \Delta \right| \gg \frac{\Gamma}{2} \approx 2\pi \times 10^6 - 10^8 Hz. \tag{11.40}$$

Therefore, the approximation in Eq. 11.27 is valid for detunings much larger

---

[1] See Table C.4 and pp. 40 and 41 in [15]
[2] See Table C.4 in [15]
[3] See Table C.1 in [15].

than typical excited state hyperfine splittings ($\approx 2\pi \times 100$ MHz). Finally, we have thus far ignored the effects of using light polarizations other than linear. This polarization dependence is discussed in Section 11.5.

Therefore, for detunings that are large compared to the excited state hyperfine splitting but small compared to the fine structure splitting, and for linearly polarized light, the position-dependent electric dipole potential energy for an atom in a hyperfine ground state F can be expressed as [16]

$$U(\mathbf{r}) = \beta \frac{\hbar\Gamma}{8} \frac{\Gamma}{\Delta} \frac{|E(\mathbf{r})|^2}{|E_s|^2}, \tag{11.41}$$

where $\mathbf{r}$ is the location of the point of interest, β is the line-strength factor of the atomic transition [2, 17], Γ is the spontaneous decay rate, E_s is the saturation field of the transition, and Δ is the laser detuning from the transition from F to the center of the excited state hyperfine manifold.

The particular cold neutral atoms used for this chapter are rubidium atoms, and the specific transition of interest is the $D2$ transition $\left(^2S_{1/2} \rightarrow ^2P_{3/2}\right)$, which has a line-strength factor of $\beta = 2/3$. Including the line-strength factor, the atom trapping potential energy becomes

$$U(\mathbf{r}) = \frac{2}{3} \frac{\hbar\Gamma}{8} \frac{\Gamma}{\Delta} \frac{|E(\mathbf{r})|^2}{|E_s|^2}. \tag{11.42}$$

For cold rubidium atoms and 780-nm laser light, the saturation electric field for this transition is $E_s = 111.5$ V/m [15] (or an intensity of 1.65 mW/cm^2), and the spontaneous decay rate is $\Gamma = 2\pi \times 6$ MHz [15].

The magnitude of the optical potential energy of the atom is directly proportional to the intensity of the laser field. The sign of the potential energy is determined by the sign of the detuning. If $\omega < \omega_o$, the energy of a photon in the laser field is less than the energy of the atomic transition, and the potential energy is negative. Potential energy minima thus occur at locations of maximum intensity. The atom is drawn towards regions of higher intensity light and red-detuned traps (RDT) are created. If the detuning is positive, or $\omega > \omega_o$, the atoms are drawn towards regions of lower intensity light and, consequently, blue-detuned traps (BDT) are created.

11.3 Diffracted Light Just beyond a Circular Aperture

In this section we present the approach of using localized high and low intensity variations found in the near-field diffraction patterns of light incident upon a circular aperture for the trapping of cold neutral atoms using the dipole force. An advantage of this method over many currently employed approaches resides in the experimental simplicity of using a single unfocused (or loosely focused) laser beam incident upon a diffracting aperture. The method presented here is extendable to more complex diffraction masks consisting of a series of circular holes in one or two dimensions and of varying diameters. Such diffraction masks would result in either one- or two-dimensional arrays of dipole traps of varying dimensions, trap depths, and distances from the aperture plane.

11.3.1 Choosing a Propagation Model, HVDT

When choosing which diffraction beam propagation model to use for a given problem there are a couple of questions to ask yourself: "How close to the aperture is my closest point of interest?" and "Which diffraction model is valid for that region of space?"

For our investigation in this section, we are proposing to use the intricate light pattern of maxima and minima located just beyond the diffracting aperture. To allow the model to be used for the extreme case where the apertures are on the same distance scale as the wavelength of light, we need a model that would be valid for regions up to and even possibly including the plane of the diffracting aperture. Thus, Hertz vector diffraction theory, HVDT, would be appropriate.[4]

Light distributions within and beyond a diffracting aperture can be accurately modeled using vector diffraction theory invoking the Hertz vector formalism [18, 19], in which the polarization potential, or the Hertz vector $\mathbf{\Pi}$, of the light field incident upon the aperture is assumed to be known in the aperture plane.

The electric and magnetic fields for any point within the aperture plane and beyond can be calculated using a single vector potential function $\mathbf{\Pi}$:

$$\mathbf{E} = k^2\mathbf{\Pi} + \nabla\left(\nabla \cdot \mathbf{\Pi}\right),\tag{11.43}$$

and

$$\mathbf{H} = -\frac{k^2}{i\omega\mu_o}\nabla \times \mathbf{\Pi} = ik\sqrt{\frac{\epsilon_o}{\mu_o}}\nabla \times \mathbf{\Pi},\tag{11.44}$$

where k is the wavenumber, ω is the angular frequency of the light, and μ_o and ϵ_o are the permeability and permittivity of free space, respectively. It has been assumed that the time-dependence of the fields follows $e^{-i\omega t}$.

If the incident field is a uniform plane wave linearly polarized, say along the x-axis, the electromagnetic field components at the point of interest can be calculated from the x-component of the Hertz vector only [18], where

$$\Pi_x(x,y,z) = \frac{iE_o}{2\pi k}\int\int\frac{e^{-ik\rho}}{\rho}dx_o dy_0,\tag{11.45}$$

where E_o is the electric field amplitude of the incident light, and ρ is the

---

[4]In this section we present only the fundamentals of HVDT necessary to describe the calculated light fields. For a complete theoretical treatment of HVDT see Section 7.5.

distance from the integration point, (x_o, y_o, z_o), in the aperture plane to the point of interest (x, y, z). The integration is performed over the open area of the aperture in the aperture plane. Once the Hertz vector has been calculated for the point of interest, and **E** and **H** have been determined using Eqs. 11.43 and 11.44, the electromagnetic field intensity is given by the Poynting vector,

$$\mathbf{S} = Re\left(\mathbf{E} \times \mathbf{H}^*\right). \tag{11.46}$$

An important parameter in the diffraction of light by a circular aperture is the ratio of the radius of the aperture, a, to the wavelength of light, λ. For aperture radius to wavelength ratios greater than 1, oscillations exist in the diffracted field intensity for points between the aperture plane and an axial distance of a^2/λ (see Ref. [19] for a more detailed discussion). In the aperture plane, the number of oscillations in the radial direction perpendicular to the polarization direction from the center of the aperture to the edge will be equal to a/λ, as will be the number of oscillations observed along the center axis of the aperture in the axial direction [19].

For distances larger than an aperture radius away from the aperture plane, calculations for the on-axis intensity distributions can be performed using the simpler Kirchhoff vector diffraction method [19]. Using Kirchhoff vector diffraction theory, an analytical expression for the on-axis electromagnetic fields can be derived. Due to cylindrical symmetry, the transverse components of the Poynting vector are zero for points along the axis of the aperture, and the z-component of the on-axis Poynting vector simplifies to

$$S_z\left(0, 0, z\right) = S_0 \left(\frac{1 + 2\left(\frac{z}{a}\right)^2}{1 + \left(\frac{z}{a}\right)^2} - \frac{2\frac{z}{a}}{\sqrt{1 + \left(\frac{z}{a}\right)^2}} \right.$$
$$\left. \times \cos\left[kz \left(\sqrt{1 + \left(\frac{a}{z}\right)^2} - 1 \right) \right] \right), \tag{11.47}$$

where $S_o = \sqrt{\frac{\epsilon_o}{\mu_o}}|E_o|^2$ is the intensity of light incident upon the aperture.

From Eq. 11.47, the number and the axial locations, $z(m)$, of the maxima and minima of $S_z(0, 0, z)$ follow:

$$z\left(m\right) = \frac{a^2}{\lambda}\left(\frac{1}{m} - \frac{m\lambda^2}{4a^2}\right), \tag{11.48}$$

where m is an integer from 1 to $2a/\lambda$. Odd integer values of m correspond to on-axis intensity maxima, and even integer values correspond to on-axis minima; for example, $z(1)$ is the position of the on-axis maximum farthest from the aperture plane. For larger ratios of a/λ the farthest away on-axis maxima have relative intensities equal to approximately four times the light intensity incident upon the aperture, and the last few minima have absolute values very close to zero.

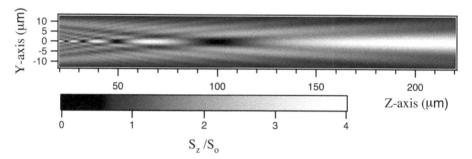

FIGURE 11.1
Calculated image plot of the z-component of the Poynting vector versus y and z for plane-wave, x-polarized, 780-nm light incident upon a 25-μm aperture [11]. The relative intensity is normalized to the intensity incident upon the aperture. (Reprinted figure with permission from G. S. Gillen, S. Guha, and K. Christandl, Phys. Rev. A, 73, 013409, 2006. Copyright 2006 by the American Physical Society.)

Figure 11.1 is an illustration using a two-dimensional image plot of the calculated light field distributions of the z-component of the Poynting vector versus the radial and axial positions (y and z) for 780-nm light incident upon a 25-μm diameter aperture and polarized in the x-direction using the Hertz vector formalism, Eqs. 11.43–11.46. The scale of Figure 11.1 is normalized to the uniform light intensity incident upon the aperture. Beyond a distance of a^2/λ, or 200 μm, the beam profile has a single, central and radially symmetric maximum that smoothly expands in width and decreases in amplitude for larger and larger axial distances, similar to the behavior of a diverging laser beam. Between the aperture plane and distance of $z(m = 1)$ an intricate diffraction pattern exists with localized regions of light and dark intensities. Figure 11.2 is a graph of the on-axis intensity of the diffracted laser light as a function of the distance from the aperture for the same axial range and parameters as in Figure 11.1. Figure 11.2 illustrates the localization of high and low intensity levels along the central axis of beam propagation. The two points of interest are the on-axis positions of approximately 66 μm and 100 μm, or the locations of the second to last on-axis maximum and the last on-axis minimum. These localized intensity distributions correspond to possible optical dipole neutral atom traps for red and blue-detuned laser light, respectively.

Although numerous regions of localized high and low intensity light distributions exist closer to the aperture, only the two possible traps at locations of $z(m = 2)$ and $z(m = 3)$ are considered for further investigation here for three reasons. First, as the trap location approaches the aperture (higher m values) the theoretical traps become smaller, more tightly spaced, and more tightly confined. Although these trap locations are theoretically appealing, these relative trap locations would present a higher level of difficulty to probe

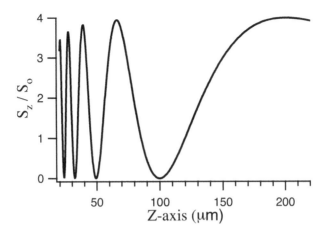

FIGURE 11.2

Calculated on-axis intensity of the z-component of the Poynting vector versus the axial distance from the aperture plane for the same parameters as Figure 11.1 [11]. (Reprinted figure with permission from G. S. Gillen, S. Guha, and K. Christandl, Phys. Rev. A, 73, 013409, 2006. Copyright 2006 by the American Physical Society.)

and address the trapped atoms experimentally. For example, the spacing between the on-axis blue-detuned traps of $m = 10$ and $m = 12$ is only 3.7 μm; whereas the spacing between the $m = 2$ and $m = 4$ traps are separated by 48 μm. Additionally, for m values greater than 4, the cross-section of the intensity profiles indicate that there is a central ellipsoidal trap surrounded by one or more toroidal traps of comparable depths. The simplest experimental approach to address atoms trapped by diffracted laser light would be from an angle perpendicular to the axis of beam propagation. Thus, the toroidal traps surrounding the ellipsoidal traps would make addressing the central on-axis trap difficult.

11.3.2 Calculated Light Fields and Atom Trap Potentials

As illustrated in Figure 11.1, the near-field distribution of light diffracted by a circular aperture contains numerous localized spots of alternating high and low laser intensities. In this section, two regions are of interest for optically trapping neutral atoms. One region is a strong candidate for the trapping of atoms using blue-detuned laser light, and the other is of interest for trapping atoms using red-detuned laser light. The primary spot of interest here for a blue-detuned trap (BDT) for trapping neutral atoms with blue-detuned laser light is the on-axis minimum farthest from the aperture (located at $z = 100$ μm in Figure 11.1). Using 780-nm laser light and a 25-μm diameter pinhole, the approximate size of this trap at full-width at half-maximum (FWHM) is roughly 5 μm by 50 μm, using the nearest intensity maxima in the radial and the axial directions. The primary spot of interest for this work for a red-detuned trap (RDT) for trapping atoms using red-detuned laser light is the second to last on-axis maximum observed in Figure 11.1, located at $z = 66$ μm. The approximate FWHM dimensions of this site are roughly 2.5 μm by 21 μm. Equation 11.47 can be used to evaluate the axial confinement of the traps. Unfortunately, analytical solutions to the diffracted intensity distributions can only be obtained for the special case of on-axis locations; see Sect. 7.7. Radial distributions must be numerically evaluated for each axial location of interest.

For cold ^{85}Rb atoms and 780-nm light, $\Gamma = 2\pi\times$ 6 MHz and $E_s = 111.15$ $\frac{V}{m}$, equation 11.42 can be written in a more convenient form as

$$\frac{U(r,z)}{S_o} - 9710 \frac{\Gamma}{\Delta} \left(\frac{S_z(r,z)}{S_o} \right) \left(\frac{\mu K}{W/cm^2} \right). \qquad (11.49)$$

Equation 11.49 gives the position-dependent dipole trap potential in micro-Kelvin per unit intensity incident upon the aperture (in units of Watts/cm^2) for the normalized field intensity distributions (S_z/S_o) calculated in Figures 11.1–11.3(a). It is noted here that care must be taken in selection of the detuning and incident intensity parameters for red-detuned dipole traps due to the fact that the dipole energy minimums occur at intensity maximums. Selection of appropriate parameters is necessary to avoid complications due to photon absorption, perturbation to the average potential from the excited state populations, trap heating, and the spontaneous force. In this investigation, we consider a detuning of $\Delta = 10^3\Gamma$ from the F=2 hyperfine ground state for blue-detuned light, and $\Delta = -10^4\Gamma$ for red-detuned light.

Even by visual inspection of Figure 11.1, it is evident that the minimum escape route for either an RDT or a BDT trapped atom is not going to be along the axis of beam propagation, or in the radial direction, but rather along a diagonal path over the lowest potential well wall. These minimum escape routes for a trapped atom to escape from either trap are illustrated in Figure 11.3.2(a), and comprise a three-dimensional conical surface. Using Hertz vector diffraction theory, the position-dependent dipole interaction potential along the minimum escape cone for ^{85}Rb atoms in red-detuned light ($\Delta = -10^4\Gamma$) is displayed in Figure 11.3.2(b). Figure 11.3.2(c) is the position-dependent potential for blue-detuned light ($\Delta = 10^3\Gamma$ from the F=2 hyperfine ground state) along a minimum escape path in the x-z plane. The vertical scales of Figures 11.3.2(b) and (c) have been converted into dipole trap depth in units of μK per W/cm^2 of uniform intensity incident upon the aperture using Eq. 11.49.

The RDT and BDT trap depths calculated and illustrated in Figure 11.3.2 scale linearly with the intensity incident upon the aperture. Trap depths in the mK range are obtainable with sub-Watt level cw lasers. For example, if a 1 mK trap depth is desired, a BDT trap would require an incident intensity of 116 W/cm^2, and an RDT trap would require 364 W/cm^2. These incident intensities can be achieved if a 57 mW or a 178 mW cw laser beam is focused

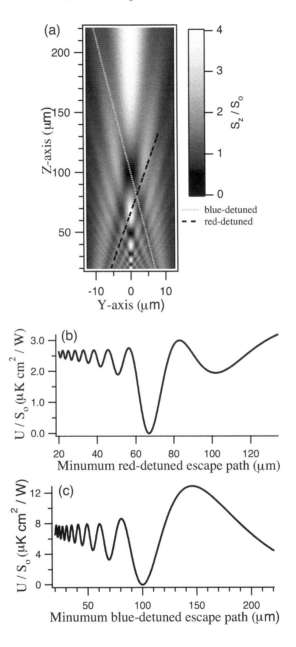

FIGURE 11.3
(a) Illustration of the minimum escape path from the on-axis dipole trap for red and blue-detuned laser light. Figures (b) and (c), are the dipole trap depths (in μK per W/cm^2 of laser intensity incident upon the aperture) for ^{85}Rb with $\Delta = -10^4\Gamma$ and $\Delta = 10^3\Gamma$ versus position along the red and blue-detuned minimum escape paths, respectively [11]. (Reprinted figure with permission from G. S. Gillen, S. Guha, and K. Christandl, Phys. Rev. A, 73, 013409, 2006. Copyright 2006 by the American Physical Society.)

down to a Gaussian e^{-1} half-width of 125 μm and incident upon the 25-μm diameter aperture, for BDT and RDT traps, respectively. The resulting radial, f_r, and axial, f_z, trap vibrational frequencies for a 1-mK BDT trap using 780-nm light and a 25-μm aperture would be $f_r = 39$ kHz and $f_z = 5.4$ kHz. The corresponding RDT trap frequencies would be $f_r = 60$ kHz, and $f_z = 6.7$ kHz.

11.4 Projection of Diffraction Patterns

Using diffracted laser light for the trapping of cold atoms, described in the previous section, allows for an experimental setup with: (1) a fairly simple optical setup comprised of an unfocused, or loosely focused, laser beam and a diffracting aperture or mask, and (2) the ability to use the same optical setup for both RDTs and BDTs by changing only the detuning of the laser. However, there are two issues that make experimental implementation of using diffracted laser light to create red-detuned and blue-detuned optical dipole traps challenging. First, the location of the diffraction patterns is very close to the diffracting aperture. Experimentally, this would mean that the diffracting aperture would have to be placed within the vacuum chamber so that cold atoms could be loaded into the diffraction pattern locations. The diffracting aperture would need to be included with the design and construction of the vacuum chamber and the optical system for the chosen atom cooling technique. For example, one commonly used technique is a magneto-optical trap (MOT). Experimentally, adjustment or replacement of the diffracting aperture or mask would require a significantly time-consuming process of: venting and disassembly of the vacuum chamber, adjustment/replacement, cleaning the vacuum chamber, baking the vacuum chamber, and waiting for it to reach the ultra-high vacuum (UHV) level required for atomic trapping. The second main experimental challenge for implementation of diffraction-based atom trapping is the fact that the diffraction pattern just beyond the aperture (and consequently the trap size, shape, trap frequencies, etc.) is a fixed function of the chosen aperture. Adjustment of the trap properties would require the replacement of the diffracting aperture or mask and all of the complexities involved in doing so.

In this section, we present an optical method, a mathematical model, and computational results for projecting the diffraction pattern located just beyond a diffracting aperture to another location away from optical components; i.e., inside a MOT cloud of cold atoms in a UHV chamber. This method eliminates the two experimental challenges of using diffracted light to trap atoms. First, by projecting the pattern away from the diffracting aperture the need to place the aperture inside the vacuum chamber is eliminated and all optical components would be located outside of the chamber where they can be easily aligned, adjusted, and/or replaced. Second, it is demonstrated here that adjustment of the aperture-to-lens distance allows for control and adjustment of the size and depth of the optical traps for a fixed diffracting aperture. If a different aperture is desired, exchanging it is a quick and straightforward process, as it is located outside of the vacuum chamber. The method presented here is only for the projection of a single aperture yielding a single RDT or BDT site within the MOT cloud. However, the method is extendable for the projection of a more complex diffracting mask of a one- or two-dimensional array of apertures yielding a projected array of trap sites.

11.4.1 Choosing a Beam Propagation Model, Fresnel Diffraction

Choosing a Model

In general, the difference between various established diffraction models is which mathematical approximations can be assumed in order to decrease both the mathematical complexity and computational times. The location of the point of interest with respect to the plane or surface area of the diffracting aperture to be integrated determines which mathematical approximations are valid. The closer the point of interest is to the diffracting aperture, the fewer the number of valid mathematical approximations, and hence the more complicated the diffraction model. One model that employs very few approximations (i.e., an infinitely thin and perfectly conducting aperture) and whose region of validity includes all points within the aperture plane and beyond is Hertz vector diffraction theory (HVDT) [18, 19]. Therefore, HVDT is the diffraction model used to first investigate which locations of the pattern are suitable trap sites to project through the lens. The results are depicted in Figure 11.4. The parameters used to calculate Figure 11.4 are an aperture radius of 50 μm, and a laser wavelength of 780 nm. It should be noted here that for a circular aperture of radius a and a laser wavelength of λ there will be a/λ on-axis bright spots, and $a/\lambda - 1$ on-axis dark spots [20]. For the purposes of this investigation, the dark spot farthest from the aperture is referred to as the *primary BDT*, and the second to last bright spot farthest from the aperture is referred to as the *primary RDT* as noted in Figure 11.4. Of all of the on-axis maxima and minima these two regions are chosen for investigation because they are the largest of each, which makes them easier to identify, load with cold atoms, and probe the atoms trapped within them in initial experiments. Both the primary BDT and the primary RDT lie within the region of validity of Fresnel approximations [20], or

$$z_1^3 \gg \frac{\pi}{4\lambda} \left[(x_1 - x_0)^2 + (y_1 - y_0)^2 \right]^2, \tag{11.50}$$

where x_1, y_1, and z_1 are the coordinates of the point of interest, and x_0 and y_0 are the locations of an integration point in the aperture plane. Therefore, the diffraction model used to calculate the projected diffraction patterns is the

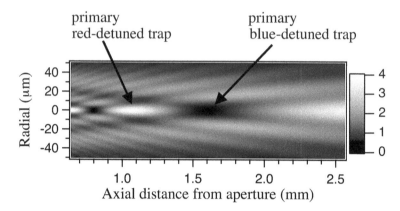

primary
red-detuned trap

primary
blue-detuned trap

FIGURE 11.4
Intensity distribution pattern just beyond a metallic circular aperture with a radius of 50 μm, and a laser wavelength of 780 nm, normalized to the intensity incident upon the aperture [21]. The locations of the localized bright region for the primary red-detuned trap site and the localized dark region for the primary blue-detuned trap site are noted. (Reprinted figure with permission from K. Gillen-Christandl and G. D. Gillen, Phys. Rev. A, 82, 063420, 2010. Copyright 2010 by the American Physical Society.)

Fresnel integral diffraction model. For a complete discussion of the Fresnel integral model see Sect. 7.1.

Diffraction and Beam Propagation Model

Figure 11.5 illustrates the theoretical setup for this investigation where a single lens of focal length f is placed a distance L from a circular aperture of radius a to project the diffraction pattern from just behind the metallic aperture into a MOT cloud of cold atoms located within an optics-free region inside of the UHV chamber.

For the theoretical setup depicted in Figure 11.5, the Fresnel approximation for the diffracted fields at a point in the near-field diffraction region (Eq. 4-17 of Ref. [22]) can be expressed as

$$E_1(x_1, y_1, z_1) = \frac{ke^{ikz_1}}{i2\pi z_1} \exp\left[i\frac{k}{2z_1}\left(x_1^2 + y_1^2\right)\right]$$
$$\times \int\int_{-\infty}^{\infty} E_{z_0} \exp\left[i\frac{k}{2z_1}\left(x_0^2 + y_0^2\right)\right]$$
$$\times \exp\left[-i\frac{k}{z_1}\left(x_1 x_0 + y_1 y_0\right)\right] \mathrm{d}x_0\,\mathrm{d}y_0, \qquad (11.51)$$

where k is the wavenumber, or $2\pi/\lambda$, (x_1, y_1, z_1) is the location of the point

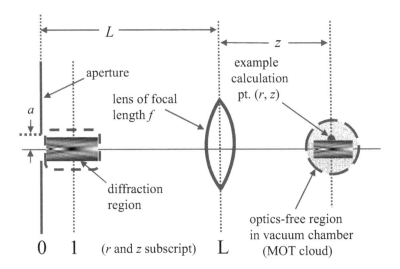

FIGURE 11.5
Theoretical setup for projecting near-field diffraction patterns to an optics-free location within a MOT cloud of cold atoms [21]. The key parameters are the aperture radius, a, the aperture to lens distance, L, the focal length of the lens, f, and the axial distance between the lens and the point of interest, z. Plane 0 is the axial location of the aperture and plane 1 is an axial plane within the diffraction region. (Reprinted figure with permission from K. Gillen-Christandl and G. D. Gillen, Phys. Rev. A, 82, 063420, 2010. Copyright 2010 by the American Physical Society.)

of interest in the diffraction region, (x_0, y_0, z_0) is an integration point in the aperture, and E_{z_0} is the distribution of the incident electric field within the open aperture area. Here, E_{z_0} is assumed to be a constant to represent an incident plane wave (i.e., a loosely focused beam with a width $\gg a$). For the diffraction of light by a circular aperture the net light fields within the aperture plane do not have cylindrical symmetry as a result of scattering effects of the aperture rim due to the vector field polarization direction [18, 19]. However, for points with axial distances greater than the aperture radius, the diffracted light fields do exhibit cylindrical symmetry [19]. Converting Eq. 11.51 to cylindrical coordinates and assuming cylindrical symmetry, the field beyond the aperture in the Fresnel diffraction region becomes

$$E_1\left(r_1, z_1\right) = \frac{ke^{ikz_1}}{iz_1}\exp\left[i\frac{kr_1^2}{2z_1}\right]$$

$$\times \int_0^a E_{z_0} J_0\left(\frac{kr_0 r_1}{z_1}\right)\exp\left[i\frac{kr_0^2}{2z_1}\right] r_0\,dr_0, \tag{11.52}$$

where J_0 is a Bessel function of the first kind of order zero.

The effect of the lens on the electric field can be approximated using a phase transformation of the light fields incident upon the lens [22]. For the lenses and optical setups used in this investigation, the paraxial thin lens phase transformation can be used. The electric field leaving the lens, E_L, in the cylindrical coordinate system of the lens, (r_L, z_L), the origin of which is at the center of the lens, is related to the field incident upon the lens, E_1, in the diffraction plane coordinate system, (r_1, z_1), by

$$E_L\left(r_L, z_L = 0\right) = E_1\left(r_1, z_1 = L\right)t\left(r_L\right), \tag{11.53}$$

where L is the axial distance between the diffracting aperture and the lens, and $t\left(r_L\right)$ is the thin lens phase transformation as a function of radial distance from the optical axis. The thin lens phase transformation for a lens of focal length f is given by Eq. (5-10) of Ref. [22], or

$$t\left(r_L\right) = \exp\left[-i\frac{kr_L^2}{2f}\right], \tag{11.54}$$

using cylindrical coordinates with cylindrical symmetry. Substituting Eqs. 11.52 and 11.54 into 11.53, the field at a single point in space immediately after the lens becomes

$$E_L\left(r_L, 0\right) = \frac{ke^{ikL}}{iL}\exp\left[i\frac{kr_L^2}{2L}\right]\int_0^a E_{z_0} J_0\left(\frac{kr_0 r_L}{L}\right)$$

$$\times \exp\left[i\frac{kr_0^2}{2L}\right]\exp\left[-i\frac{kr_L^2}{2f}\right] r_0\,dr_0, \tag{11.55}$$

where r_L is the radial location of a point within the plane of the center of the lens. To obtain the projected electric field at the point of interest in the optics-free region of the MOT cloud of cold atoms it is necessary to integrate the electric field just after the lens over the open area of the lens and propagate the fields to the point of interest. Using the Fresnel diffraction integral, the field at the point of interest, (r, z), where the coordinate system has the same origin as the lens coordinate system, becomes a function of the field distribution just after the lens

$$E(r, z) = \frac{ke^{ikz}}{iz} \exp\left[i\frac{kr^2}{2z}\right] \int_0^R E_L(r_L, L)$$

$$\times J_0\left(\frac{kr_L r}{z}\right) \exp\left[i\frac{kr_L^2}{2z}\right] r_L \, dr_L, \tag{11.56}$$

where the integration is over the open area of the lens and R is the radius of the lens. Substitution of Eq. 11.55 into Eq. 11.56 yields

$$E(r, z) = -\frac{k^2 e^{ik(z+L)}}{Lz} \exp\left(i\frac{kr^2}{2z}\right)$$

$$\times \int_0^R \left[\int_0^a E_{z_0} J_0\left(\frac{kr_0 r_L}{L}\right) \exp\left(i\frac{kr_0^2}{2L}\right) r_0 \, dr_0\right]$$

$$\times J_0\left(\frac{kr_L r}{z}\right) \exp\left[i\frac{k}{2}\left(\frac{1}{L} + \frac{1}{z} - \frac{1}{f}\right) r_L^2\right] r_L \, dr_L. \tag{11.57}$$

If the lens size is significantly larger than the diffraction pattern incident upon the lens then the limit of the dr_L integral can be assumed to be infinity. The integral of the lens plane can then be evaluated explicitly and Eq. 11.57 can be simplified to

$$E(r, z) = \frac{dke^{ik(z+L)}}{iLz} \exp\left[i\frac{k}{2z}\left(1 - \frac{d}{z}\right) r^2\right]$$

$$\times \int_0^a E_{z_0} J_0\left(\frac{kr_0 r d}{Lz}\right)$$

$$\times \exp\left[i\frac{k}{2L}\left(1 - \frac{d}{L}\right) r_0^2\right] r_0 \, dr_0, \tag{11.58}$$

where the parameter d is defined to be

$$d = \left[\frac{1}{L} + \frac{1}{z} - \frac{1}{f}\right]^{-1}. \tag{11.59}$$

Equation 11.58 is the general integral result of this section yielding the electric field at the point of interest in the optics-free region within the MOT cloud. It should be noted here that the axial distance from the lens that forces the value of d to approach infinity is the axial location of the image of the aperture plane according to geometrical optics.

On-Axis Fields

While the general integral result of Eq. 11.58 is complete and yields the scalar light field for any point within the desired projection region, it can be cumbersome to quickly determine where the projected light distributions can be found. The mathematics of the general integral result of Eq. 11.58 simplify for on-axis locations, $r = 0$, and the integral can be directly evaluated yielding the more convenient analytical form

$$
E\left(0, z\right) = -\frac{E_{z_0} d}{z\left(1 - \frac{d}{L}\right)} e^{ik(z+L)}
$$
$$
\times \left[\exp\left(i\frac{ka^2}{2L}\left[1 - \frac{d}{L}\right]\right) - 1\right], \tag{11.60}
$$

which is dependent upon the wavenumber of the light, k, the aperture radius, a, the distance between the aperture and the lens, L, and the focal length of the lens, f. The on-axis projected scalar field of Eq. 11.60 only exhibits localized maxima and minima around the region of the location of the imaged near-field pattern according to geometrical optics. Equation 11.60 can be evaluated to obtain the locations of the on-axis maxima and minima. The exact locations of on-axis maxima and minima occur when

$$
z = f\frac{1 - nL\frac{\lambda}{a^2}}{1 - n\left(L - f\right)\frac{\lambda}{a^2}}, \tag{11.61}
$$

where n is a non-zero integer. On-axis maxima correspond to when n is an odd integer, and on-axis minima correspond to when n is an even integer. On-axis maxima represent locations of possible projected red-detuned dipole atom traps, and on-axis minima represent locations of possible projected blue-detuned dipole atom traps. Equation 11.61 can also be written in the more familiar style of a thin lens equation in geometrical optics, or

$$
\frac{1}{z} + \frac{1}{L - \frac{a^2}{n\lambda}} = \frac{1}{f}. \tag{11.62}
$$

For the two on-axis locations of interest for this investigation, a value of $n = 2$ corresponds to the projection of the primary BDT, and a value of $n = 3$ corresponds to the projection of the primary RDT.

11.4.2 Projection Calculations

Intensity Calculations

The method introduced in this section allows for the remote control and manipulation of the optical dipole traps within the vacuum chamber by placing the diffracting aperture and projection lens outside of the vacuum chamber. Without breaking vacuum, the placement, radial and axial sizes of the trap sites, as well as the trap depths can be adjusted and controlled for a given laser frequency by changing any one of three optical parameters: the aperture radius, a, the focal length of the lens, f, and the aperture-to-lens distance, L.

Figure 11.6 illustrates some of the control over the projected diffraction pattern by adjustment of only the parameter L. All of the image plots for Figure 11.6 are for a wavelength of 780 nm, an aperture radius of 25 microns, and a focal length of 40 mm. The numerical intensity scale of each part of Figure 11.6 is normalized to the intensity of the laser light incident upon the diffracting aperture. Figure 11.6(a) is a calculation of the fields just beyond the aperture using HVDT. Figure 11.6(b) is an image plot of a projected diffraction pattern, using Eq. 11.58, and approximately equal in size to that of the original. To project a diffraction pattern equal to that of the original pattern in physical size and relative intensity an aperture-to-lens distance of just over twice the focal length is chosen, or $L = 80.4$ mm. This particular choice forces the size of the projected pattern to approximately equal that of the original, and forces the distance of the projected BDT from the lens to equal exactly the distance of the original BDT to the lens. The relative intensity of the projected pattern is equal to that of the original pattern where the intensity of the RDT is 4× that of the incident laser intensity. Increasing the aperture-to-lens distance beyond $2f$ has two effects upon the projected pattern: the entire projected pattern will be smaller than the original and, consequently, the relative intensity of the bright spots will be amplified with respect to the incident laser intensity, as illustrated in Figure 11.6(c) and (e) where L is set to be 125 mm. Conversely, by adjusting the aperture-to-lens distance to values less than $2f$, the projected pattern will be larger than the original and have a lower overall intensity, as illustrated in Figure 11.6(d) where L is set to be 68 mm. Both Figures 11.6(c) and (d) are intentionally set to the same radial scaling and axial scaling as 11.6(a) to better illustrate the effects of changing L on the overall projected pattern size. Figure 11.6(e) is an expanded view of the smaller projection of the diffraction pattern, part (c). Note here the differences in radial and axial sizes of the projected traps, as compared to the original trap sites of part (a). In addition, the peak intensity of the projected RDT is now 17.9× the incident laser intensity, whereas that of Figure 11.6 (d) is only 1.93× the incident laser intensity.

With the diffracting aperture located outside of the vacuum chamber, exchanging the aperture for one with a different radius is a straightforward process. Changing the size of the aperture radius effects the radial and axial scalings of the near-field diffraction pattern differently; the radial dimensions of the pattern scale linearly with the aperture radius, whereas the axial scaling of the pattern is dependent upon the square of the aperture radius [20]. For circular apertures where the radius is much larger than the wavelength of the laser light such as those of Figure 11.6, where $a/\lambda = 32$, the trap sites tend to be much longer in the axial dimension than the radial dimension. Because of the different dependencies of the radial and axial scalings on the aperture radius, the aspect ratio of the radial size to the axial size increases as the

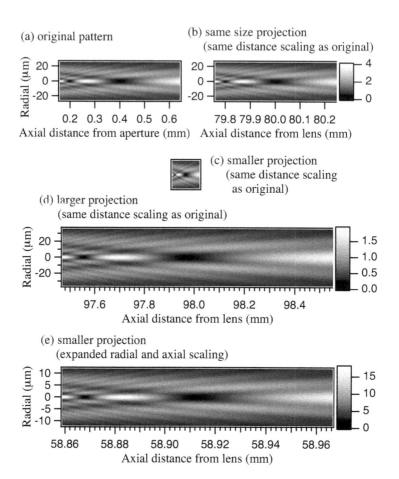

FIGURE 11.6
Diffraction patterns for an aperture with a radius of 25 microns, where (a) is
the original pattern just beyond the diffracting aperture, (b) is a projection
with the aperture-to-lens distance, L, chosen such that the projection is approximately the same size as the original, (c) is a projection with L chosen
such that the projection is smaller than the original, and (d) is a projection
with L chosen such that the projection is larger than the original [21]. Parts
(b)–(d) are all set to the same distance scaling as (a) for direct comparison
of the size differences, and part (e) is an expansion of part (c). All numerical intensity scaling values are normalized to the intensity incident upon the
aperture. (Reprinted figure with permission from K. Gillen-Christandl and
G. D. Gillen, Phys. Rev. A, 82, 063420, 2010. Copyright 2010 by the American Physical Society.)

size of the diffracting aperture decreases. It should be noted here that care must be taken when choosing smaller apertures to remain within the region of validity of the Fresnel approximations used in this model, where it is assumed that $a \gg \lambda$. With the assumptions used for this model and the locations of the primary RDT and primary BDT, a lower limit on the aperture size would be an aperture size of about 5 microns, where $a/\lambda = 6.25$. Figure 11.7(a) is an image plot of the intensity pattern just beyond a 5 micron radius circular aperture using HVDT. Sub-micron sized traps, or "nanotraps," can be created by using our projection technique. As L becomes larger than $2f$ the projected pattern becomes smaller than the original diffraction pattern. Fig 11.7(b) is an illustration of the creation of a nanotrap. First, we start with Fig 11.7(a), or the smallest original diffraction pattern within the region of validity of the projection model used here, and then we set $L > 2f$. For this particular illustration, f is chosen to be a 40-mm lens, and L is chosen to be 125 mm, which yields RDT and BDT axial trap dimensions of ≈ 1 micron, and radial trap dimensions of < 1 micron.

Trapping Potential Energy Calculations

Conversion of the intensity distributions of Figures 11.6 and 11.7 to potential energy wells for trapping cold neutral atoms requires a choice of the laser detuning and an incident laser intensity. Additionally, for detunings not large compared to the hyperfine ground state splitting, a choice of a hyperfine ground state of the atom is required. Using the same detuning as our previous work [11] ($\Delta = -10^4\Gamma$ for RDT sites and $\Delta = 10^3\Gamma$ from the $F{=}2$ hyperfine ground state of ^{85}Rb for BDT sites), disturbances of the possible atom traps due to photon absorption, population of excited states, and the spontaneous force are negligible. Including the saturation field for the chosen transition, the trap potential energy, Eq. 11.42, can be expressed as a function of the chosen detuning, and normalized to the intensity incident upon the aperture, or

$$\frac{U(r,z)}{S_o} = 9710\frac{\Gamma}{\Delta}\left(\frac{S_z(r,z)}{S_o}\right)\frac{\mu K}{W/cm^2}, \qquad (11.63)$$

where the term S_z/S_o is the normalized intensity displayed in the numerical intensity scaling of Figures 11.6 and 11.7.

Properties of Calculated Traps

Tables 11.1 and 11.2 give the trap depths and trap dimensions for all of the diffraction patterns illustrated in Figures 11.6 and 11.7. The depth of each trap site is taken to be the difference between the minimum of the trap potential well and the lowest potential barrier to getting out of an RDT or a BDT along the escape path. For a complete discussion on escape paths from each type of trap, see Sec. 11.3.2. The radial and axial dimensions reported for each diffraction pattern are defined as the approximate width and length of each

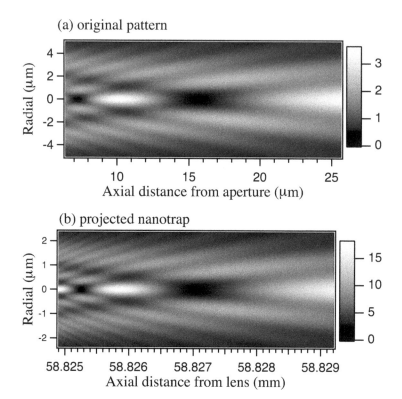

(a) original pattern

(b) projected nanotrap

FIGURE 11.7
Original and projected intensity patterns for the creation of a nanotrap [21].
The aperture radius is 5 microns, the focal length of the lens is 40 mm, and
the aperture to lens distance is set to 125 mm. Both intensity scaling values
are normalized to the intensity incident upon the aperture. (Reprinted figure
with permission from K. Gillen-Christandl and G. D. Gillen, Phys. Rev. A,
82, 063420, 2010. Copyright 2010 by the American Physical Society.)

Diffraction pattern	Figure number	RDT trap depth per 100 W/cm^2 (μK)	Radial Width (μm)	Axial length (μm)
original $a = 25\ \mu$m	11.6(a)	277	6.03	120
projected $a = 25\ \mu$m, $L = 80.4$ mm	11.6(b)	280	6.03	120
projected $a = 25\ \mu$m, $L = 125$ mm	11.6(c) & (e)	1250	2.84	26.7
projected $a = 25\ \mu$m, $L = 68$ mm	11.6(d)	134	8.88	252
original $a = 5\ \mu$m	11.7(a)	241	1.31	4.89
projected nanotrap $a = 5\ \mu$m, $L = 125$ mm	11.7(b)	1260	0.59	1.08

TABLE 11.1
Properties of the primary red-detuned trap (RDT) for various diffraction patterns [21]. For all patterns, the lens focal length is 40 mm, the detuning parameter is set to $\Delta = -10^4\,\Gamma$, and the trap depth is given per $100\,\mathrm{W/cm^2}$ of light intensity incident upon the aperture. For reference, a 17 mW laser focused to a spot diameter of $\sim 150\,\mu$m gives an incident intensity of $\sim 100\,\mathrm{W/cm^2}$. (Reprinted table with permission from K. Gillen-Christandl and G. D. Gillen, Phys. Rev. A, 82, 063420, 2010. Copyright 2010 by the American Physical Society.)

of the traps within the equipotential energy surface equal to the trap depth. Tables 11.1 and 11.2 quantitatively illustrate the effects of changing either the distance between the lens and the diffracting aperture, or changing the size of the diffracting aperture on the depth, dimensions, and axial-radial aspect ratio of the trap sites.

11.5 Polarization-Dependent Atomic Dipole Traps

In Sections 11.3 and 11.4 we investigated the interaction of light fields with the dipole potential to create atom trap sites dependent upon the *scalar* intensity of the light field. In this section we present an approach for trapping

Diffraction pattern	Figure number	BDT trap depth per 100 W/cm^2 (μK)	Radial Width (μm)	Axial length (μm)
original $a = 25~\mu$m	11.6(a)	820	9.05	125
projected $a = 25~\mu$m, $L = 80.4$ mm	11.6(b)	823	9.05	125
projected $a = 25~\mu$m, $L = 125$ mm	11.6(c) & (e)	3680	4.26	27.9
projected $a = 25~\mu$m, $L = 68$ mm	11.6(d)	393	13.1	264
original $a = 5~\mu$m	11.7(a)	818	1.86	5.24
projected nanotrap $a = 5~\mu$m, $L = 125$ mm	11.7(b)	3700	0.85	1.11

TABLE 11.2
Properties of the primary blue-detuned trap (BDT) for various diffraction patterns [21]. For all patterns, the lens focal length is 40 mm, the detuning parameter is set to $\Delta = 10^3\,\Gamma$, and the trap depth is given per 100 W/cm^2 of light intensity incident upon the aperture. (Reprinted table with permission from K. Gillen-Christandl and G. D. Gillen, Phys. Rev. A, 82, 063420, 2010. Copyright 2010 by the American Physical Society.)

atoms where the properties of the atom trap sites depend upon the *vector* nature of the propagating light field; i.e., the polarization. By manipulation and control of the polarization of the diffracted light field one can trap atoms in a particular magnetic substate. We present this approach for trapping atoms with one particular application in mind: a two-qubit gate for neutral atom quantum computing. For this particular application there are a few necessary steps:

- the ability to trap two atoms in two different trap sites located next to each other,

- the ability to move the atoms together to create an entangled quantum state in the shortest possible time, and

- the ability to move the atoms apart again.

For the purposes of this section we will demonstrate that polarization-dependent diffraction patterns of two different laser beams incident upon the same aperture at different angles have the ability to trap and move adjacent atoms together and apart without losing the atoms from the traps.

11.5.1 Theory of the Polarization Dependence of the Optical Dipole Trapping Potential Energy

An electric field such as that from a laser induces an electric dipole moment in a neutral atom. In general, this induced dipole moment depends on the polarization of the laser light, as well as on the hyperfine level and Zeeman magnetic substate of the atom. The induced dipole moment due to a certain electric field is determined by the polarizability of the atom. The interaction of the induced dipole with the electric field of the laser light results in a potential energy and its associated force, which can trap the atom in regions of high or low light intensity. The classical treatment of this interaction was shown in Section 11.2. The semi-classical equivalent of Eq. 11.14 is the potential energy operator for the light-atom interaction, given by [23]

$$\hat{U}(\mathbf{r}) = -\frac{1}{4}\mathbf{E}_0^*(\mathbf{r})\hat{\alpha}\mathbf{E}_0(\mathbf{r}). \tag{11.64}$$

Here, $\mathbf{E}_0(\mathbf{r})$ is the complex amplitude for an electric field written in the form $\mathbf{E}(\mathbf{r}, t) = \text{Re}(\mathbf{E}_0(\mathbf{r})e^{-i\omega t})$, $\hat{\alpha}$ is the atomic polarizability tensor (for detailed discussions see [23, 24]), and ω is the angular frequency of the laser light. Alternatively, the electric field is often written in its Fourier series form with positive and negative frequency components, $\mathbf{E}^{(+)} = \mathbf{E}_0^*/2$ and $\mathbf{E}^{(-)} = \mathbf{E}_0/2$, respectively,

$$\mathbf{E}(\mathbf{r}, t) = \mathbf{E}^{(+)}(\mathbf{r})e^{-i\omega t} + \mathbf{E}^{(-)}(\mathbf{r})e^{i\omega t}. \tag{11.65}$$

The corresponding expression for the dipole potential energy operator is then

$$\hat{U}(\mathbf{r}) = -\mathbf{E}^{(+)}(\mathbf{r})\hat{\alpha}\mathbf{E}^{(-)}(\mathbf{r}). \tag{11.66}$$

As derived in [25] and Appendix A of [14], for an alkali atom, the polarizability tensor components in the spherical basis are

$$\hat{\alpha}_{q',q} = (-1)^{q'} \sum_{F'} [\alpha_{0,F'F} f_{F'F}$$

$$\times \sum_{m} (c^{F,1,F'}_{m+q-q',q',m+q} c^{F,1,F'}_{m,q,m+q} \tag{11.67}$$

$$\times |F, m + q - q'\rangle\langle F, m|)],$$

where $q', q = \pm 1, 0$ stands for the spherical basis components, $\alpha_{0,F'F}$ is the characteristic polarizability scalar, $f_{F'F}$ is the relative oscillator strength of the $F \rightarrow F'$ hyperfine transition, and the c's are the Clebsch-Gordan coefficients for the $F, m \rightarrow F', m + q$ dipole transition, and related to the $F', m + q \rightarrow F, m + q - q'$ dipole transition (see Appendix A of [14]). The relative oscillator strength of an $F \to F'$ transition is

$$f_{F'F} = (2J' + 1)(2F + 1) \left| \left\{ \begin{matrix} F' & I & J' \\ J & 1 & F \end{matrix} \right\} \right|^2, \tag{11.68}$$

where the curly braces signify the six-J symbol, and I is the quantum number for the nuclear spin of the atom. The characteristic polarizability scalar is given by

$$\alpha_{0,F'F} = -\frac{|\langle J'||d||J\rangle|^2}{\hbar \Delta_{F'F}}. \tag{11.69}$$

Here, $\Delta_{F'F}$ is the angular frequency detuning from the $F \to F'$ transition, and $\langle J'||d||J\rangle$ is the reduced dipole matrix element for the $J \to J'$ fine structure transition.

There are three common normalization conventions for the reduced dipole matrix element. A comparison of the three conventions, as well as an example for unit conversion of the reduced dipole matrix elements, are given in Appendix B of [14].

In this section, we used the following relation (with the same normalization as [26]) for the polarizability scalar of an alkali atom to calculate the reduced matrix element [23]:

$$\alpha_{0,F'F} = -\frac{3\lambda^3}{32\pi^3} \frac{\Gamma}{\Delta_{F'F}}. \tag{11.70}$$

This equation is for the wavelength λ in cm and gives $\alpha_{0,F'F}$ in cgs units. In our work, we use SI units throughout our code, so we are also listing the SI version of this equation (i.e. with λ in units of meters),

$$\alpha_{0,F'F} = -\frac{3\lambda^3}{32\pi^3} \frac{\Gamma}{\Delta_{F'F}} \cdot 1.11 \times 10^{-10} \frac{Jm^2}{V^2}. \tag{11.71}$$

In this investigation, we only present the diabatic potentials [27], which are the diagonal components of the potential energy operator in the F, m_F basis. This is appropriate, because we plan on trapping pre-cooled atoms in these traps, which will remain at the bottom of the potential energy wells, rather than traveling through the wells. For the configurations that involve movement of the traps, we need to consider two aspects of the speed of this motion. First, it must be slow compared to the trap frequency in order to reduce atom loss or state disturbance during motion. On the other hand, the motion should occur fast enough to reduce the probability of Raman transitions that could flip atoms into a different magnetic substate if the two atom potential energies are very similar (e.g. when two traps are fully overlapped). Such Raman transitions can change the state of an atom into either an untrapped state, or a state trapped by a different well (see Section 11.5.2). As long as the Raman transition probability remains sufficiently low, we can use the diabatic potentials to describe the atom dynamics in our traps. To calculate the diabatic potential energy for a certain F, m_F state, we calculate the expectation value of the potential energy operator as follows:

$$U_{F,m_F} = \langle F, m_F | \hat{U} | F, m_F \rangle. \tag{11.72}$$

Plugging in the dipole energy operator (Eq. (11.64)) explicitly, we get

$$
\begin{aligned}
U_{F,m_F} = &-\frac{1}{4} \sum_{q',q} E_{0q'}^* E_{0q} \hat{\alpha}_{q',q} \\
= &-\frac{1}{4} \sum_{q',q} \{ (-1)^{q'} E_{0q'}^* E_{0q} \sum_{F'} [\alpha_{0,F'F} f_{F'F} \\
&\times \sum_m (c_{m+q-q',q',m+q}^{F,1,F'} c_{m,q,m+q}^{F,1,F'} \\
&\times \langle F, m_F | F, m+q-q' \rangle \langle F, m | F, m_F \rangle)] \}.
\end{aligned}
\tag{11.73}
$$

Here, the $E_{0q',q}$ with $q',q = \pm 1, 0$ are the spherical components of the electric field amplitude, corresponding to right and left circular light polarization, $\sigma^\pm$, and linear light polarization, π, respectively.

Exploiting orthonormality, this expression simplifies to

$$
\begin{aligned}
U_{F,m_F} = &-\frac{1}{4} \sum_q (-1)^q |E_{0q}|^2 \\
&\times \sum_{F'} \alpha_{0,F'F} f_{F'F} (c_{m_F,q,m_F+q}^{F,1,F'})^2.
\end{aligned}
\tag{11.74}
$$

The electric field distributions for the diffraction pattern immediately behind a circular aperture were obtained using Hertz vector diffraction theory [18, 19]. Calculations using the model yield the Cartesian components of the electric field. To find the spherical components we use the spherical unit vectors [15]

$$
\begin{aligned}
\mathbf{e}_{-1} &= \frac{1}{\sqrt{2}} (\mathbf{e}_x - i\mathbf{e}_y) \\
\mathbf{e}_0 &= \mathbf{e}_z \\
\mathbf{e}_{+1} &= -\frac{1}{\sqrt{2}} (\mathbf{e}_x + i\mathbf{e}_y).
\end{aligned}
\tag{11.75}
$$

Here, $\mathbf{e}_x$, $\mathbf{e}_y$, and $\mathbf{e}_z$ are the Cartesian unit vectors. From this we find the spherical components of the complex amplitude of the electric field defined by

$$
\begin{aligned}
\mathbf{E}_0 &= E_{0x}\mathbf{e}_x + E_{0y}\mathbf{e}_y + E_{0z}\mathbf{e}_z \\
&= E_{0-1}\mathbf{e}_{-1} + E_{00}\mathbf{e}_0 + E_{0+1}\mathbf{e}_{+1}.
\end{aligned}
\tag{11.76}
$$

Here E_{0j} for $j = x, y, z$ are the Cartesian components of the electric field amplitude. The spherical and Cartesian components are related by

$$E_{0-1} = \frac{1}{\sqrt{2}} (E_{0x} + iE_{0y})$$
$$E_{00} = E_{0z} \qquad\qquad (11.77)$$
$$E_{0+1} = \frac{1}{\sqrt{2}} (-E_{0x} + iE_{0y}).$$

These spherical components were then plugged into Eq. (11.74).

Equations (11.68), (11.71), (11.74), and (11.77) were used to calculate the computational results presented in the next section. The computations were performed using a code that will take any arbitrary electric field distribution in Cartesian coordinates, decompose it into its spherical components, and then calculate both the diabatic and adiabatic potentials for any given detuning $\Delta \ll \Delta_{\mathrm{fs}}$, where Δ_{fs} denotes the fine structure splitting of the excited state. Note that the detuning can be made arbitrarily small, including smaller than the excited state hyperfine splitting. For larger detunings, the expressions simplify tremendously, as only the fine structure splitting needs to be considered [2, 25]. Also note that the E_{0j} are complex, so both the real and imaginary components must be supplied for this calculation. Appendix C of [14] shows the details of our specific electric field configurations (single laser beam incident at an angle and pair of oppositely circularly polarized laser beams incident at an angle). The code has two variable input parameters: the laser detuning Δ from the transition $F \rightarrow$ maximum F', which is contained in $\Delta_{F'F}$, and the laser intensity I_0. The electric field amplitude at the aperture for each of the incident beams starts out normalized to 1. To change this to meaningful units, we insert the scaling due to intensity, and convert the units of the potential energy from J to mK as follows:

$$U_{F,m_F}(\mathrm{mK}) = 1000 \frac{2}{3k_{\mathrm{B}}} \frac{2I_0}{\epsilon_0 c} U_{F,m_F}. \qquad (11.78)$$

Here, the factor of 1000 is for converting K to mK, the factor of $\frac{2}{3k_{\mathrm{B}}}$ is for conversion from J to K, and the factor of $\frac{2I_0}{\epsilon_0 c}$ is for inserting physical units for the electric field. For an intensity in $\mathrm{W/m^2}$, this yields electric field units of $\mathrm{V/m}$. k_{B} is Boltzmann's constant, ϵ_0 is the permittivity of free space, and c is the speed of light. This is how we obtained the numerical results presented in the next section.

11.5.2 Computational Results for Atom Traps beyond a Circular Aperture

Movable Atomic Dipole Traps

Consider the diffraction pattern resulting from a laser beam incident on a circular aperture with a diameter of 25 μm at an angle of γ=0.055 rad as shown in Figure 11.8. Depending on the laser detuning, three-dimensional (3D) atom traps will form either in the localized bright spots or dark spots of this diffraction pattern very close to the aperture. Figure 11.8(c) shows the diabatic trapping potential energy (calculated from the electric field distribution using Eq. (11.74)) for the $F=1$, $m_F=0$ magnetic substate of ^{87}Rb, for a laser intensity of 364 W/cm^2, and a laser detuning of -10,000 Γ (red detuning) from the Rb D2 transition (λ=780 nm). The D2 linewidth of Rb is Γ=2π×6 MHz [15]. Atoms are trapped in the bright spots on the laser beam axis. Similarly, Figure 11.8(d) shows the diabatic trapping potential energy for the $F=1$, $m_F=0$ magnetic substate of ^{87}Rb, for a laser intensity of 116 W/cm^2, and a laser detuning of 1,000 Γ (blue detuning) from the Rb D2 transition. Here, localized atom traps form in the dark spots on the laser beam axis.

We analyzed the properties of the traps formed farthest from the aperture (z=67 μm for the farthest, well-localized bright spot for the red-detuned case, and z=100 μm for the blue-detuned case) and compared them to the normal incidence case. We chose these traps because they are the biggest and most easily accessible for initial experiments. The traps formed closer to the aperture are also viable, and, in fact, advantageous for quantum computing as they have larger trap frequencies. To determine the trap frequencies, we approximated the bottom of the (non-harmonic) traps with a harmonic oscillator potential energy well. The values of the trap frequencies obtained depend on the fit range used. In this investigation, we chose a fit range of the bottom 200 μK of the well, unless otherwise stated. Other trap properties of relevance for quantum information applications are the size of the motional harmonic oscillator ground state wavefunction along a spatial dimension j,

$$\beta_j = \sqrt{\frac{\hbar}{2\pi f_j m_{\mathrm{Rb}}}}, \tag{11.79}$$

for the $1/e$ half width of the probability density, and the energy difference $\Delta U_j \equiv h f_j / k_{\mathrm{B}}$ between two motional states of the potential energy well. Here, f_j is the trap frequency along spatial dimension j, and m_{Rb} is the mass of one ^{87}Rb atom. We denote the spatial dimensions of the trap with indices $j = r_x$ for the radial dimension along x in the coordinate system of Figure 11.8(a), $j = r_{yz}$ for the radial dimension in the y-z plane, and $j =$ axial for the axial dimension of the trap. In addition, the coherence of qubits in dipole traps is often limited by the scattering rate of trap photons. For blue-detuned traps with zero intensity at the bottom, the scattering rate for a ground state atom (averaged over the extent of the wavefunction) can be written as [28]

$$\eta = \frac{\pi}{2}(f_{r_x} + f_{r_{yz}} + f_{\mathrm{axial}})\frac{\Gamma}{\Delta}. \tag{11.80}$$

For red-detuned traps, a conservative estimate of the scattering rate is the peak scattering rate [2]

$$\eta = \frac{\Gamma}{\Delta\hbar}U_{\mathrm{min}}, \tag{11.81}$$

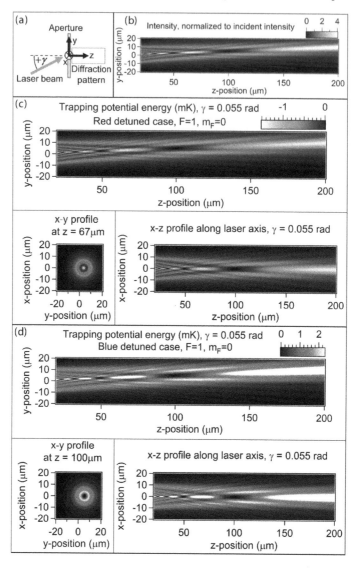

FIGURE 11.8

Diabatic trapping potential energy for a single laser beam (σ^+ polarization was used) incident on a circular aperture with a diameter of 25 μm at an angle of γ=0.055 rad [14]. (a) Diagram of setup. (b) Intensity pattern. (c) Trapping potential energy for the light-polarization-independent F=1, m_F=0 magnetic substate of ^{87}Rb, for a laser intensity of 364 W/cm^2, and a laser detuning of -10,000 Γ. (d) Trapping potential energy for the F=1, m_F=0 magnetic substate of ^{87}Rb, for a laser intensity of 116 W/cm^2, and a laser detuning of 1,000 Γ. (Reprinted figure with permission from K. Gillen-Christandl and B. D. Copsey, Phys. Rev. A, 83, 023408, 2011. Copyright 2011 by the American Physical Society.)

where $U_{\min}$ is the potential energy at the intensity peak of the trap. We determined the trap depth ΔU_{trap} by finding the peak potential energy of the path of weakest confinement ("escape path") with respect to the potential energy of the bottom of the well.

We determined that the trap properties calculated in [11] for normal incidence stay largely intact when the laser is incident at an angle. For comparisons to [11], we must mention that the trap frequencies cited there are for a harmonic fit range of 1 mK, whereas in this section we cite trap frequencies for 200 μK, which we deemed the relevant range for an atom sample pre-cooled in a magneto-optical trap. The corresponding normal incidence frequencies for the red detuned example are a radial trap frequency of $f_{r_x} = f_{r_{yz}} = 71$ kHz, and an axial trap frequency of $f_{\text{axial}} = 6.9$ kHz. The other trap properties are: $\beta_{r_x} = \beta_{r_{yz}} = 40$ nm, $\beta_{\text{axial}} = 130$ nm, $\Delta U_{r_x} = \Delta U_{r_{yz}} = 3.4$ μK, $\Delta U_{\text{axial}} = 0.33$ μK. The trap photon scattering rate is 4.3 kHz. The trap depth is 1 mK. Similarly, for blue-detuned light at normal incidence we have $f_{r_x} = f_{r_{yz}} = 28$ kHz, $f_{\text{axial}} = 5.6$ kHz, $\beta_{r_x} = \beta_{r_{yz}} = 64$ nm, $\beta_{\text{axial}} = 140$ nm, $\Delta U_{r_x} = \Delta U_{r_{yz}} = 1.3$ μK, and $\Delta U_{\text{axial}} = 0.27$ μK. For the blue detuned traps, the deviation from a harmonic potential energy well is particularly pronounced, with the bottom being very flat. To fully describe these traps, we performed fits for a fit range of 20 μK, yielding fits valid only for very low temperature atoms (< 1 μK) such as for atoms loaded from a Bose-Einstein condensate. The radial trap frequency for the bottom of the well for normal incidence is approximately 10 kHz, so it is comparable to the axial trap frequency (as are the other properties, $\beta_{r_x} = \beta_{r_{yz}} = 0.11$ μm, $\Delta U_{r_x} = \Delta U_{r_{yz}} = 0.48$ μK). A conservative estimate (using the larger trap frequencies) for the scattering rate is 15 Hz. The trap depth is 1 mK.

Figure 11.9 shows the trapping potential energy curves along the laser beam direction, as well as along the direction of weakest confinement for both the red and blue detuned examples listed above. For a beam incident at an angle of $\gamma = 0.055$ rad, for the red detuned example (laser beam intensity of 364 W/cm^2, laser detuning of -10,000 Γ) we find a trap depth of 1 mK, $f_{r_x} = f_{r_{yz}} = 74$ kHz, $f_{\text{axial}} = 6.8$ kHz, $\beta_{r_x} = \beta_{r_{yz}} = 40$ nm, $\beta_{\text{axial}} = 130$ nm, $\Delta U_{r_x} = \Delta U_{r_{yz}} = 3.6$ μK, and $\Delta U_{\text{axial}} = 0.33$ μK. The average scattering rate is 4.3 kHz. These results are very similar to those for normal incidence. For the blue detuned example (laser beam intensity of 116 W/cm^2, laser detuning of 1,000 Γ) we find a trap depth of 0.9 mK, $f_{r_x} = f_{r_{yz}} = 26$ kHz, $f_{\text{axial}} = 5.5$ kHz, $\beta_{r_x} = \beta_{r_{yz}} = 67$ nm, $\beta_{\text{axial}} = 150$ nm, $\Delta U_{r_x} = \Delta U_{r_{yz}} = 1.3$ μK, and $\Delta U_{\text{axial}} = 0.26$ μK. The radial frequencies describing the behavior of the bottom of the blue well are $f_{r_x} = 5.6$ kHz and $f_{r_{yz}} = 6.7$ kHz, with $\beta_{r_x} = 140$ nm, $\beta_{r_{yz}} = 130$ nm, $\Delta U_{r_x} = 0.27$ μK, and $\Delta U_{r_{yz}} = 0.32$ μK. At this level, we are starting to notice differences between the untilted x-direction and the dimension with tilt (y-z plane). Again, these results are very close to the normal incidence values. Therefore, the traps stay intact upon tilting.

Thus, by tilting the incident laser beam, an atom trapped at a bright spot or dark spot can be moved. We propose using this to bring two qubits together

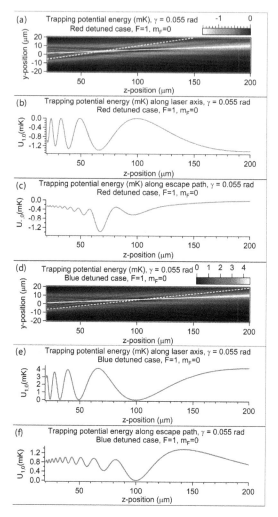

FIGURE 11.9
Diabatic trapping potential energy along the laser axis and along the paths of weakest confinement ("escape paths") for red detuned and blue detuned traps, for a single laser beam (σ^+ polarization was used) incident at $\gamma=0.055$ rad, for the light-polarization-independent $F=1$, $m_F=0$ magnetic substate of ^{87}Rb [14]. (a) Axial path (solid line) and escape path (dashed line) for a laser intensity of 364 W/cm^2, and a laser detuning of -10,000 Γ. (b) Trapping potential energy along the laser axis for red detuned trap (solid line in (a)). (c) Trapping potential energy along the escape path for red detuned trap (dashed line in (a), weakest trap direction). (d) Axial path (solid line) and escape path (dashed line) for a laser intensity of 116 W/cm^2, and a laser detuning of 1,000 Γ. (e) Trapping potential energy along the laser axis for blue detuned trap (solid line in (b)). (f) Trapping potential energy along the escape path for blue detuned trap (dashed line in (b), weakest trap direction). (Reprinted figure with permission from K. Gillen-Christandl and B. D. Copsey, Phys. Rev. A, 83, 023408, 2011. Copyright 2011 by the American Physical Society.)

and apart by employing two laser beams at an angle, as shown in Figure 11.10. One atom is placed in each of the two bright spot traps (for red-detuned light) or dark spot traps (for blue-detuned light). The laser beams are then tilted together to overlap the atoms for two-qubit quantum operations, and tilted apart to separate the atoms.

There are several issues with this approach. When overlapping the wells, there is a significant probability for the atoms to tunnel between the traps and switch places. This is detrimental for quantum computing. In addition, for blue-detuned traps, the wall from one trap will push the atom out of the other trap. Both of these issues can be addressed by exploiting the light polarization dependence of the trapping potential energy for atoms in different magnetic substates, as discussed in the next section.

Atomic Trapping Potential Energy for Different Magnetic Substates

Due to the dependence of the trapping potential energy on the light polarization and the magnetic substate of a trapped atom as outlined in Section 11.5.1, atoms in different magnetic substates placed in the same light pattern have a different trapping potential energy curve. Consider the configuration in Figure 11.10. A right-circularly polarized (σ^+) laser beam and a left-circularly polarized (σ^-) laser beam are incident on a circular aperture at angles γ and $-\gamma$, respectively. Figure 11.10(b) shows the intensity pattern for an incident angle of $\gamma=0.055$ rad. We chose this angle as an example, as the trap sites examined here are well separated for this angle. Figure 11.10(c)-(j) show the diabatic trapping potential energy for the eight magnetic substates of the ^{87}Rb hyperfine ground state manifold, for the red-detuned example. The results are for a laser detuning of $\Delta=-10,000\ \Gamma$ from the $F=1 \rightarrow F' = 3$ transition and from the $F=2 \rightarrow F' = 3$ transition, respectively. Table 11.3 summarizes the trap properties for both the red-detuned trap formed by the well-localized bright spot farthest from the aperture ($z=67\ \mu$m) and the blue-detuned trap formed by the dark spot farthest from the aperture ($z=100\ \mu$m). Note that only the 200 μK fit results are shown. For blue-detuned traps, the properties of the bottom of the well are better approximated with a 20 μK fit. For the x-direction, this yields trap properties comparable to the axial trap properties listed. For the y-z plane, a miniwell is formed at the bottom of the major well, and has trap properties comparable to the radial trap properties listed in the table. The polarization dependence of the potential energy for each magnetic substate is evident. For both examples, we chose a detuning that is large compared to the excited state hyperfine splitting, thus there is very little dependence of the detuning on F' during the F' summation in Eq. (11.74). Also, the results shown are for the same detuning from the F to F' transition for both $F=1$ and $F=2$, so the differences in the potential energies are mostly due to the magnetic substate and the light polarization.

One consequence of this polarization dependence of the potential energy for

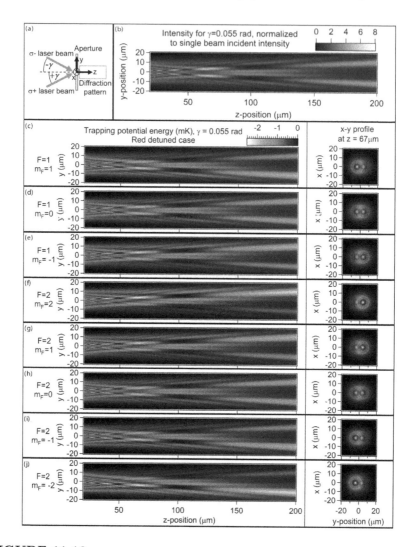

FIGURE 11.10

Diabatic trapping potential energy for two laser beams of opposite circular polarization incident on a circular aperture with a diameter of 25 μm at an angle of $\gamma=0.055$ rad [14]. (a) Diagram of setup. (b) Intensity pattern, normalized to the incident intensity of one incident circularly polarized laser beam. (c)-(j) Trapping potential energy of the intensity pattern in (b) for the eight magnetic substates of the hyperfine groundstate manifold in ^{87}Rb, for a laser intensity of 364 W/cm^2, and a laser detuning of -10,000 Γ from the transitions from the respective F states. (Reprinted figure with permission from K. Gillen-Christandl and B. D. Copsey, Phys. Rev. A, 83, 023408, 2011. Copyright 2011 by the American Physical Society.)

State F, m_F	Red detuned					Blue detuned				
	f (kHz)	β (nm)	ΔU_j (μK)	$\frac{\eta}{2\pi}$ (kHz)	ΔU_{trap} (mK)	f (kHz)	β (nm)	ΔU_j (μK)	$\frac{\eta}{2\pi}$ (Hz)	ΔU_{trap} (mK)
1, 1	83, 86, 8.1	37, 37, 120	4, 4.1, 0.39	6.7	1.2	27, 37, 6	65, 56, 140	1.3, 1.8, 0.29	18	1.2
1, 0	74, 79, 7.5	40, 38, 120	3.6, 3.8, 0.36	6.0	0.98	26, 35, 5.3	67, 58, 150	1.2, 1.7, 0.26	16	1.0
1, −1	83, 86, 8.1	37, 37, 120	4, 4.1, 0.39	6.7	1.2	27, 37, 6	65, 56, 140	1.3, 1.8, 0.29	18	1.2
2, 2	90, 92, 8.7	36, 36, 120	4.4, 4.4, 0.42	7.3	1.5	29, 41, 6.8	63, 53, 130	1.4, 2, 0.33	19	1.5
2, 1	83, 86, 8.1	38, 37, 120	4, 4.1, 0.39	6.7	1.2	28, 38, 6.2	65, 55, 140	1.3, 1.8, 0.3	18	1.3
2, 0	74, 79, 7.5	40, 38, 120	3.6, 3.8, 0.36	5.9	0.98	26, 35, 5.4	67, 58, 150	1.2, 1.7, 0.26	16	1.1
2, −1	83, 86, 8.1	38, 37, 120	4, 4.1, 0.39	6.7	1.2	28, 38, 6.2	65, 55, 140	1.3, 1.8, 0.3	18	1.3
2, −2	90, 92, 8.7	36, 36, 120	4.4, 4.4, 0.42	7.3	1.5	29, 41, 6.8	63, 53, 130	1.4, 2, 0.33	19	1.5

TABLE 11.3

Trap properties for all F, m_F hyperfine ground states in ^{87}Rb for the red detuned ($I_0=364$ W/cm^2, $\Delta=-10{,}000\,\Gamma$) and blue detuned ($I_0=116$ W/cm^2, $\Delta=1{,}000\,\Gamma$) examples [14]. For trap frequencies, ground state sizes, and energy differences, values are given in the order r_x, r_{yz}, axial. ΔU_{trap} denotes the trap depth. (Reprinted table with permission from K. Gillen-Christandl and B. D. Copsey, Phys. Rev. A, 83, 023408, 2011. Copyright 2011 by the American Physical Society.)

use in quantum computing is the following. As can be seen in Figure 11.10(c), we find that an atom (qubit) in the $F=1$, $m_F=1$ substate experiences strong confinement in all dimensions in the bright spot from the σ^- beam (i.e. the bottom or left bright spot in the y-z or x-y profile, respectively), at $z=67$ μm from the aperture. While the bright spot due to the σ^+ beam (top or right bright spot in the y-z or x-y profile, respectively) is also confined in all dimensions, the confinement is significantly weaker due to the polarization dependence of the potential energy. Therefore, a $F=1$, $m_F=1$ atom, seeking the location of lowest potential energy is trapped in the σ^- bright spot. Similarly, as visible in Figure 11.10(e), an atom in the $F=1$, $m_F=-1$ substate is trapped in the σ^+ bright spot. Consequently, both atoms (qubits) can be stored in separate locations within the same light pattern, shown in Figure 11.10(b). A similar polarization dependence effect has been successfully demonstrated in 3D optical lattices [29].

11.5.3 Bringing Atom Traps Together and Apart for Two-Qubit Operations

Red-Detuned Diffraction Trap

Figure 11.11 shows the trapping potential energy plots for an atom in the $F=1$, $m_F=1$ substate, trapped in the σ^- beam, and an atom in the $F=1$, $m_F=-1$ substate, trapped in the σ^+ beam of Figure 11.10(a), for several angles. Figure 11.11(a) depicts the intensity pattern created by the setup from Figure 11.10(a) for several angles, and Figure 11.11(b) shows the potential energy profile along the y-direction, at the location of the primary red-detuned trap, $z=67$ μm, for a pair of 364 W/cm^2 laser beams with opposite circular polarization, tuned 10,000 linewidths to the red of the ^{87}Rb D2 transition.

Figure 11.11(b) demonstrates that for the red detuned case the two traps move together continuously as the lasers are tilted to normal incidence, at which point the two traps are completely overlapped. This process can be reversed by tilting the laser beams apart. Each atom will be most probable to follow its trap as there is a difference in trapping strength between the two traps, due to the light polarization dependence. For instance, an atom in state $F=1$, $m_F=1$, trapped in the primary bright spot of the normal incidence configuration, follows the solid potential energy curve towards the negative y-direction in Figure 11.11(b) as the angle γ of the two beams is slowly increased. By the same means, an atom in $F=1$, $m_F=-1$ remains in the dashed potential energy minimum, moving towards the positive y-direction as the two beams are separated. In this way, we can bring two atoms together and apart without experiencing significant trap or tunneling losses.

The major source of tunneling in this setup is due to trap photon Raman transitions when the potential energy curves for both atoms (i.e. both states) cross, for example when the wells are completely overlapped. The probability

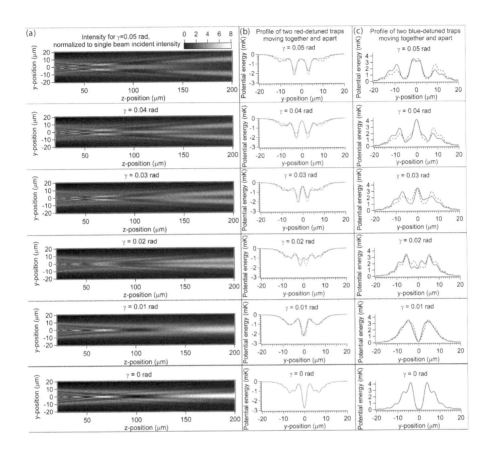

FIGURE 11.11

Bringing two traps together. (a) Column showing the intensity pattern, normalized to the incident intensity of one incident circularly polarized laser beam, for several incident angles γ. (b) Column showing the diabatic potential energy profile along the y-direction at z=67 μm for the F=1, m_F=1 (solid line) and m_F=-1 (dashed line) magnetic substates of ^{87}Rb trapped in the intensity pattern from (a) for several incident angles γ, for a laser intensity of 364 W/cm^2, and a laser detuning of -10,000 Γ. (c) Column showing the diabatic potential energy profile along y-direction at z=100 μm for the F=1, m_F=1 (solid line) and m_F=-1 (dashed line) magnetic substates of ^{87}Rb trapped in the intensity pattern from (a) for several incident angles γ, for a laser intensity of 116 W/cm^2, and a laser detuning of 1,000 Γ. (Reprinted figure with permission from K. Gillen-Christandl and B. D. Copsey, Phys. Rev. A, 83, 023408, 2011. Copyright 2011 by the American Physical Society.)

of such a transition can be reduced by performing this operation significantly faster than the Raman scattering rate.

Blue-Detuned Diffraction Trap

Figure 11.11(c) shows the corresponding trapping potential energy profiles at the location of the primary blue-detuned trap, $z=100$ μm, for a pair of 116 W/cm^2 laser beams with opposite circular polarization, tuned 1,000 linewidths to the blue of the ^{87}Rb D2 transition.

Since for blue-detuned light atoms are trapped in dark spots, it may generally be desirable to use blue-detuned traps for quantum computing to ensure long decoherence times. As illustrated in Figure 11.11(a), as the two beams are moved together and apart, an intensity wall moves through the dark spot of each beam. However, as can be seen in Figure 11.11(c), the associated potential energy of the intensity wall is not large enough to push the atoms out of the trap. Thus, the two atoms can still be overlapped completely, without switching wells, as the σ^- well traps the $F=1$, $m_F=1$ atom more strongly than the σ^+ well, and vice versa.

But the potential wall will influence the motion of the atom. For quantum computing, this disturbance must be kept negligible or must be reversible to avoid deterioration of the computation. If this is not possible, the advantages of trapping the atoms in locations of low light intensity are erased by the disturbance due to this potential energy wall, and trapping in the bright spots with sufficiently large detuning may be preferable.

Acknowledgments

Work contained in Sections 11.2, 11.4, and 11.5 was supported by the National Science Foundation under Grant Number PHY-0855524 and the Department of the Navy, Office of Naval Research, Grant Numbers N00014-06-1-1111, N00014-07-1-1152, and N00014-08-1-1209. Any opinions, findings, and conclusions or recommendations expressed in this material are those of the authors and do not necessarily reflect the views of the National Science Foundation or the Office of Naval Research.

Bibliography

[1] J. E. Bjorkholm, R.R. Freeman, A. Ashkin, and D. B. Pearson, Phys. Rev. Lett. **41**, 1361 (1978).

[2] R. Grimm, M. Weidemüller, and Y. B. Ovchinnikov, Adv. At. Mol. Opt. Phys. **42**, 95 (2000).

[3] N. Friedman, A. Kaplan, and N. Davidson, Adv. At. Mol. Opt. Phys. **48**, 99 (2002).

[4] S. Chu, J. E. Bjorkholm, A. Ashkin, and A. Cable, Phys. Rev. Lett. **57**, 314 (1986).

[5] P. Verkerk, B. Lounis, C. Salomon, C. Cohen-Tannoudji, J.-Y. Courtois, G. Grynberg, Phys. Rev. Lett. **68**, 3861 (1992).

[6] A. Hemmerich and T. W. Hänsch, Phys. Rev. Lett. **70**, 410 (1993).

[7] G. Grynberg, B. Lounis, P. Verkerk, J.-Y. Courtois, and C. Salomon, Phys. Rev. Lett. **70**, 2249 (1993).

[8] Y. B. Ovchinnikov, I. Manek, A. I. Sidorov, G. Wasik, and R. Grimm, Europhys. Lett. **43**, 510 (1998).

[9] T. Kuga, Y. Torii, N. Shiokawa, T. Hirano, Y. Shimizu, and H. Sasada, Phys. Rev. Lett. **78**, 4713 (1997).

[10] V. V. Klimov and V. S. Letokhov, Opt. Commun. **121** 130 (1995).

[11] G. D. Gillen, S. Guha, and K. Christandl, Phys. Rev. A **73**, 013409 (2006).

[12] T. N. Bandi, V. G. Minogin, and S. N. Chormaic, Phys. Rev. A **78**, 013410 (2008).

[13] L. Chen and J. Yin, Phys. Rev. A **80**, 065401 (2009).

[14] K. Gillen-Christandl and B. D. Copsey, Phys. Rev. A **83**, 023408 (2011).

[15] H. J. Metcalf and P. van der Straten, *Laser cooling and trapping*, (Springer, New York, NY, 1999).

[16] A. H. Barnett, S. P. Smith, M. Olshanii, K. S. Johnson, A. W. Adams, and M. Prentiss, Phys. Rev. A **61**, 023608 (2000).

[17] L. J. Curtis, *Atomic structure and lifetimes: a conceptual approach*, (Cambridge University Press, Cambridge, UK, 2003). Chapter 6.

[18] G. Bekefi, J. Appl. Phys. **24**, 1123 (1953).

[19] S. Guha and G. D. Gillen, Opt. Express **13**, 1424 (2005).

[20] G. D. Gillen and S. Guha, Am. J. Phys. **72**, 1195 (2004).

[21] K. Gillen-Christandl and G. D. Gillen, Phys. Rev. A **82**, 063420 (2010).

[22] J. W. Goodman, *Introduction to Fourier optics*, (Roberts & Company Publishers, Englewood, CO, 2005).

[23] I. H. Deutsch, P. S. Jessen, Opt. Commun. **283**, 681 (2010).

[24] J. M. Geremia, J. K. Stockton, H. Mabuchi, Phys. Rev. A **73**, 042112 (2006).

[25] I. H. Deutsch, P. S. Jessen, Phys. Rev. A **57**, 1972 (1998).

[26] M. E. Rose, *Elementary theory of angular momentum* (John Wiley & Sons, New York, 1957), p. 88.

[27] P. S. Jessen, I. H. Deutsch, Adv. At. Mol. Opt. Phys **37**, 95 (1996).

[28] K. Christandl, Ph.D. thesis, The Ohio State University, 2005.

[29] O. Mandel, M. Greiner, A. Widera, T. Rom, T. W. Hänsch, I. Bloch, Phys. Rev. Lett. **91**, 010407 (2003).

A

Complex Phase Notation, Engineer's vs. Physicist's

A.1 Sinusoidal Waves

This book presents mathaematical models for the propagation of light through various apertures, optical elements, and linear and nonlinear media. Light is an electromagnetic wave where both the electric and magnetic fields are sinusoidal in terms of both position and time. The simplest mathematical representation of sinusoidal waves is using the trigonometric functions sin and cos. Choosing the cosine function, the general expression for a sinusoidal wave can be written as

$$f(z,t) = A_o \cos [k(z - vt) + \phi], \tag{A.1}$$

where the *phase* of the wave is everything inside the trigonometric function argument, or $[k(z - vt) + \phi]$.

Equation A.1 represents a sinusoidal wave traveling in the $+z$-direction with an amplitude of A_o, a wavenumber of k, and travelling with a speed of v. The phase constant, ϕ, determines the value of the oscillating function at the origin at a time of $t = o$. For example, if we wanted our oscillation wave to look like a cosine function (where the value of the function is a maximum at the origin at $t = 0$) then we would set our phase constant equal to zero. If we wanted our oscillating wave to look like a sine function (where the value of the function at the origin is zero with a positive slope at $t = 0$) then we would set our phase constant to be equal to $-\pi/2$ or $+3\pi/2$. In other words, if we delay a cosine function by a factor of $+3\pi/2$ we have created a sine function. It is due to this reason that ϕ is also known as the "phase delay."

Equation A.1 can also be written as

$$f(z,t) = A_o \cos [kz - \omega t + \phi], \tag{A.2}$$

where we have multiplied the wavenumber into the expression $(z - vt)$ and used the relationship between the wavenumber, wave speed, and angular frequency of the oscillation, ω,

$$\frac{\omega}{k} = v. \tag{A.3}$$

We should also note here that the values of k and ω are assumed to be positive. We have also previously stated that Eqs. A.2 and A.2 represent a wave traveling in the $+z$-direction. Without this being explicitly stated, the direction of propagation of the wave can be determined from the signs of the position and time terms of the phase. In the case of Eq. A.2 that is the kz and ωt terms:

- if kz and ωt have *opposite* signs the wave is traveling in the *positive* z direction, and

- if kz and ωt have *the same* sign, the wave is traveling in the *negative* z direction.

A.2   Complex Notation Using Euler's Formulas

"Complex" notation simply means that we are using complex numbers, functions, or phases. A complex number is defined to be a number that can be represented as a sum of two numbers, a and b, in the form

$$z = a + ib, \tag{A.4}$$

where i is the fundamental imaginary number

$$i = \sqrt{-1}, \tag{A.5}$$

a is the *real* part of z, and b is the *imaginary* part of z. The real and imaginary parts of z can also be written as

$$a = \mathrm{Re}\,[z] \tag{A.6}$$

and

$$b = \mathrm{Im}\,[z]. \tag{A.7}$$

Similarly, complex functions or phases are simply functions of complex numbers and/or complex variables.

Euler's formulas allow the user to express oscillatory trigonometric functions as an exponent of the natural number e as either

$$e^{i\theta} = \cos\theta + i\sin\theta, \tag{A.8}$$

or

$$e^{-i\theta} = \cos\theta - i\sin\theta. \tag{A.9}$$

Euler's equations can also be rewritten in the form where the trigonometric fuctions are expressed as complex exponentials:

$$\cos\theta = \frac{e^{i\theta} + e^{-i\theta}}{2}, \tag{A.10}$$

and

$$\sin \theta = \frac{e^{i\theta} - e^{-i\theta}}{2}. \tag{A.11}$$

A simpler way of extracting the trigonometric functions from complex exponentials is to apply Eqs. A.6 and A.7 to Eq. A.8, or

$$\cos \theta = \text{Re} \left[e^{i\theta} \right] \tag{A.12}$$

and

$$\sin \theta = \text{Im} \left[e^{i\theta} \right]. \tag{A.13}$$

Finally, the traveling wave described by Eq. A.2 can now be expressed as

$$f(z,t) = A_o \text{Re} \left[e^{i(kz - \omega t + \phi)} \right]. \tag{A.14}$$

When expressing traveling waves using Euler's form of complex exponentials the function for the wave is more conveniently written simply as

$$f(z,t) = A_o e^{i(kz - \omega t + \phi)} \tag{A.15}$$

where it is *understood* that the *actual* wave is only the real part of Eq. A.15.

Now, many students when they first encounter Euler's formulas and begin representing traveling waves as complex exponentials have the reaction, "This just seems like *a lot* of extra work! Why would you even want to go through all of the work to take a simple trig funtion, convert it into a complex function, put it into an exponent, and then have to find the real part of this complex exponential just to get back to where we started? That doesn't make any sense to me." This is quite the valid question.

First, if we start the problem using complex notations and stay in complex notation throughout the entire problem we do not have to deal with converting the waves back and forth between trigonometric functions and complex exponentials. However, the *major advantage* of using complex exponetial notation for waves is that performing mathematical functions on the waves (i.e., adding waves, multiplying waves, finding the phase, etc.) is much, much easier than dealing with sines and cosines and vast tables of trigonometric identites.[1]

Let us assume that we do not want to have to write "$+\phi$" in the phase of the complex exponential every time we write the function for the wave. Using complex notation, it is fairly straightforward to absorb this part of the phase into the amplitude due to the fact that a sum of phase terms in an exponent is mathematically the same as a product of exponential functions. In other words, separating out the phase constant is achieved by

$$f(z,t) = A_o e^{i(kz - \omega t + \phi)} = A_o e^{i(kz - \omega t)} e^{i\phi} = A_o e^{i\phi} e^{i(kz - \omega t)} \tag{A.16}$$

---

[1] For an example of how much easier it is to simply add two oscillating waves using complex exponentials than it is to use trigonometric functions, see Example 2 of Chapter 8 of D. J. Griffiths, *Introduction to Electrodynamics*, Second Edition (Prentice-Hall, Inc., Englewood Cliffs, New Jersey, 1989.)

or

$$f(z,t) = Ae^{i(kz-\omega t)}, \tag{A.17}$$

where A is now the complex amplitude

$$A = A_o e^{i\phi}, \tag{A.18}$$

which contains both the amplitude of the actual wave and the initial phase delay of the wave.

A.3   Engineer's vs. Physicist's Notation

When comparing the complex exponential functions representing sinusoidal waves typically used by engineers to those typically used by physicists, there are two subtle differences.

First, the fundamental imaginary number, $\sqrt{-1}$, is typically found to be represented in physics textbooks as an i. In engineering textbooks, the fundamental imaginary number can be found to be represented by a j.

The second, and slightly more subtle, difference between engineering notation and physicist notation comes from a choice of representing the time-dependence of the oscillation:

- Engineer's choice for the time-dependence $\longrightarrow$ $e^{i\omega t}$

- Physicist's choice for the time-dependence $\longrightarrow$ $e^{-i\omega t}$.

Recall that it is the relationship between the position-dependent and time-dependent terms in the phase that determine the direction of propagation. For a wave traveling in the $+z$-direction, the two phase terms need to have opposite signs.

Using the engineer's choice for the time-dependence, a function representing a wave traveling in the $+z$-direction can be written as a product of complex exponetials having opposite signs for the position and time-dependence, or

$$f(z,t) = Ae^{i\omega t}e^{i(-kz)}, \tag{A.19}$$

which can be re-written as

$$f(z,t) = Ae^{-i(kz-\omega t)} \quad \text{engineer's notation.} \tag{A.20}$$

Similarly, using the physicist's choice for the time dependence, the function for the same wave takes the form

$$f(z,t) = Ae^{-i\omega t}e^{ikz}, \tag{A.21}$$

Complex function	Engineering notation	Physicist notation
chosen time-dependence	$e^{i\omega t}$	$e^{-i\omega t}$
plane wave	$Ae^{-i(kz-\omega t)}$	$Ae^{i(kz-\omega t)}$
Green's function	$\dfrac{e^{-ik\rho}}{\rho}$	$\dfrac{e^{ik\rho}}{\rho}$
E from **Π**	$\mathbf{E} = k^2\mathbf{\Pi} + \nabla\left(\nabla\cdot\mathbf{\Pi}\right)$	same
H from **Π**	$\mathbf{H} = -\dfrac{k^2}{i\omega\mu_o}\nabla\times\mathbf{\Pi}$	$\mathbf{H} = \dfrac{k^2}{i\omega\mu_o}\nabla\times\mathbf{\Pi}$

TABLE A.1
Similarities and differences between engineering complex notation and physicist complex notation for various mathematical functions.

which can be re-written as

$$f\left(z,t\right) = Ae^{i(kz-\omega t)} \quad \text{physicist's notation.} \tag{A.22}$$

Even though Eqs. A.20 and A.22 are mathematically different, they *do* represent the same *physical* wave by analyzing Eqs. A.8 and A.9 and utilizing the fact that the actual wave is only the real part of the complex exponential. The actual waves for each notation can be written as

$$\text{Re}\left[Ae^{-i(kz-\omega t)}\right] = A_o\cos\left[-\left(kz-\omega t\right)\right] \quad \text{engineer's notation,} \tag{A.23}$$

and

$$\text{Re}\left[Ae^{i(kz-\omega t)}\right] = A_o\cos\left[kz-\omega t\right] \quad \text{physicist's notation.} \tag{A.24}$$

Recalling the fact that $\cos\left(-\theta\right) = \cos\theta$, we see that the two notation styles do represent the same physical wave.

Expanding beyond the simple traveling wave described here, Table A.1 illustrates how the similarities and differences between the two notations manifest themselves for various mathematical functions.

Topic/model	Section(s)/Chapter(s)	Style
Light & EM waves	Ch. 1– 5	physicist's
Gaussian beams	Ch. 6	physicist's
Scalar diffraction theories	Sec. 7.1– 7.3	physicist's
Vector diffraction, apertures	Sec. 7.4 – 7.7	engineer's
Vector diffraction, curved surfaces	Ch. 9	engineer's
Vector diffraction, Gaussian beams	Ch. 10	engineer's

TABLE A.2
Complex notation styles used within this book according to topic, section, or chapter.

A.4   Use of Engineer's and Physicist's Complex Notation in This Book

In this book we use both engineering and physicist complex notation. However, within each topic, or beam propagation model, we are consistent with using only a single complex notation style. Whether engineer's or physicist's complex notation is used for a particular topic of the book depends upon the style of notation used in other books, or references, from which the model in this book is derived or closely resembles. Table A.2 clarifies which topics, or sections, of this book use which style of complex notation.

A.5   Some Commonly Used Electrodynamics and Optics Books

I. These commonly used, or cited, books use *engineer's complex notation*:

- G. Bekefi, "Diffraction of electromagnetic waves by an aperture in a large screen," J. App. Phys. **24**, 1123–1130 (1953).[2]

---

[2]Although this paper by Bekefi is neither a book, nor a commonly used citation, it is included in this list as it is the basis for much of the vector diffraction theory used in Chapters 7, 8, and 10 of this book.

- D. Marcuse, *Light Transmission Optics*, Second Edition (Van Nostrand Reinhold Company, New York, 1982.)

- A. Siegman, *Lasers* (University Science Books, Mill Valley, CA 1986.)

- W. T. Sifvast, *Laser Fundamentals*, Second Edition (Cambridge University Press, New York, NY, 2004.)

- A. Yariv, *Optical Electronics*, Fourth Edition (Saunders College Publishing, Chicago, IL, 1991.)

- A. Yariv, *Quantum Electronics*, Third Edition (John Wiley & Sons, New York, 1989.)

II. These commonly used, or cited, books use ***physicist's complex notation***:

- M. Born and E. Wolf, *Principles of Optics*, Seventh Edition (Cambridge University Press, Cambridge, UK, 1999.)

- R. W. Boyd, *Nonlinear Optics*, Second Edition (Academic Press, New York, 2003.)

- J. W. Goodman, *Introduction to Fourier Optics*, Third Edition (Roberts & Company Publishers, Englewood, CO, 2005.)

- D. J. Griffiths, *Introduction to Electrodynamics*, Third Edition (Prentice Hall, Inc., Upper Saddle River, NJ, 1999.)

- E. Hecht, *Optics*, Fourth Edition (Addison Wesley, New York, 2002.)

- J. D. Jackson, *Classical Electrodynamics*, Third Edition (John Wiley & Sons, New York, 1999.)

- R. K. Luneburg, *Mathematical Theory of Optics* (University of California Press, Berkely, CA, 1964.)

- F. L. Pedrotti, L. S. Pedrotti, and L. M. Pedrotti, *Introduction to Optics*, (Pearson Prentice Hall, Upper Saddle River, NJ, 2007.)

- J. A. Stratton, *Electromagnetic Theory* (McGraw-Hill Book Company, Inc., New York, 1941.)

Index

Milton Keynes UK
Ingram Content Group UK Ltd.
UKHW031140141024
449569UK00024B/1179